Marine Clastic Sedimentology

Concepts and Case Studies

Marine Clastic Sedimentology
Concepts and Case Studies

A volume in memory of C. Tarquin Teale

edited by

J. K. Leggett

Imperial College
Royal School of Mines
London, UK

G. G. Zuffa

Istituto di Geologia e Paleontologia
dell'Università
Bologna, Italy

Graham & Trotman
A member of the Kluwer Academic Publishers Group
LONDON/DORDRECHT/BOSTON

First published in 1987 by

Graham & Trotman Limited
Sterling House
66 Wilton Road
London SW1V IDE
UK

Graham & Trotman Inc.
Kluwer Academic Publishers Group
101 Philip Drive
Assinippi Park
Norwell, MA 02061
USA

British Library Cataloguing in Publication Data

Marine clastic sedimentology : concepts
and case studies : a volume in memory of
C. Tarquin Teale.
1. Marine sediments 2. Clastic sediments
I. Leggett, Jeremy K. II. Zuffa, G.G.
III. Teale, C. Tarquin
551.46'083 GC380.15

ISBN-13: 978-94-010-7954-9 e-ISBN-13: 978-94-009-3241-8
DOI: 10.1007/978-94-009-3241-8

LCCCN 87-16787

Softcover reprint of the hadrcover 1st edition 1987

Typeset in Great Britain by Santype International, Salisbury, Wilts., England

Contents

Contributors

C. CURTIS, Department of Geology, The University, Mappin Street, Sheffield S09 5NH, UK

I. HAYDEN, Department of Geology, Imperial College, Royal School of Mines, Prince Consort Road, London SW7 2BP, UK

A. KEMP, Department of Oceanography, The University, Highfield, Southampton S09 5NH, UK

E. MUTTI, Istituto di Geologia, Università degli studi di Parma, Via J. F. Kennedy 4, 43100 Parma, Italy

K. MYERS, Department of Geology, Imperial College, Royal School of Mines, Prince Consort Road, London SW7 2BP, UK

W. NORMARK, USGS, 345 Middlefield Road, Menlo Park, California 94025, USA

K. PICKERING, Department of Geology, The University, University Road, Leicester LE1 7RH, UK

M. SANTANTONIO, Servizio Geologico d'Italia, Largo S. Susanna 13, 00187 Rome, Italy

R. SMITH, Department of Earth Sciences, Downing Street, Cambridge CB2 3EQ, UK

T. TEALE, deceased

A. THICKPENNY, Lomond Associates, 10 Craigendoran Avenue, Helensburgh, Glasgow G84 7AZ, UK

J. YOUNG, Department of Geology, Imperial College, Royal School of Mines, Prince Consort Road, London SW7 2BP, UK

G. ZUFFA, Istituto di Geologia e Paleontologia dell'Università, Via Zamboni 63, 40127 Bologna, Italy

Preface

Tarquin Teale, a sedimentology/stratigraphy postgraduate student at the Royal School of Mines, was killed in a road accident south of Rome on 17 October 1985. Premature death is a form of tragedy which can make havoc of the ordered progress which we try to impose on our lives. As parents, relatives and friends, we all know this, and yet somehow when it touches our own world there is no consolation to be found anywhere. In Tarquin's case the enormity of the loss felt by those of us who knew him can barely be expressed in words.

Tarquin had everything which we aspire to. His fellow graduate students envied his dramatic progress in research. We his advisors, in appreciating this progress, marvelled at how refreshingly rare it was to see such precocious talent combined with such a caring, modest and well-balanced personality. He was destined for the highest honours in geoscience and there is no doubt that he would have lived a life, had he been granted the chance, which would have spread colour, intellectual insight and goodness.

At Sheffield University he gained the reputation of being one of the finest undergraduates to ever pass through that Department. Yet he also managed to be active in a number of extra-curricular fields, was Secretary of the student Geology Society, and ran regularly for the university in cross-country, track, and orienteering events. One of his tutors, Jack Soper, commented in a written reference that Tarquin combined "... high intelligence with a degree of dedication unusual in a young person with such a wide range of other interests. His work capacity far exceeds that of the average student. He has flair and originality and has great research potential ..." Tarquin left Sheffield in the summer of 1983 with a First Class Honours Degree, the Fearnsides Prize for fieldwork, and the Laverick-Webster Prize for the best undergraduate performance of the year.

And so he joined Imperial College. His task was to investigate the sedimentary-tectonic history of the Longobucco sequence, a Jurassic rift-basin succession in NE Calabria. Exposed in mountainous and remote country, the rocks in question had been little-studied. They were believed to record rifting on the "African" margin of Tethys during the particularly interesting, but little-understood, Mesozoic history of the Calabrian arc. The project grew from a special relationship which exists between Imperial College and the Dipartimento di Scienze della Terra of the University of Calabria, where one of us (GZ) was then head of department. We both knew the difficulties awaiting the student who, under our joint supervision, was to undertake the project, but neither of us came near to predicting the scientific rewards.

The fact that Gianni Zuffa was ill when Tarquin started his research meant that he commenced his first long field season in a very foreign environment and culture almost alone, no easy task for such a sociable person. However, by the end of the season he had managed to do much basic groundwork, and started on his long trail of geological discoveries and reinterpretations. Moreover, Tarquin managed not just to live in Longobucco but to become an accepted member of the village community. Those of us who were lucky enough to visit him in his field area will never forget walking with him down the crowded *piazza* during *passagiata* hour. Scores of people would greet him with enthusiasm and animation. Their regard and affection for him was obvious, as was his for them. He was truly at home with his friends there in the *cantine* of Longobucco.

He made friends wherever he went, but in Italy this called for more than just his personality. He acquired a thorough knowledge of not only standard Italian, but also the *dialletto* spoken in rural Calabria. I once went to visit Tarquin in Longobucco with an educated North Italian geologist. He had difficulty understanding the locals' dialect, and it was soon evident that, apart from being compatriots, he and they had little in common. Later he said to me while Tarquin was in full flight, "Tarquinio speaks Italian like a peasant". It was a great compliment.

In his subsequent field seasons Tarquin completed the mapping of the entire Longobucco Basin, some 330 square kilometres, at a scale of 1:10,000. This he complemented with some 2 km of sedimentary logs. In addition, his collaboration with Italian palaeontologist Massimo Santantonio led to extension of his research into previously unstudied Jurassic pelagic sequences outside the Longobucco basin.

Tarquin was a first class field geologist, with a knack for finding key outcrops. His abilities and dedication produced rapid results. He was already publishing on three major aspects of his research. With

Massimo and two London-based nannofossil workers—Jeremy Young and Paul Bown—he had produced a complete revision of the stratigraphy and biostratigraphy of the Longobucco Group. They proved that the sequence was entirely Liassic in age rather than ranging up to Cretaceous. This effectively stood previous reconstructions of the Calabrian arc on their heads.

His identification, with Emiliano Mutti, of house-sized olistoliths of shelf sediments enclosed within turbidites of the Longobucco Group led inevitably to a reassessment of the basin configuration and the role of syn-sedimentary extensional tectonics. He went beyond this, however, producing an elegant model for the mode of emplacement of these olistoliths, with the aid of modern analogs and field evidence. This work, published in this volume, we believe will stand as a valuable contribution to sedimentary geology.

Tarquin had also nearly completed a major paper, with Massimo, on the Caloveto Group pelagic sediments. From these isolated patches of condensed sequences they were able to reconstruct a fascinating sequence of Jurassic environments. This again had considerable implications for the tectonic evolution of the region, and this volume contains a paper summarizing that work.

Tarquin had mastered not just the basics but the details of several sub-disciplines of our science, and so his talents did not end with sedimentology. He had become increasingly involved in structural interpretation. For the Longobucco basin he had assembled evidence for syn-sedimentary rifting, subsequent compressional telescoping of the unit and associated, previously unidentified, strike-slip deformation. His final field project, with Steve Knott from Oxford, had been on the structural evolution of the Calabrian arc.

In addition, the journal *Sedimentology* had also just accepted a long article on bentonites in the Welsh Borderland. These he had discovered during his undergraduate mapping, and characteristically he pursued the topic vigorously, with Alan Spiers of Sheffield.

With his wide-ranging and deep knowledge of geology Tarquin was much in demand at Imperial College as a demonstrator, on field trips and in practicals. Through these, his active involvement in the De La Bêche club, and his never-failing charm and consideration for others, he became unquestionably one of the best known, and well-loved, members of the Geology Department at Imperial. He was equally popular at Calabria University, where he had worked with several members of the staff, and seemed to know virtually every student. From here he had led a party of fifty of the most eminent geoscientists in sedimentary geology to Longobucco, after a major international conference in 1984.

Tarquin seemed destined to shoot to the top of whichever branch of geology he chose to invest his talents. In addition to his published work, he had assembled an encyclopaedic knowledge of the sedimentology, stratigraphy, ichnology and palaeogeography of the Longobucco basin. He had unravelled its structural history and greatly contributed to knowledge of the structural development of the Calabrian Arc. This research, which was intended to be his Ph.D thesis, seemed certain to be an outstanding piece of work. Following this Tarquin had hoped to return to Italy to do full-time post-doctoral research with Emiliano Mutti.

Those of us who worked with or knew him were devastated by his loss and we resolved to provide a deeply symbolic memorial to him: one which would be lasting and useful, so that we may gather some small solace every time we think, over the years, of his tragically early departure. This book is part of that memorial. The other part is the Tarquin Teale Memorial Fund, which each year will pay for an outstanding British graduate student to do sedimentological fieldwork in Italy, or for an outstanding Italian graduate student to do fieldwork in Britain. All the royalties from this book will go towards that fund.

Tarquin, this is your book.

Jeremy Leggett
Gian Gaspare Zuffa

Introduction

The geologist, perhaps more than any other breed of natural scientist, loves to try and impose order on the many natural processes which arouse his/her interest. Models and classification schemes abound in all the many sub-disciplines of geoscience. Who of suitable vintage can forget, for example, the many types of "geosyncline" which in the mid-1960s purported to classify the multifarious sedimentary basins of the planet, and their ancient equivalents in the mountain belts?

Plate tectonics has been a fertile breeding ground for models. Many models which have sprung from the "mobilist" philosophies behind plate tectonics have splendidly stood the tests of an ever-expanding data-base. Recently, however, it has become clear that one of the most important models in sedimentary geology has not.

Submarine fans, large cones of clastic detritus which accumulate in great thicknesses on the floors of sedimentary basins of all scales, are particularly important because of their petroleum potential. Once buried, and slightly deformed, their muddier portions often provide a source for hydrocarbons, and their sandier portions often provide reservoirs. In consequence, commercial interest in submarine fans, modern and ancient, has been intense over the years.

One of the most sought-after consultants on this topic has been Emiliano Mutti of the University of Parma. This is because in 1972, together with his colleague Franco Ricci Lucchi, he published a predictive model for sand-body geometry in submarine fans. It was based on a marriage of knowledge from ancient submarine-fan strata superbly exposed in the Italian Apennines with what little was then known of the physiography of modern fans. Since then the Mutti/Ricci Lucchi model has been applied enthusiastically all over the world. Its success was guaranteed in that it predicted successfully "ahead-of-the-drill" enough times to keep oil company sedimentologists happy.

But as his work continued, Mutti grew progressively less happy. In particular, he was uneasy with the increasing complexity being revealed in modern fans by the research of oceanographer-geologists such as William Normark of the USGS. In addition, his commercial work began to reveal increasingly sophisticated seismic reflection profiles which showed unexpected patterns. Finally last year, he decided to team up with Normark to refute his own, by most yardsticks hugely successful, early work. The philosophy he now advances is that there are too many complex interacting factors which control submarine fan sedimentation (sea-level, tectonics, sediment supply, etc.) to enable us to arrive at predictive models until we know much more about both modern and ancient examples.

In February 1986, nearly four hundred geologists were privileged to see Normark and Mutti in London, presenting their arguments for a more anarchistic approach to submarine fan sedimentation. The occasion was a symposium organized by the De La Bêche Club, the student geologists' society of Imperial College. The symposium was organized by the students to provide a memorial to a graduate member of their community who had been killed in a road accident in Italy, Tarquin Teale.

Tarquin had worked in the field with both Mutti and Normark prior to his tragically premature death. His research, complete and all but written-up when he was killed driving to the airport returning home from his field area, concerned both submarine fans and the evolution of mountain belts—those of Calabria in southern Italy. He was respected by all geologists who knew him, both as a precociously talented sedimentologist, and as a warm and genuine person with time for all, irrespective of status, race or sex. Most particularly he was admired for his capacity for painstaking work in the field. At the Imperial College meeting Mutti referred to this quality of Tarquin's. Given a natural span of time and experience, his work would no doubt have added significantly to our progress in developing reliable predictive modelling in submarine fan environments. The same, of course, applies to unravelling the mountain belts.

The ten papers in Tarquin's Memorial Volume fall into four groups. The first two concern models of deep-water clastic sedimentation. We believe that the long paper by Mutti and Normark, based on their presentation at the DLB symposium, will become a landmark in the history of sedimentary geology. Gianni Zuffa's paper builds on the theme of how to read geological history from deep-water clastics by interpreting their petrography.

The next five papers provide case histories: from the Tethyan rocks of Calabria, and the Lower Palaeozoic Appalachian-Caledonian belt. The two Tethyan papers summarize the main discoveries which Tarquin Teale made in the Longobucco Basin. We believe they will be of interest to all those concerned with sedimentary basins and their tectonic settings. The first, jointly researched with Massimo Santantonio, describes how detailed biostratigraphic and facies work can be utilized to work out the relationship between sedimentation and tectonics in a rift basin. The second describes huge allochtonous blocks which fell from the edge of the basin as it evolved. This case study, we believe, will be of great interest to those involved in sediment dynamics and the processes involved in the emplacement of olistoliths.

The Appalachian-Caledonian case-histories are fine examples of what can be done with techniques for facies modelling (such as those articulated by Mutti and Normark) when diligent sedimentary geologists are faced with complex ancient terranes. The Welsh Basin was situated on the south side of the Iapetus Ocean during Early Palaeozoic times. Ru Smith describes aspects of its Silurian fill. The Southern Uplands, most geologists seem to believe, was a trench on the northern side of the Iapetus.

Next we move on to the mudrocks, the unromantic component of basin-fill and Charles Curtis gives us a masterly review of the geochemical processes which govern the diagenesis of these economically-important rocks. Alan Kemp describes how detailed facies analysis, coupled once again with diligent biostratigraphy, can be used to unravel the changing geometry of its fill during Silurian times. Andrew Thickpenny describes a Lower Palaeozoic case-history: the Scandinavian Alum Shales. Keith Myers describes his important new technique for relating gamma-ray logs of mudrock successions to original sedimentary processes.

The palaeogeographic position of basins in an orogenic belt such as that built during, and after, Iapetus closed is not always clear. Kevin Pickering reconstructs the history of a foreland basin from the Quebec Appalachians, and compares the history of that basin with a more-complicated Appalachian basin in Newfoundland.

All the contributors to this volume knew Tarquin Teale well; the editors were his supervisors. Emiliano Mutti and Bill Normark went into the field with Tarquin at various times for guided tours in the Longobucco Basin. That they chose to contribute their important new paper here, and not to a journal such as the *American Association Of Petroleum Geologists' Bulletin*, demonstrates their regard for Tarquin.

Massimo Santantonio and Jeremy Young were collaborators of Tarquin's. Andy Thickpenny and Keith Myers were fellow graduate students at Imperial. Ru Smith, Alan Kemp and Kevin Pickering knew Tarquin from his presentations, and conviviality, at the annual British Sedimentological Research Group meetings. Charles Curtis taught him as an undergraduate. We all miss him.

There is an old maxim that the more you research something the more complicated it becomes. In the mountain belts and on the basin floors it is certainly true: we have moved on from the days of the simple cartoon models of the early- and mid-1970s. Tarquin Teale would have been closely involved at the cutting edge in helping modern sedimentology move on still further.

Jeremy Leggett
Gian Gaspare Zuffa

Chapter 1

Comparing Examples of Modern and Ancient Turbidite Systems: Problems and Concepts

Emiliano Mutti, University of Parma, 43100 Parma, Italy, and
William R. Normark, US Geological Survey, Menlo Park, California 94025, USA

ABSTRACT

A useful comparison of modern and ancient submarine fans can be based only on well-understood and thoroughly mapped systems. In addition, the examples selected for comparison must represent depositional systems similar in such characteristics as type of basin, size of sediment source, physical and temporal scales, and stage of development. Many fan sedimentation models presently in use do not meet these criteria.

A conceptual framework for comparing modern with ancient, ancient with ancient, and modern with modern turbidite systems defines our approach to the problems involved. To attempt to select similar depositional systems, we define four basic types of turbidite basins based on size, mobility of the crust, effects of syndepositional tectonic activity, and volume of sediment available in the source areas. The difference in physical scale and the great dissimilarities in types of data available are particularly important in the comparison of modern and ancient deposits. Comparisons can be done for basin-fill sequences or complexes (1st order), for individual fan systems (2nd order), for stages of growth within an individual system (3rd order), or for the scale of specific elements (facies associations and component substages) within a system, e.g. lobes, channel deposits, overbank deposits (4th order). Valid comparison requires that the field mapping is sufficient to recognize the stage of development of the system.

To facilitate application of this conceptual framework, working definitions of individual fan elements attempt to provide criteria applicable to both modern and ancient settings. These elements are channels, overbank deposits, lobes, channel/lobe transition features, and scours (major erosional nonchannel features). *Derived* characteristics, such as fan divisions and sedimentation models, are considered as secondary points only used as necessary for discussion. The use of morphologic terms to describe ancient deposits is also qualified. The primary emphasis remains on detailed, complete field work, both on land and at sea, to provide factual characterizations of the sediment and rock assemblages to ensure that similar features are being compared in terms of both temporal and physical scales.

INTRODUCTION

The recent review of 28 separate modern and ancient submarine fans provided by the international COMFAN (COMmittee on FANs) meeting in 1982 (Bouma *et al.*, 1985a; Normark *et al.*, 1985) has helped identify major problems in developing generalized fan models. Developing generalizations (models) for understanding the growth of submarine turbidite systems, however, is as complicated as attempting to judge the wide variety of wines in the world by a single set of criteria. What is required is a clear definition of the common characteristics of the objects being compared as well as an appreciation of the wide range of variables that influence their development. (This applies to wines as well as to submarine fans.) Many of the existing fan models have not adequately dealt with the following limitations:

1. Not all turbidite systems have formed submarine fan sequences. The characteristic turbidite facies can be developed in a variety of restricted basins and in tectonically active regions that do not permit the development of complete fan systems.

2. It is necessary to recognize the major factors that control the growth of submarine fan complexes. As a minimum, these factors include (a) the type of crust on which the turbidite system is formed, (b) the longevity of the sediment source(s) and the rate of sediment supply, and (c) the local tectonic and global sea-level controls on sediment supply.

3. Comparison of modern and ancient fans, or modern with modern and ancient with ancient, cannot provide valid results unless the scale of observations is similar. These scale problems include both time—the length of time over which the turbidite system develops—and spatial dimensions.

4. Studies of modern and ancient turbidite systems are based on different types of data that provide different degrees of resolution and reflect differing physical attributes of the deposits.

5. There are complications caused by the terminology used to describe turbidite systems. These complications result from (a) the lack of accepted definitions of features in the mapping of either modern or ancient systems and (b) the use of morphologic terms to describe ancient turbidite sequences where, except for local channels, little original morphologic relief is preserved.

The problems of comparing turbidite systems are compounded by attempts to modify or develop depositional models based on poorly or incompletely mapped systems. For ancient fans, many of the difficulties in mapping stem from a lack of adequate exposure of the sedimentary sequences; for modern fans, the sheer size of the deposits and the lack of adequate sampling and imaging capabilities severely restrict our ability to map the systems. Therefore, we do not attempt to review all the various models or examples of turbidite systems to reconcile the present inherent discrepancies. Rather, we suggest a moratorium on model building until adequate field work has resolved many of the questions posed. Unfortunately, we shall need to use some model concepts to introduce and describe our stages. The purpose is not to promote or fully develop such models, however, as they are discussed elsewhere (Mutti, 1985). We also attempt to define the criteria to minimize the problems and to suggest some standardization of terms, wherever possible developing compatible definitions for modern and ancient turbidite systems.

Most of the work on both modern and ancient turbidite depositional systems over the past 15 years has been directed primarily to understand (1) facies and related depositional processes (see summaries in Carter, 1975; Lowe, 1972, 1979, 1982; Middleton and Hampton, 1973; Mutti and Ricci Lucchi, 1972, 1975; Nardin *et al.*, 1979, Walker, 1966, 1978; Walker and Mutti, 1973) and (2) the factors that control turbidite deposition and the growth of submarine turbidite systems (see, for example, summaries in Howell and Normark, 1982, Nelson and Nilsen, 1974; Normark, 1978; Pickering, 1982b; Stow *et al.*, 1985; Walker, 1978). In this paper, we emphasize the importance of framing turbidite deposits within accurate patterns of stratigraphic correlations established within the deposits. This stratigraphic framework, which is the prerequisite for comparison of turbidite depositional features, will be considered at different temporal and spatial scales, from that of seismic stratigraphy down to that of facies and facies association packages.

The examples of turbidite depositional systems that we will discuss in the following pages are, therefore, mainly selected from those with which we are most familiar, to ensure that these scale factors are correctly understood. In this way, we hope to show that valid comparisons of modern fans and ancient systems can provide very useful insights into the understanding of turbidite successions.

TYPES OF TURBIDITE BASINS

The morphology, internal structure, and facies associations of submarine fans are primarily controlled by the long-term stability of the basin and the volume of sediment supplied to the depositional area. The largest submarine fan complexes form over tens of millions of years and require not only a relatively stable basin but also a long-term supply of large volumes of sediment. It is probably within these long-lived systems that the effects of *global* sea-level changes (Vail *et al.*, 1977) are most pronounced. Turbidite basins formed in active tectonic settings tend to be short lived, especially in cases where both the source area and depositional basin are affected by crustal movements. In these short-lived systems, local tectonic control greatly influences the development of the turbidite deposit.

The major factor controlling the stability (or, conversely, the mobility) of the turbidite basin is the type of crust underlying the basin. Long-lived submarine fans are generally on oceanic crust, and the size of the fan reflects the amount of sediment supply and not the confines of the basin. Turbidite basins formed on continental crust, especially along active continental margins, tend to be short lived, even where sedimentation rates are relatively high.

The crustal setting of turbidite deposits must be considered for comparisons of modern and/or ancient examples as well as the effects of sediment supply. In reviewing the major turbidite deposits that have been described in the literature (see Barnes and Normark, 1985), we suggest that there are at least four main types of basin. For most turbidite systems mentioned in this section, the summary by Barnes and Normark (1985) provides references and tabulation of data.

Type A

Type A basins are those formed on oceanic crust with large, long-lived sediment source(s) and little or no tectonic activity. These conditions are generally found on mid-plate margins in the Atlantic and Indian Oceans. The largest fan complexes are all found offshore from major river systems and have been building for several million years to as much as 20 million years. Examples of this type include the Bengal, Indus, Amazon, Laurentian, and Mississippi Fans, which are all fed by large rivers (Bouma *et al.*, 1985b; Damuth and Flood, 1985; Emmel and Curray, 1985; Kolla and Coumes, 1985; Piper *et al.*, 1985). Small fans can also form within this type of basin wherever the source area is restricted, e.g. the four-million-year-old San Lucas Fan (off the Baja California peninsula; Normark, 1970a). Because large pieces of oceanic crust are not generally obducted or overthrust along continental margins, good ancient examples of these large deep-ocean fan complexes are rarely available.

Type B

Type B basins are formed on oceanic crust with relatively long-lived sediment source(s) but with tectonic activity in the source/basin transition area. These basins are also long lived, some exceeding 20 million years, but they generally have lower rates of sediment supply than do Type A basins, although the supply is sufficient to overcome the effects of tectonic activity. Two subtypes are common: (a) transform margins, where examples include the Monterey and Delgada fans on oceanic crust off central California (Normark, 1985; Normark and Gutmacher, 1985); and (b) subduction margins, in examples where the rate of sediment supply greatly exceeds the rate at which the fan is accreted to the margin. Examples of this type include the Astoria and Nitinat fans of the north-east Pacific (Nelson, 1985) and probably the Cretaceous fan of the Gottero Sandstone in north-western Italy (Nilsen and Abbate, 1985).

The mid-Tertiary-age Zodiac Fan in the Gulf of Alaska is a large, long-lived fan on the Pacific lithospheric plate; its supply of sediment has been cut off by the modern Aleutian trench (Stevenson *et al.*, 1983). Plate motion reconstructions show that when the fan was growing it was a Type B.

Some turbidite basins exhibit extensive zones of diapirism along the margin that can disrupt submarine fan development (e.g. Rhone Fan; Droz and Bellaiche, 1985); this disruption, however, generally does not prevent the development of long-lived systems so we do not include it as a separate subtype.

Type C

Type C basins are formed on continental crust with relatively large and long-lived sediment supply but with structural control of basin configuration and duration. Where continuing tectonic activity over several million years along the margin results in a migrating foreland basin, a succession of turbidite systems results. In the early stages of thrust propagation, relative and global sea-level variations are probably the main factor in controlling turbidite accumulation in basin development. Further orogenic shortening results in the narrowing and fragmentation of foreland basins; during this stage, uplift and emersion associated with frontal and lateral ramp areas of moving thrust sheets induce self-edge and/or intrabasinal instability processes that will lead to failures of unconsolidated or poorly consolidated sediment of shallow and/or deep-marine origin. Under these circumstances, tectonic control becomes of primary importance in turbidite deposition, regardless of global sea-level variations. The best-known examples of this type of basin are the Eocene Hecho Group turbidites of the south-central Pyrenees, Spain (Mutti, 1985; Mutti *et al.*, 1985), and the classic Oligocene to Pliocene foredeep sandy flysch of the northern Apennines of Italy (Ricci Lucchi and Ori, 1985). Few data are publically available from large modern examples of this type.

Type D

Type D basins are those formed on continental crust where continuing tectonic activity results in relatively rapid changes in basin shape and in short-lived sediment sources. The resulting turbidite deposits form in relatively short time-spans (10^4–10^5 years), are dominantly coarse-grained channel and lobe sequences, and are volumetrically much smaller than turbidite fill of the other three basin types. These small, actively deforming and short-lived basins form on all active margins: during early rifting phases of an opening ocean, along active transform margins, and during final stages of continental collision.

Modern examples include fans of the California Continental Borderland, e.g. Navy, La Jolla, and Redondo Fans in transform-controlled basins (Bachman and Graham, 1985; Haner, 1971; Normark *et al.*, 1979). Ancient examples are also common, and include the Tertiary Piedmont Basin of north-western Italy, an area of post-collision rifting (Cazzola *et al.*, 1985); the Upper Cretaceous Aren Sandstone of south-central Pyrenees in Spain that occurs within a foreland basin (Simó and Puigdefabregas, 1985), and the Jurassic deposits of Greenland (Surlyk, 1978) that formed during early stages of rifting.

In providing examples of modern and ancient submarine fans that fall into the four types of basins, we do *not* imply that these fans fit generalizations or 'models' for fan growth. In addition to the type of basin and degree of tectonism, it is necessary to consider the effects of scale differences and types of observations (data sets) used to map the fans before attempting comparisons.

SCALE FACTORS

A geologically significant comparison of modern and ancient turbidite deposits requires that the comparison be made at similar spatial and temporal scales. Figure 1 shows the spatial scale of features that comprise turbidite basin deposits and of the types of observations used to map the deposits. Figure 2 introduces the terminology we suggest for comparison of turbidite systems, and relates depositional events to both temporal and spatial scales.

Turbidite complexes ('suites' of Ricci Lucchi, 1975), systems, stages, and substages form the basis for the classification shown in Fig. 2. These units have been amply discussed in previous papers (Mutti, 1985; Mutti *et al.*, 1985), and will be briefly reviewed in a following section. Although these distinctions have been studied primarily in ancient sequences thus far, we think their recognition in modern fans would probably greatly improve the possibilities for significant comparative studies in the future.

A *turbidite complex* refers to a basin-fill succession and is composed of several turbidite systems that are stacked one upon the other.

A *turbidite system* is a body of genetically related mass-flow sediments deposited in stratigraphic continuity. Systems are commonly bounded, above and below, by mudstones (in many cases reflecting high stands of sea level) or by submarine erosional unconformities. As such, a turbidite system is part of a depositional sequence in the sense of Vail *et al.* (1977) and will commonly occur at the base of a sequence initiated during a relative low stand of sea level.

The definition of a turbidite system as given here is basically *sensu stricto*, and a later section will expand upon this usage. This *sensu stricto* usage should not be confused with that in the COMFAN discussions (Bouma *et al.*, 1985a) which is *sensu lato* in that 'related turbidite systems' are meant to include all turbidite deposits, many of which do not constitute recognizable submarine fans or represent short-lived episodes of deposition.

A *turbidite stage* consists of facies associations and erosional surfaces that formed during a specific period of growth within a system (*sensu stricto*). A *substage*, where more specific associations can be recognized, encompasses time-equivalent facies and erosional surfaces within a facies association.

The framework of Fig. 2 allows consideration of turbidite deposits within five main orders of scale that are discussed below in descending order of magnitude.

First order

First-order features are at the scale of entire basin-fills of successions of turbidites and thus encompass periods of time of several millions of years and longer. These features compare in significance to large fan complexes in that they are generally built up by the stacking of individual turbidite systems (*sensu stricto*) in the same long-lived depocenter. First-order complex successions, that may reach volumes up to several hundreds of thousands of km^3, can be split into major depositional sequences in the sense of Vail *et al.* (1977) through both seismic reflection techniques and through detailed stratigraphic mapping. These sequences allow the framing of turbidite sediments within relative variations of sea level (Mutti, 1985). Figure 3 shows an example of a first-order turbidite succession and its subdivision in

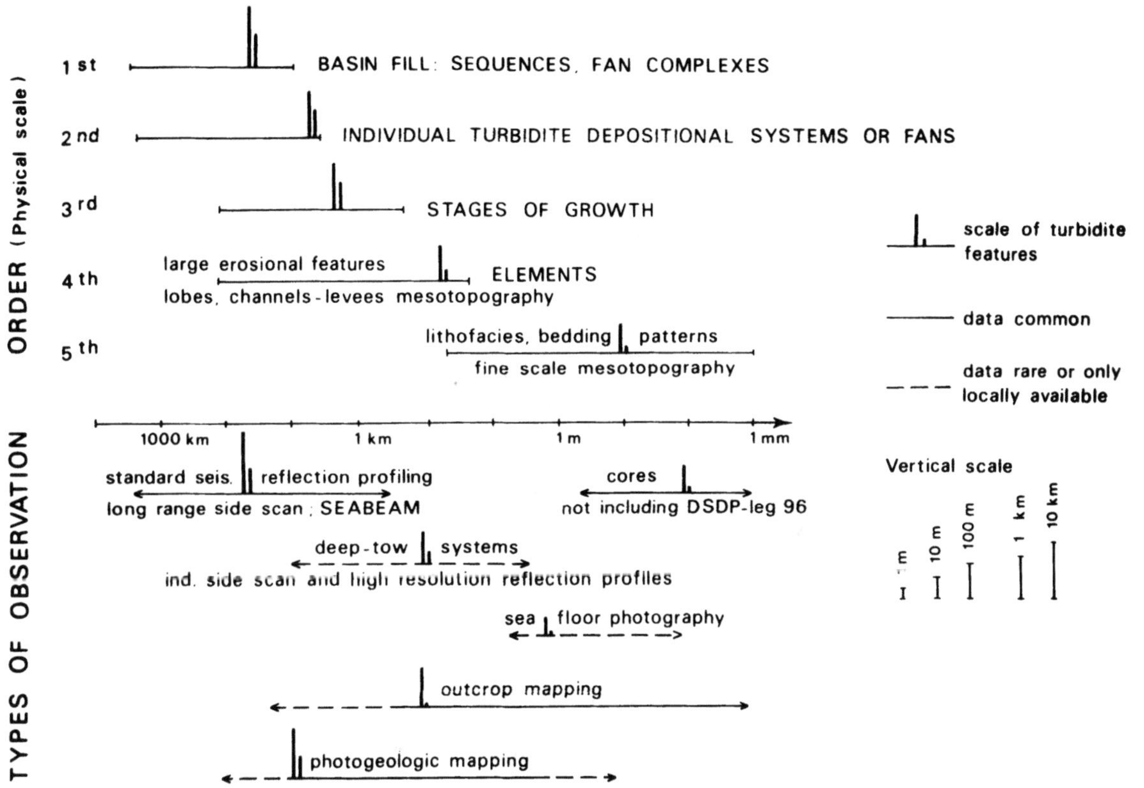

Fig. 1. Physical scale comparisons for vertical and horizontal dimensions of turbidite basin-fill and depositional features (upper), and of typical observations—outcrops, cores, geophysical records (lower). All multibeam echo-sounding systems are represented by the term SEABEAM in this diagram. The vertical scale for core samples does not include Deep Sea Drilling Program cores (Leg 96 is the Mississippi Fan sampling: Bouma *et al*. 1985c).

EVENTS	*TERMINOLOGY*	*HIERARCHY*	*DURATION (years)*
DEPOSITION AFFECTED BY MAJOR BREAKS IN SEDIMENTATION (UNCONFORMITIES) PRODUCED BY LONG-TERM SEA LEVEL VARIATIONS AND TECTONIC ACTIVITY	TURBIDITE COMPLEX	1st ORDER	x 10^6 to 10^7
SHORT-TERM SEA LEVEL VARIATIONS AND TECTONIC ACTIVITY PRODUCE CHANGES IN SEDIMENTATION BUT NO SIGNIFICANT BREAKS (UNCONFORMITIES)	TURBIDITE SYSTEM	2nd ORDER	x 10^5 – 10^6
	TURBIDITE STAGE	3rd ORDER	x 10^4 – 10^5
HIGH-FREQUENCY CHANGES IN DEPOSITIONAL AND EROSIONAL PROCESSES OF POORLY UNDERSTOOD ORIGIN	TURBIDITE FACIES ASSOCIATION AND COMPONENT SUB-STAGES	4th ORDER	x 10^3 – 10^4
"NORMAL" SMALL-SCALE EROSION AND DEPOSITION	BEDS AND THEIR FEATURES	5th ORDER	VIRTUALLY INSTANTANEOUS FEATURES

Fig. 2. Types of events characterizing turbidite systems and their physical scale subdivisions (re: Fig. 1) showing associated time scale for depositional units.

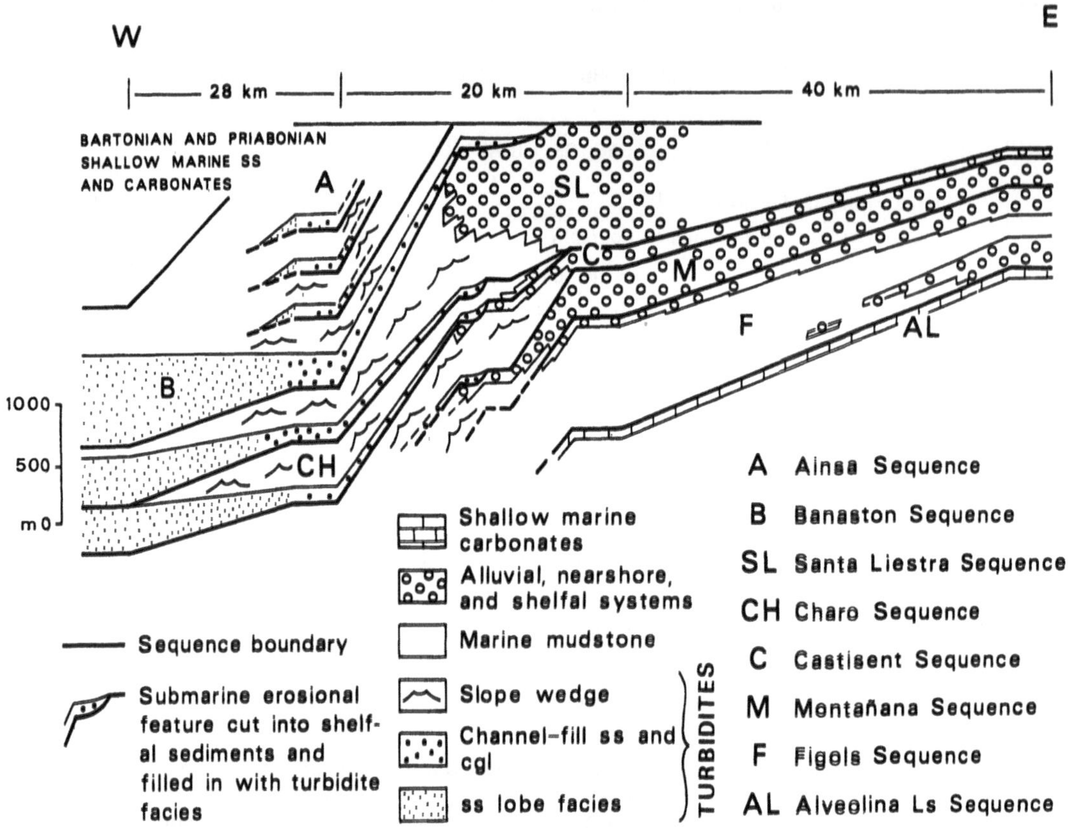

Fig. 3. Stratigraphic cross-section parallel to elongate axis of sedimentary fill in an Eocene foreland basin in the south-central Pyrenees, Spain. The turbidite complex fills the western, deeper sector of the basin and formed in about 10–12 million years between early and late middle Eocene. Six turbidite depositional systems are recognized. The three stratigraphically lower systems are examples of Type 1 deposits (Mutti, 1985) in which the bulk of the sand is deposited in the lobes. The three systems that are stratigraphically higher are smaller, and the sandstone facies are found predominantly as channel-fill and immediately adjacent lobe deposits (Mutti Type II, 1985). Thick overbank wedges form the upper part of each system along the eastern side of the basin. The basal boundary of most systems is correlative with a marked stratigraphic unconformity in the shelf-edge region. Modified from Mutti *et al.*, 1985.

depositional sequences from the Eocene Hecho Group basin, north-eastern Spain.

Second order

Second-order features, or turbidite depositional systems (*sensu stricto*), are bodies of turbidite sediments that occur within individual depositional sequences and are typically separated vertically by highstand mud facies. Most commonly, these systems encompass spans of geologic time measurable in the order of 10^5–10^6 years, and roughly match the physical scale of formations and/or members of conventional lithostratigraphic usage. Such units can therefore be mapped at appropriate scales in both surface and subsurface studies and in modern basins. Depending upon local conditions of sediment supply and basin size and configuration, such systems may reach thicknesses of several hundreds of meters, and lengths of as much as a thousand kilometers. Both thickness and length may be considerably affected by basin size.

Turbidite depositional systems (Mutti, 1985) have the same geologic significance as an individual fan-lobe deposit (Bouma *et al.*, 1985b), i.e. a package of turbidites that forms during a period of fan activity and is bounded above and below by fine-grained facies produced during phases of fan deactivation and/or by submarine unconformities (Feely *et al.*, 1985). Most commonly, each system develops through channel-fill deposits that are replaced in a down-current direction by non-channelized facies. Examples of ancient turbidite systems are those shown in Fig. 3; conceptually similar units are the

Pleistocene fan lobes of the Mississippi Fan complex (Bouma *et al.*, 1985b; Feely *et al.*, 1985).

Third order

Third-order features define stages within the development of an individual turbidite system (Mutti, 1985). In modern fans, third-order features are represented by distinct morphometric units such as low-relief lobes, levee–channel complexes, and large-scale erosional features. In ancient turbidites, third-order features are defined by facies associations or large-scale erosional surfaces that reflect different stages in the development of the system (Fig. 2).

As discussed in more detail later, these different stages may develop in an ordered vertical sequence that is apparently dictated by relative sea-level variations, or as independent features of growth phases of one or more stages. Third-order features can also be recognized on seismic-reflection profiles provided that the facies association related to a specific stage develops thicknesses of several tens of meters or more, and can thus be expressed as a distinct seismic facies. The length of time involved in the deposition of facies associations that reflect different stages of evolution within each system is probably of the order of 10^4–10^5 years.

The identification of third-order features in exposed ancient sequences generally depends on the extent and quality of the exposures; accurate time-correlation patterns are required. Large erosional features that are critical to an identification of stage deposits are easily missed in the field and can generally be recognized only through careful air-photographic mapping after a first phase of field reconnaissance (Mutti *et al.*, 1985).

Fourth order

Fourth-order features are here defined as erosional or depositional features that develop within an individual stage in the growth of a system and are characteristic of specific depositional environments. Most commonly these features match the concepts of channel-fill and lobe facies associations, as derived from ancient deposits, i.e. features developed at the scale of a few tens of meters down to a few meters in vertical scale and whose channelized or non-channelized geometry is commonly identifiable at the scale of subaerial exposures with lengths of a few hundreds to several tens of meters. The same geometrical characteristics are commonly recognized also at the scale of mesotopography in recent settings (Normark *et al.*, 1979).

Fifth order

Fifth-order features go down to the scale of 'common' exposures, e.g. fine-scale mesotopography, cores, and samples in modern deposits. In ancient rocks, such features include lithofacies, bedding patterns, cut-and-fill features, mud-draped scours, and internal structure and texture of beds, i.e. depositional or erosional features developed at the scale of few meters down to millimeters. Except for mudstone units that can be deposited over periods of 10^3 years, such features are the product of events that represent deposition over 10^0–10^3 years, or are virtually instantaneous.

Confusion among the five orders of temporal and physical scales noted above leads inevitably to equivocal comparisons between ancient and modern turbidite systems (*sensu lato*), as well as between modern and modern, and ancient and ancient. Figure 1 summarizes the scale problems encountered in analyzing modern and ancient turbidite deposits.

DIFFERENCES IN TYPES OF OBSERVATIONS

In addition to the scale problems described in the previous section, the comparison of modern and ancient turbidite deposits is further complicated by a relative lack of common data sets. Facies associations for ancient fans are based upon the types of observations that are least available to marine geologists, i.e. details of bed thickness, internal structure, and vertical succession of sand beds.

Modern fans

The study of modern fans provides a view of the overall shape of the deposit and of the larger morphological features, e.g. major channels, levees, and large slide deposits. The bathymetry for most fans is an interpolation between individual wide-beam, echo-sounding lines spaced from hundreds of meters to tens of kilometers apart. Thus, without narrow-beam or deep-towed sounding systems, bathymetric features generally less than one kilometer in horizontal extent are poorly defined or, in most cases, not recognized (Normark *et al.*, 1978, 1979). Long-range, side-scanning sonar and deep-towed, side-scan systems help to define the shape of intrachannel features and the mesotopography, respectively, but these types of data are only now becoming available. At the smallest scale, sea-floor photographs can provide some insight into relief less than five meters across on modern fans.

Seismic-reflection profiling systems provide much of our data base for modern turbidite deposits. High-energy, deep-penetration, seismic-reflection profiles are useful for basin-scale studies. Total thickness and horizontal extent of basin-fill as well as major unconformities or lithologic contacts can be adequately mapped, but with the same limitations resulting from interpolation between profiles that apply to surface morphology. These systems, however, cannot generally resolve surfaces less than several tens of meters apart. High-resolution, seismic-reflection profiles (both 3.5 kHz surface-ship and deep-towed systems) can provide resolution of surfaces separated by as little as two meters, but only in silt and clay. Thick sand beds (>50 cm) are not generally penetrated by these low-power systems. Thus, the internal structures of sandy sections less than thirty to fifty meters thick are commonly not resolved by any seismic-reflection technique.

For purposes of comparing reflection profile data with measured sections from ancient turbidites, it is necessary (1) to correct the reflection data to true vertical scale, an operation that requires knowledge of the sound velocity in the modern turbidites, and then (2) to correct for the uncompacted thickness of modern muddy sediments. Much of the existing literature on fan models does not take into account these factors.

Sediment sampling on modern fans can provide some direct observations of the turbidite units. Most of the available samples, however, are short (<10 m) and commonly show distortion caused by coring, especially where the core diameter is less than 10 cm. Sand beds are not adequately sampled—many cores from modern turbidites are stopped by thicker sand beds. Deep Sea Drilling Project (DSDP) drilling on the Mississippi Fan (Bouma *et al.*, 1985c) provides one of the few exceptions for samples from modern turbidites. Direct (visual) lateral correlation between adjacent cores is not possible, thus wavy bedding, amalgamation, and pinch-out are examples of bedding features not observed in modern turbidites over scales exceeding core width (maximum of about 0.5 m in box cores).

The major problems in acquisition of data from modern fans are (1) lack of sufficient data on structure and bed thickness of sandy intervals because of the problems in obtaining long cores, (2) lack of morphologic data at 'outcrop' scale (100 m or less in horizontal extent), and (3) a lack of detail from the larger fan systems because of the ship time required for adequate mapping of turbidite basins of 200 km or more in lateral dimensions.

Ancient turbidite systems

In many respects, the study of ancient turbidite deposits provides information in those areas where the study of modern fans is weakest. Detailed outcrop studies generally allow reconstruction of precise vertical successions of beds, facies, and facies associations through sediment thicknesses commonly up to several tens of meters. However, the limitation is that lateral continuity is rarely seen, thus inhibiting reconstruction of the original geometry and of the spatial and temporal relations with adjacent sediments. As a result, outcrop studies provide a wealth of local information through a specific time interval, but the sedimentary evolution that can be reconstructed for such intervals can only rarely be compared with that of adjacent areas. Thus, the problem in most cases is one of lack of precise correlation patterns. Marker beds are generally rare or only of local significance; and precise biostratigraphic correlations are commonly prevented by high rates of sedimentation.

Visual correlation of individual beds across and between outcrops, particularly when integrated with paleocurrent data, also allows effective sampling that can demonstrate lateral changes in grain size, sorting, composition, and internal structures related to change in flow conditions within the depositing current. Morphologic features within individual outcrops are comparable in scale to the mesotopography on modern fans (Normark *et al.*, 1979). The recognition of these features is generally limited to a single section with no knowledge of the horizontal shape. This can result in a misidentification of the cause of the relief seen in section only. The most reliably identified morphologic relief is within the sand beds, as compaction of the shale intervals during lithification commonly distorts the shape and can even invert the relief where sandy sections are adjacent to shales.

The major problems in the acquisition of data from exposed ancient turbidite deposits are generally: (1) no knowledge of the morphology of the deposit except for very local relief; in general, even levee relief is not preserved; (2) limited outcrops and structural deformation of the basin preclude the determination of the extent and shape of the turbidite deposit; and (3) the nature of the source (provenance and type, e.g. canyon, delta, fan-delta) must be deduced from the sediments themselves. The problems noted above can only be overcome in relatively well-exposed sequences through detailed mapping and stratigraphic analysis. In some cases, small turbidite systems (e.g. Cazzola and Rigazio, 1983; Cazzola *et al.*, 1985) or significant portions of large systems (e.g. Enos, 1969; Hirayama and Nakajima, 1977; Mutti *et al.*, 1978; Ricci Lucchi and Valmori, 1980; Tokuhashi, 1979) can be significantly reconstructed in their vertical and lateral facies distribution patterns. The approach is, however, extremely time consuming and it requires extensive and careful field work; this is probably the reason why detailed analyses of ancient turbidite systems are so rarely available in the literature. It should also be mentioned that the examples quoted above refer primarily to non-channelized turbidite facies, or lobes in some usage, that greatly facilitate physical correlations because of their lateral extent over significant distances. In well-exposed sequences, integration of conventional field techniques with air-photographic mapping allows a considerable improvement in both correlation patterns and in the recognition of large-scale erosional features that are commonly missed because of scale problems in outcrop studies. An example of such an integrated approach is the study of the Eocene Hecho Group basin, in south-central Pyrenees (Mutti, 1985; Mutti *et al.*, 1985), where particularly well-exposed turbidite deposits allow the recognition of several huge (extending up to 20 km in length and with widths in excess of 5 km) submarine erosional features that are generally initiated during periods of tectonic instability. The largest channel features recognizable at the outcrop scale are generally in the order of 0.5–1 km in width.

The considerable improvement of techniques for subsurface basin analysis over the past 20 years, specifically reflection-seismic profiles and well logs, has significantly contributed to a better definition of stratigraphic position, geometry, gross facies variations, and internal correlation patterns of many turbidite systems (see recent summaries in Mitchum, 1985; Tillman and Ali, 1982; and Vail *et al.*, 1977). Subsurface data have thus become increasingly important for the understanding of turbidite depositional patterns. In particular, these data partly bridge the gap that exists between the different types of data sets from modern fans and exposed turbidite sequences.

Seismic stratigraphy allows the framing of turbidite systems (*sensu stricto*) within the temporal and spatial boundaries of depositional sequences, as well as the recognition and mapping of different seismic facies within systems. These different seismic facies can allow, in their turn, the recognition and the mapping of the different phases of growth of each system (e.g. Feeley *et al.*, 1985; Mitchum, 1985).

Seismic resolution is, however, generally insufficient to establish geometry and internal correlation patterns of individual facies associations and substages (fourth- and fifth-order features of Fig. 2), which can be recognized in subsurface analysis only in cores and through well-log interpretation. Examples of such studies include Siemers *et al.* (1981), Tillman and Ali (1982), and Walker (1984b, 1985). It should be noted, however, that the reliability of facies interpretations derived from well logs largely depends upon the amount of control from core analysis. Detailed correlation patterns established from well logs suffer about the same limitations as surface correlation patterns. Subsurface correlations can be very accurate if key-beds and/or laterally persistent shaly or sandstone units can be confidently traced from well to well, as in the case of some sandstone lobe sequences (e.g. Casnedi, 1981). Conversely, in channelized deposits, and more generally in all those cases where turbidite sandstone facies form thick, homogeneous units devoid of significant shaly separations (the so-called 'blocky' character of well-log terminology), detailed correlation patterns become subjective.

OPERATIONAL PROCEDURES —APPLICATION TO MAPPING

The conceptual framework that we have developed is difficult to apply in fieldwork (on rocks or at sea) without more specific discussion of the main diagnostic features within turbidite sequences. To understand the stages of growth, the major depositional and erosional events must be accurately interpreted. We recognize five primary erosional and/or depositional *elements* that can be mapped in both ancient and modern turbidite sequences: channels and their fill, overbank deposits, lobes, channel–lobe–transition deposits and morphologic features, and major erosional features (excluding channels), especially scours and shelf-edge slump scars. Large shelf-edge failures, such as those described from the Mississippi Delta by Coleman *et al.* (1983), can be very prominent pathways for sediment to turbidite systems and ultimate source areas, but they will not be discussed in this paper.

Channels

A channel is the expression of negative relief produced by confined turbidity current flow, and represents a major, long-term pathway for sediment transport. Features scoured and filled by one, or only a few

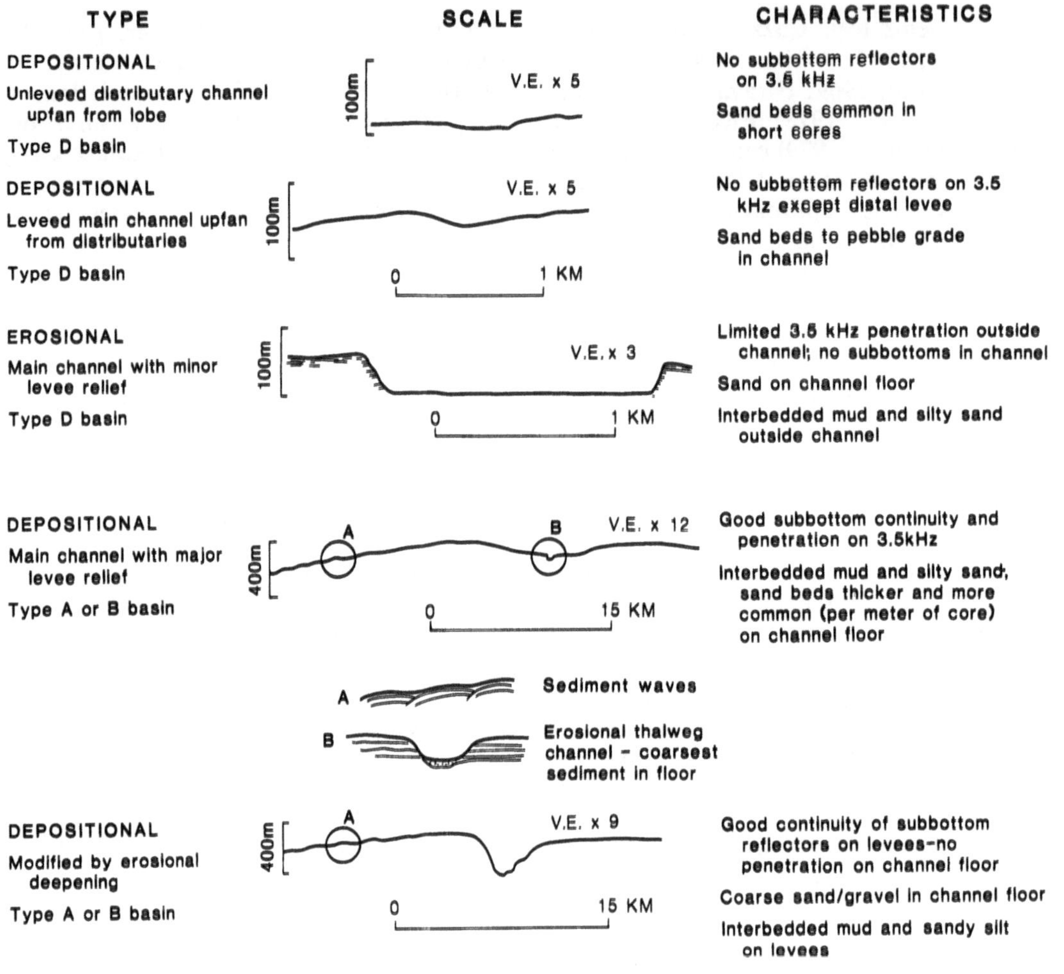

Fig. 4. Comparison of modern turbidite channels illustrating extremes in scale and geometry for depositional, erosional, and mixed channel types.

events, do not constitute channels because they do not represent a long-term conduit for sediment moving to the fan. Typically, both modern channels and ancient channel-fill sequences are associated with the coarsest sediment supplied to the system.

Modern channels

There are two main types of channels on modern submarine fans: large, leveed valleys that are the primary sediment feeder systems, and smaller, commonly unleveed channels that appear to function as distributaries (Normark, 1978). Channels on modern fans range in size from a relief of two meters (the limit of resolution using a narrow-beam, deep-tow echo-sounder) and tens of meters in width to huge features as much as a kilometer in relief and with channel floors exceeding ten kilometers in width (Fig. 4; Normark, 1978; Barnes and Normark, 1985). Our knowledge of the smaller channel systems, less than 250 m in width, is limited to only the few fans mapped with deep-tow systems or that occur in shallow water (<1000 m). By contrast, most channels described in ancient turbidites are much less than 250 m in width.

The length of channel systems on modern fans, as well as the width and depth of the larger channels (>1 km wide), are available from standard echo-sounding profiles. Both detailed deep-tow surveys and long-range, side-scan sonar profiles, however, have shown that the channel pattern (plan) may be quite sinuous and are thought to indicate meandering (Bouma *et al.*, 1985b; Damuth and Flood, 1985). Where the major channels are relatively straight or gently curving, a sinuous thalweg channel is generally present (Droz and Bellaiche, 1985; Normark, 1978). Branching distributary channel systems are commonly suggested for modern fans, but such distributary systems generally are inferred by interpolation between profile lines. Detailed deep-tow, side-scan studies suggest that true distributary systems may be rare (Normark, 1970a; Normark *et al.*, 1979). Similarly, braided channel systems have been inferred for several modern turbidite systems, but direct evidence from long-range, side-scan sonar profiles is limited (Belderson *et al.*, 1984). The flow conditions that could lead to development of meandering, braiding, or branching of submarine turbidite channels are poorly known at present.

In modern systems, channel relief can be dominantly erosional or depositional in origin or can result from a combination of both processes (Normark, 1970b). In depositional channels, relief is maintained through time under conditions of vertical aggradation of the channel floor together with overbank deposition. The greatest relief observed on modern channels is formed by a combination of deposition and erosion. Erosion can result from a number of perturbations of a fan system, including changes in sea level or in base level for deposition. In some cases, overbank deposition of fines can continue to build levee relief even though major erosion characterizes the channel floor, e.g. the Laurentian Fan (Piper *et al.*, 1985).

Although the coarsest sediment recovered from most modern fans is in the larger channels (see morphometric maps in wall chart comparison; Barnes and Normark, 1985), the lack of sufficiently long and closely spaced cores results in a substantially poorer knowledge of facies types along and across these channels, as well as of the lateral and vertical facies sequences that develop during phases of activity and abandonment. Sediments as coarse as gravels have been cored from modern channels, but cobbles cannot be sampled and sand beds thicker than a few meters are generally not completely penetrated. The relatively high amplitude and discontinuous nature of reflecting horizons underlying many modern channel floors is generally thought to indicate thick, coarse-grained sediments within the channel floor environment (Bouma *et al.*, 1985c; Normark and Gutmacher, 1985, Fig. 3; Stelting *et al.*, 1985). Thus, in some modern fans, channel sections can be relatively continuous vertical successions of coarse-grained facies and may not exhibit thinning- and fining-upward trends commonly used to infer ancient channel areas. In the Cap-Ferret Fan (Cremer *et al.*, 1985), successive sequences of coarse channel-floor sediments are overlain with finer, overbank sediments where the main channel has migrated laterally. Alternating coarse and fine intervals with a channel-fill sequence could indicate the effects of sea level on controlling sediment supply (Hess and Normark, 1976).

Ancient channels

Problems arise in understanding ancient channel-fill sequences and particularly in comparing these sequences with modern channels. From a geological standpoint, our view of the latter can be regarded as essentially instantaneous as they are 'empty' features with samples limited to the most recent events. Ancient channel-fill sequences commonly display a complex internal stratigraphy recording channel evolution with time that is expressed by repeated periods of erosion and infilling (e.g. Hein, 1982; Johnson and Walker, 1979; Kelling and Woollands, 1969; Piper, 1970; Walker, 1975). In addition, the depth, (and particularly) width, and length of these ancient channels are generally difficult to determine because of the common lack of sufficiently large exposures, or detailed correlation patterns between discontinuous exposures.

A further problem in understanding ancient channel-fill sequences is scale. The term 'channel' has been used in the literature in a very loose sense to denote features that range from small scours at the base of individual sandstone beds up to sequences of beds with an aggregate thickness of as much as several hundreds of meters that have filled in large-scale submarine erosional features.

The term 'channel-fill deposit' will be used herein to denote sediments that filled a relatively long-lived feature with common erosional characters that served as a pathway for sediment transport. Ancient channel-fill deposits may display a great variety of facies types, internal structures, and geometries; nevertheless, these deposits will typically consist of sequences of beds and/or erosional events that testify to the relatively long-lived character of the original features. True ancient channel-fill sequences occur as composite features resulting from the vertical stacking and the lateral juxtaposition of several individual channel-fill units. Such complexes rarely exceed thicknesses of a hundred meters and widths of one to two kilometers.

Ancient channel-fill sequences are subdivided into three major categories: (1) erosional, (2) depositional, and (3) mixed. The general characterization of these types of channel fill is given in Fig. 5.

Erosional channel-fill sequences are characterized by coarse-grained and scoured facies that have a sharp basal contact with an erosional surface in the channel axis. These sequences are typically composed

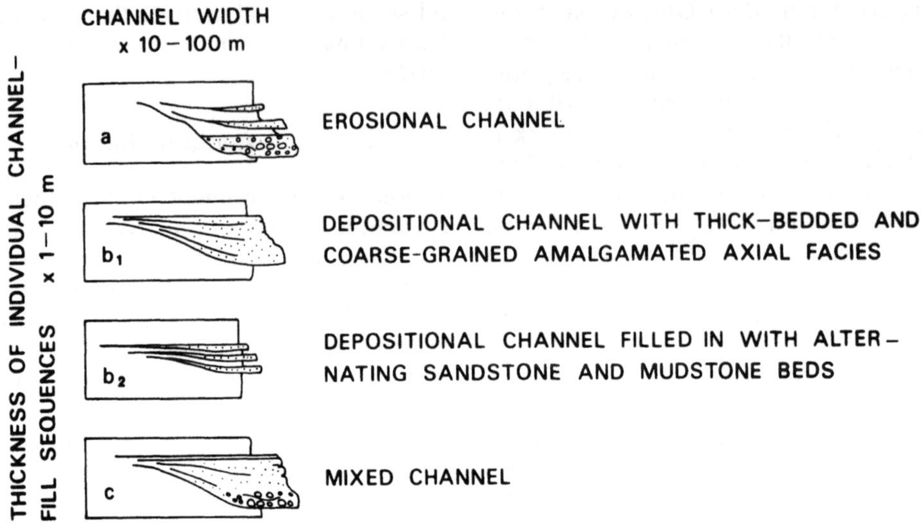

Fig. 5. Main types of channel-fill deposits observed in ancient turbidites. Vertical and horizontal scales range from several to tens of meters.

of lenticular, clast-supported conglomeratic units that are locally associated with debris flow deposits, graded pebbly mudstones, and (less commonly) crudely horizontal and/or cross-stratified alternations of coarse sandstone and granule-to-pebble conglomerate. Examples of this type of channel-fill sediment are known from many systems (e.g. Carter and Norris, 1977; Clifton, 1984; Hein and Walker, 1982; Mutti *et al.*, 1985).

The clast-supported and scoured conglomerate units suggest deposition from debris flows that evolved through entrainment of water into concentrated turbidity currents that transport substantial sediment into the basin downslope. The mud-supported conglomerates are thought to be deposited by debris flows that became frozen during down-channel movement because of insufficient flow thickness and/or gradient within the channel. Field evidence indicates that no, or volumetrically minor, overbank deposits are associated with these coarse-grained, channel-fill facies.

Depositional channel-fill sequences consist of thick, extensively scoured and amalgamated graded sandstone and pebbly sandstone beds in the axial part of broad (~100 m to 1 km) and commonly relatively shallow (5–50 m) channel features. These thick-bedded and coarse-grained axial facies are replaced toward the channel margins by progressively thinner-bedded and finer-grained deposits that are separated by mudstone partings. These finer-grained units, in turn, pass into channel-margin overbank facies (see overbank discussion). Characteristically, the bedding surfaces of depositional channel-fill sequences converge toward the channel margins, thus forming beds that pinch out in both directions normal to the sediment flow (Fig. 6). Descriptions of channel-fill sequences of this type are common in the existing literature (e.g. Mutti, 1977; Pickering, 1982a; Ricci Lucchi, 1969; Walker, 1966, 1975).

We interpret this type of channel-fill deposit to represent a stage of deposition during which many of the turbidity currents deposit much of their sediment load directly in the channel system rather than farther basin-ward (see later).

Mixed channel-fill deposits are characteristically composed of facies sequences that record an erosional stage expressed by coarse-grained, residual conglomeratic beds followed by a period of sand deposition. This sequence can occur singly or be repeated several times within the same channel-fill succession. Examples of this type of channel-fill deposits are well documented as well and include studies by Carter and Lindquist (1975), Cazzola *et al.* (1981), Dupuy *et al.* (1963), Mutti (1969), and Walker (1985).

All three types of channel-fill sequences pass upward into either of two general sediment successions. In one case, there is an abrupt transition to mudstone facies, probably resulting from a sudden deactivation of the channel. In the second case, which is more common, the channel-fill sequences are overlain by thin-bedded and fine-grained deposits associated with small channel-fill sequences developed within the thalweg of the main channel. Sandstone beds within some of these small-scale channel features, which are commonly a few tens of meters

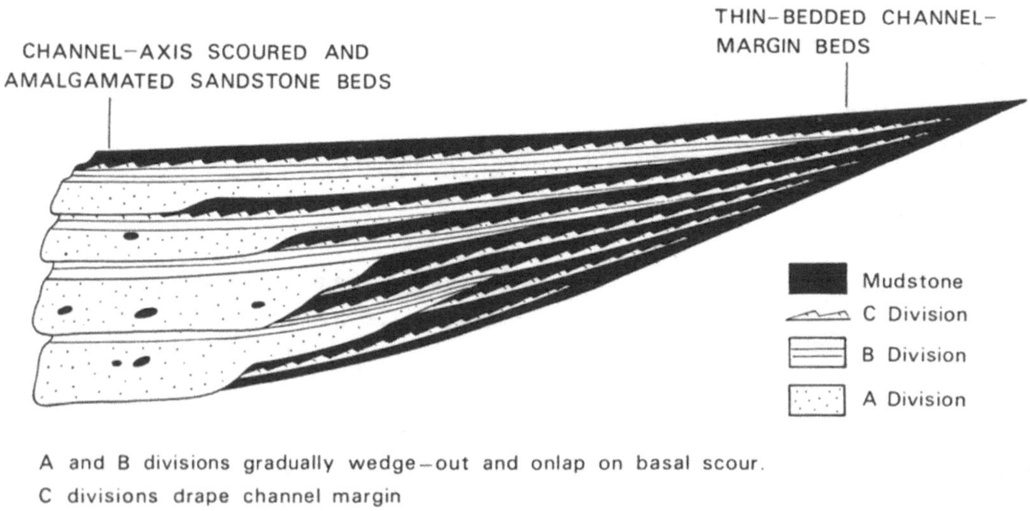

Fig. 6. Lateral relation between channel-axis and channel-margin deposits within a depositional channel. The greater number of thin-bedded units that characterize channel margins (relative to channel axis) apparently results from a higher potential for preservation of the thin beds on the margin.

wide and only a few meters deep, locally display distinct lateral accretion surfaces, indicating a meandering course (Mutti *et al.*, 1985, p. 562). This second case reflects gradual abandonment of the channel and a concomitant decrease in the volume of the flows in the channel.

Figure 7 reviews the characteristic features of channel-fill deposits that are considered diagnostic in recognition of this element of turbidite systems.

Scours

The term 'scour' is used to denote isolated, roughly equidimensional cut-and-fill features where the erosion and subsequent depositional fill are most commonly produced by the same flow. Scours vary considerably in size and geometry, ranging from small-scale flute casts to large and relatively deep holes that are often misinterpreted in field mapping because of the limited size of the exposures. In addition, when viewed in single section only, large scours may be confused with channels. More generally, scours are found as relatively small and deeply incised features at the base of sandstone and conglomerate beds within channel-fill sequences as well as broad and generally shallow erosion at the base of sandstone beds in depositional lobes (see below). In both cases, scours are typically shallower than three meters and are generally expressed by amalgamation surfaces (Fig. 8). Very commonly, and particularly in lobe sandstones, these features are associated with an abundance of outside rip-up shale clasts. The type of substratum can obviously greatly affect the geometry and characteristics of scour features.

Scours on modern fans range in size from those that can be observed in bottom photographs (several meters across, i.e. outcrop scale) to some that can be resolved by narrow-beam echo-sounding and side-scanning sonar techniques (Fig. 9). The largest scour observed on a deep-tow, side-scan record is about one kilometer wide and is floored by an extensive field of large-scale current ripples or small dunes (Hess and Normark, 1976). Regardless of their size, scour depressions on modern fans are found in areas where turbidity currents exit the channels; specifically (1) where the levee relief decreases abruptly near channel terminations and (2) where flow-stripping (Piper and Normark, 1983) most commonly occurs. Both settings are areas where the larger turbidity currents will undergo a hydraulic jump as the slope decreases and the currents also spread laterally. Scours 2–3 m deep and 10–60 m wide have been identified on sediments downslope of a delta front in Howe Sound, British Columbia (D. B. Prior, personal communication, May 1986; see Fig. 2 in Prior and Bornhold, 1986). These scours occur near a major break in slope, a highly probable area for hydraulic jumps in density currents moving down the delta front.

Filled scours have not yet been described from modern fans, probably because many are filled by coarse sediment, and as discussed earlier, deep-tow,

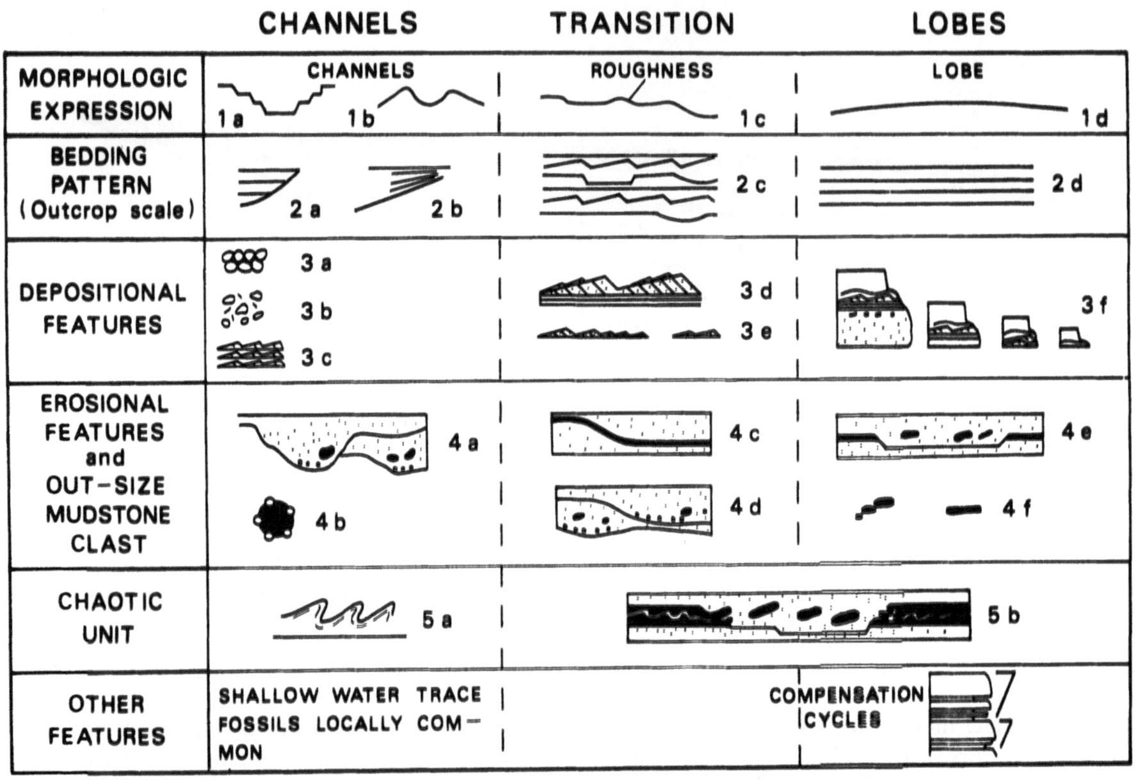

Fig. 7. Main characteristic features of channel, channel–lobe transition, and lobe deposits. 1a, erosional channel; 1b, depositional channel; 1c, zone of roughness; 1d, lobe relief; 2a, beds truncate against channel margin; 2b, beds converge toward channel edge; 2c, bedding irregularity resulting from scours and large-scale bedforms; 2d, even-parallel bedding pattern; 3a, clast-supported conglomerates (residual facies); 3b, mud-supported conglomerates (debris-flow deposits); 3c, thin-bedded overbank deposits; 3d, 3e, coarse-grained, internally stratified sandstone facies (see Fig. 15 for details); 3f, complete and base-missing Bouma sequences; 4a, deep and relatively narrow scours locally associated with stone clasts; 4b, armored mudstone clasts; 4c, mud-draped scours; 4d, broad scours locally associated with mudstone clasts; 4e, tabular scours invariably associated with mudstone rip-up clasts from underlying substratum; 4f, nests of mudstone clasts commonly showing inverse grading and 'take-off' attitude of individual clasts; 5a, slump units; 5b, impact features.

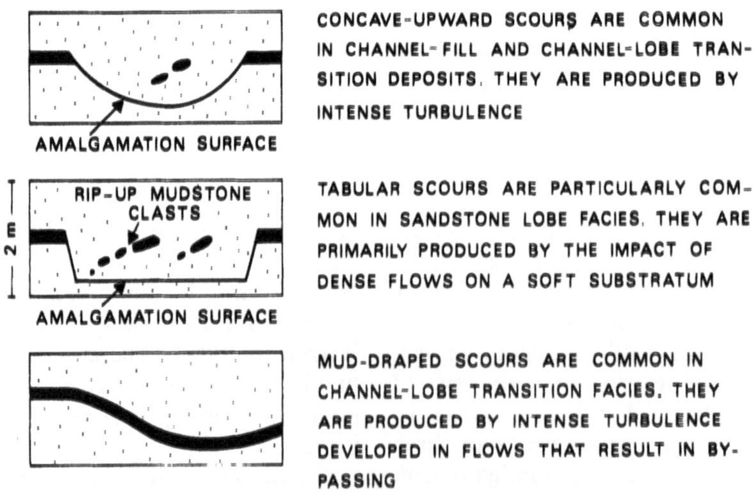

Fig. 8. Main types of scours observed in ancient turbidite beds at outcrop scale.

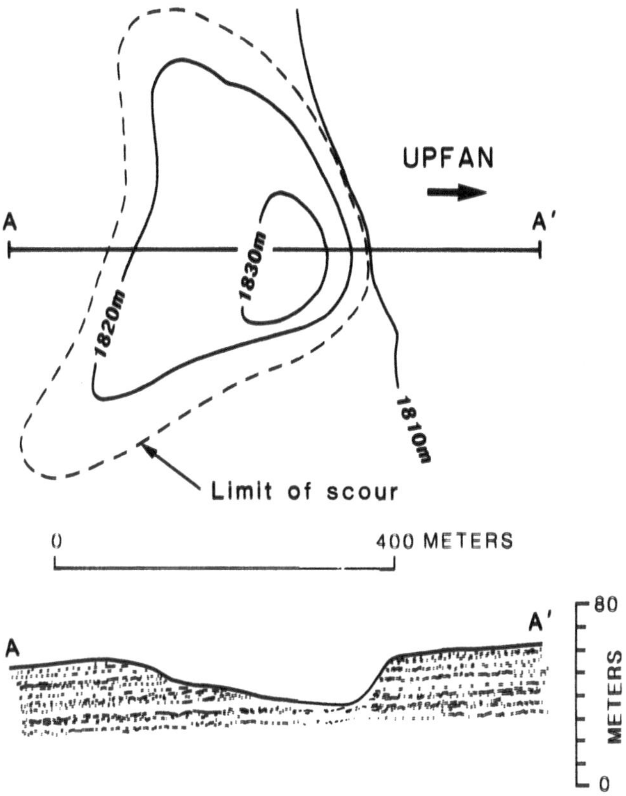

Fig. 9. Plan view and section of large scour from modern fan (Navy). Interpretive lithology based on adjacent piston core (from Normark *et al.*, 1979).

high-resolution reflection profiling cannot resolve bedding in thick sand beds, and core samples do not penetrate thick sand beds.

The term 'mud-draped scour' defines features that are very similar to normal scours in scale and geometry but show evidence that they were filled by later flows, commonly after draping of the erosional surface with mud (Fig. 8). The existence of unfilled scours on modern fans suggests that many such features will be draped by hemipelagic mud before being filled by later turbidity currents. Mud-draped scours, which clearly indicate sediment bypass for the flows that cut them, are particularly common in channel–lobe transition deposits (see below).

Overbank deposits

Overbank sediments are generally fine-grained, thin-bedded, and current-laminated deposits that result from lateral spreading from the main body of a confined turbidity current. Thus, graded mudstone units are generally volumetrically predominant. As such, overbank sediments can occur in many parts of a turbidite depositional system; there is, however, a general tendency to use the terms 'levee' and 'overbank' interchangeably. Persistent overbank deposition along the margins of an active channel can result in the construction of positive relief that is properly called a levee. Because 'levee' is a morphologic term and because the morphologic expression of a levee is rarely preserved in ancient turbidite systems, the more general term 'overbank deposits' is preferred for ancient deposits. Many smaller channels on modern fans are without recognizable levees, yet overbank deposition dominates on many areas of the fan away from the channels. Small turbidity currents can even deposit an overbank-sediment facies within a channel (related to the thalweg; e.g. Clifton, 1981).

Modern overbank deposits

On modern fans dominated by leveed-valley systems, much of the fan will consist of overbank deposits. On all fans, most of the fan surface upslope from the lobe deposits, with the exception of the channel floors, is built by overbank flows; thus, overbank deposits can form more than 50% of the fan area on some large deep-sea systems (see Barnes and Normark, 1985). The width of overbank areas underlain by demonstrable levee relief on modern fans tends to increase as the total fan area increases (Normark, 1978). On the largest fan complexes, active and buried levee-relief features commonly reach 25–40 km in width; levee–channel complexes can exceed 75 km where the fan width exceeds several hundred kilometers (Damuth and Flood, 1985; Emmel and Curray, 1985; McHargue and Webb, 1986).

Sediment waves are developed on many of the large levees on the sides of the channel where the Coriolis force causes preferential overflow (right-hand side looking down-channel in the Northern hemisphere). Wavelengths of these asymmetric bedforms commonly range from 0.25 to 2 km and wave heights up to 40 m (Normark *et al.*, 1980). These bedforms migrate upslope during growth of the levees. Sediment waves, even at a much smaller scale, do not appear to be as common on smaller fan systems.

Neither the relief of the sediment waves nor the levees themselves are likely to be recognizable in ancient turbidites. The lateral changes in thickness of individual levee beds that are only a centimeter to several tens of centimeters thick amount to a few tens of meters of cumulative difference in thickness over horizontal distances of one to ten (or more) kilometers. The scale of these features thus precludes recognition in most outcrops. The recognition of levee areas and sediment waves in ancient overbank

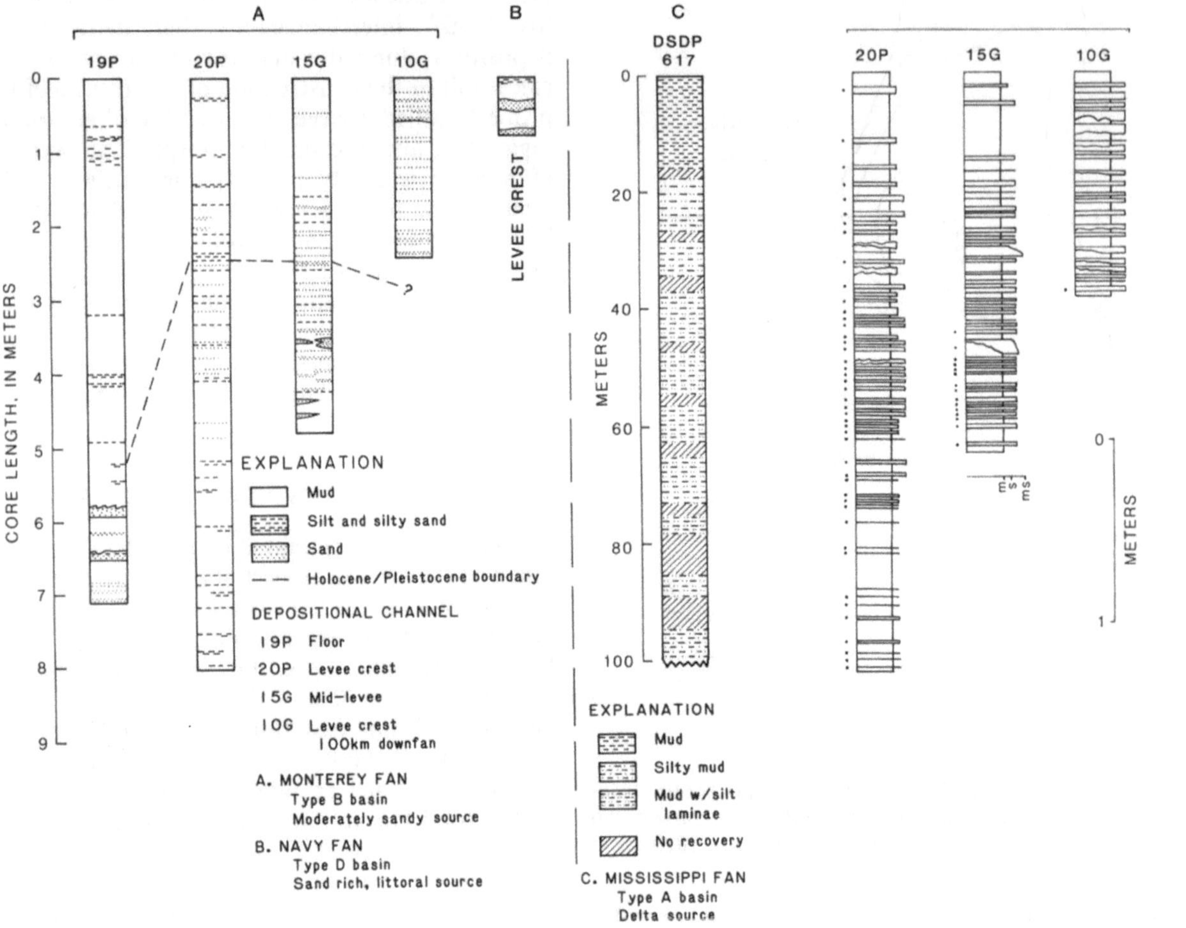

Fig. 10. Lithologic columns from overbank (levee) deposits on modern fans. Mississippi Fan section taken from Normark and Meyer (1986). The right part of the figure shows the Monterey Fan levee cores reconstructed for the effects of compaction of the muddy intervals with an effective decrease in porosity from 75% to 25%, which is common for shale. The dots along the left side of each core section indicate beds that are discontinuous within the 8-cm-wide core.

turbidites will thus probably depend on diagnostic characters (yet to be determined) within the sediments, especially the coarser beds, rather than on bedding geometries.

Samples from modern turbidite–channel levees are a bit more representative of depositional conditions than samples from channels and scours. Interbedded mud, silt, and silty sand units are common (Fig. 10). The sand units are thin, generally less than 10 cm thick, and commonly are discontinuous or show marked wedging within the narrow core sections. The DSDP cores from the Mississippi Fan provide the greatest vertical sequence for modern overbank deposits, but generally lack the sandy intervals cored on other systems (Fig. 10).

The longer cores that are recovered from modern overbank areas reflect both thinner (more easily penetrated) sand and silt beds and more mud than from adjacent channels. The greater amount of mud in the overbank cores, however, means that direct comparison of modern core sections with lithologic sections from ancient deposits can be misleading unless the effects of compaction are considered. The three Monterey Fan levee cores in Fig. 10 are reconstructed in the right part of the figure to show the effects of compaction and dewatering. In addition, the cores are shown in the same format as lithologic sections used for ancient turbidites. Many levee silt and fine-sand beds are discontinuous and have irregular upper and/or lower contacts.

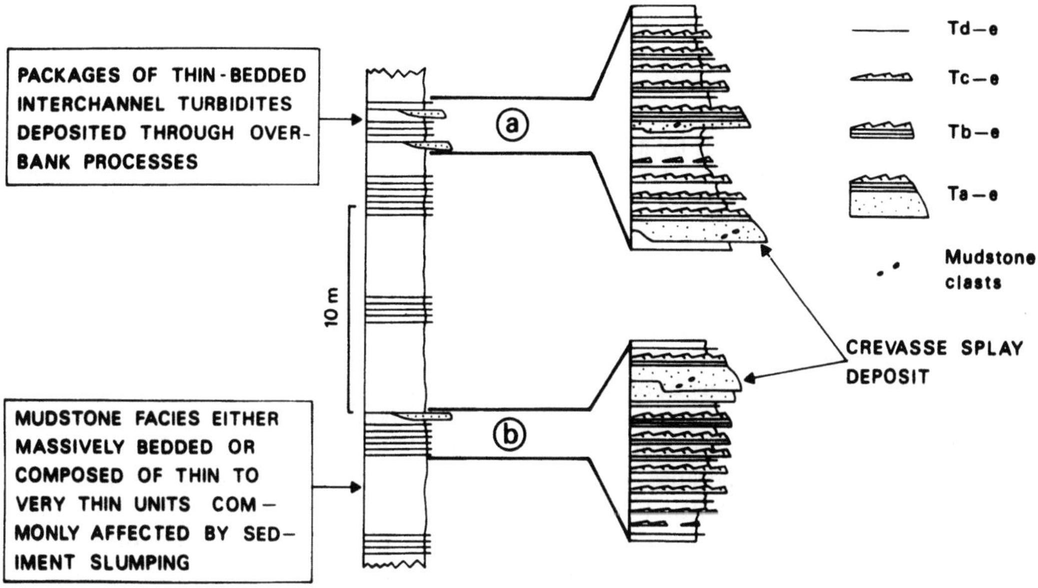

Fig. 11. Interchannel overbank thin-bedded turbidites. Both thickening- and coarsening-upward (lower part of diagram) and thinning- and fining-upward (upper part of diagram) trends are depicted, suggesting increased and decreased channel activity, respectively.

Ancient overbank deposits

Ancient overbank sediments are generally associated with a variety of channel features as well as slump-generated relief and associated deposits. Obviously, such a broad definition implies that this facies assemblage is poorly defined and not well understood. We subdivide these sediments into two main groups: (1) channel-related overbank deposits and (2) overbank wedges. The first group is one of the major elements of a turbidite system; the second group, as reflected by both scale and facies relations, is marginal to the main turbidite deposit and is more readily discussed in terms of a stage within a system (see later).

Channel-related overbank deposits consist of alternating beds of mudstone and extensively current-laminated, fine-grained sandstone. These overbank facies have been sufficiently described in several papers (see Clifton, 1981; Mutti, 1977; Pickering, 1982a; Walker, 1966, 1984a), and they occur within channels (related to the thalweg), along the edges of channels, and within interchannel areas in association with broad and shallow channelized bodies thought to be crevasse deposits (see facies details and suggested terminology in Mutti, 1977). Some of the characteristics and vertical facies sequences shown by these sediments are depicted in Fig. 11.

Mapping within the Hecho Group turbidite system in northern Spain suggests that channel-related overbank deposits are not volumetrically important during all stages within the turbidite system but primarily form during the stage in which sand is deposited within channels and/or in areas immediately beyond channel mouths. During this stage, relatively large-volume turbidity currents that mainly deposit their sediment load in the distal part of the system are recorded in both channel margin and interchannel areas. Reduction in flow volume produces a 'backstepping' of sand deposition confined within channels in the proximal part of the system; in this stage, overbank beds on channel margins can be traced into thicker and coarser beds in the channel axis. Further deactivation of the system, with progressively smaller turbidity currents, produces overbank deposits and minor associated channel-fill facies, both restricted within the main channel.

Overbank wedges are bodies of fine-grained sediment that form distinct depositional stages within the Hecho Group turbidite system (Fig. 3). These wedge-shaped bodies are as much as several hundred meters thick near the margin of the basin and gradually wedge out (basinwards) over distances of tens of kilometers into thin mudstone units. Stratigraphic relations indicate that these wedges of fine-grained turbidites are associated with active fluvio-deltaic deposition on the adjacent shelf (Mutti *et al.*, 1985).

Overbank wedges are primarily composed of thin

and subtly graded mudstone beds that alternate with discontinuous thin beds or packets of beds of fine-grained, and generally extensively current-laminated, sandstone. The bedding pattern in these sediments is extremely irregular because they are cut by abundant slump scars, and contain slump and related chaotic deposits, and a variety of channelized features. Channel-fill sediment, which typically occurs as sand-filled depositional channels only a few meters thick and generally less than a hundred meters wide is, however, a volumetrically minor component of overbank wedges. It is unlikely that these minor channels are the major pathways for the bulk of sediment in the overbank wedges.

The overbank wedges were tentatively described as channel-levee deposits mainly because of their internal geometric unconformities (Mutti, 1985), but these deposits are probably equivalent to the slope-front-fill deposits of the seismostratigraphic models of Vail *et al.* (1977) and, more generally, to slope drapes associated with active deltaic deposition on adjacent shelf areas (Pickering, 1982b). Deposition of overbank wedge sediments involves not only the obvious channel overflow processes but also a variety of slope and shelf-edge processes, including dilute, open-slope turbidity currents, surficial slumping, and sediment creep, all related to sediment instability of adjacent deltaic margins (see Coleman *et al.*, 1983).

Lobes

In general, the term 'lobe' does not have the same connotation for modern fans that it has for ancient turbidites. (1) Lobes defined for modern fans are based more on morphologic characters than on sedimentary features. (2) Modern fan lobes commonly have channels. (3) The relatively tabular and laterally extensive character of sand beds observed in ancient lobes cannot be demonstrated for modern lobes because of the lack of penetration by coring devices and by high-resolution reflection-profiling systems in these sandy sequences.

Modern lobes

On modern fans, the term 'lobe' has been used to describe a wide range, in both scale and origin, of depositional features that constitute areas of predominantly sandy sediments (Fig. 12). On Navy Fan, Normark *et al.* (1979) describe small, sandy lobes that are slightly convex-upward areas, one to two kilometers in width, with smooth, channel-less surfaces. These small lobes are fed by short distributary channels and occur on the upper 'suprafan' area. The thickest sand beds (2 m) from Navy Fan were on the lobe areas.

The suprafan feature described for several modern sand-rich fans was defined as a convex-upward bulge on the fan surface that is channelized on the inner part and smooth on the outer part (Normark, 1970a). The suprafan morphologic feature is an area of sand deposition at the mouth of the major channel crossing the upper fan and was interpreted to be a prograding delta-like body. The entire suprafan feature has been termed a lobe deposit (Normark, 1970a, 1978; Walker, 1978) that encompasses areas as much as 10–20 km across.

Large, gently sloping areas on several modern fans that are characterized by relatively sandy sediment have been identified as depositional lobes; these sandy areas on modern fans, however, are associated with leveed and unleveed, relatively widely spaced channel systems that interrupt the otherwise smooth surface of the fan. These sandy 'lobes' have widths of more than 100 km (Fig. 12).

The term 'lobe' has also been used to describe both topographic relief and seismically defined intervals on modern fans that do not represent areas of predominantly sandy deposition. Delgada Fan has a prominent south-westerly extending bulge that dominates the fan morphology. This feature is tentatively interpreted as a recently abandoned leveed valley complex; but because it formed as a prograding, convex-upward topographic feature it has been loosely referred to as a type of lobe (Normark and Gutmacher, 1985). A similar term, 'fanlobe' (Bouma *et al.*, 1985b), has been used to define system-wide sequences of sediment on the Mississippi Fan that generally encompass a major leveed-channel complex and that are defined on the basis of seismic reflectors that are correlated across the basin. A fanlobe represents an entire depositional system that includes all of the basic elements we describe in this section.

The purpose of this brief review of lobe features from modern fans is to emphasize that there is no standard usage and no limit to the scale of features that might be termed lobes. Whenever the term 'lobe' is used in a publication, one must carefully check to see exactly how the author uses the term. Otherwise, comparisons are useless.

Ancient lobes

In ancient turbidite sequences, sandstone lobes were first recognized and defined from the Miocene San

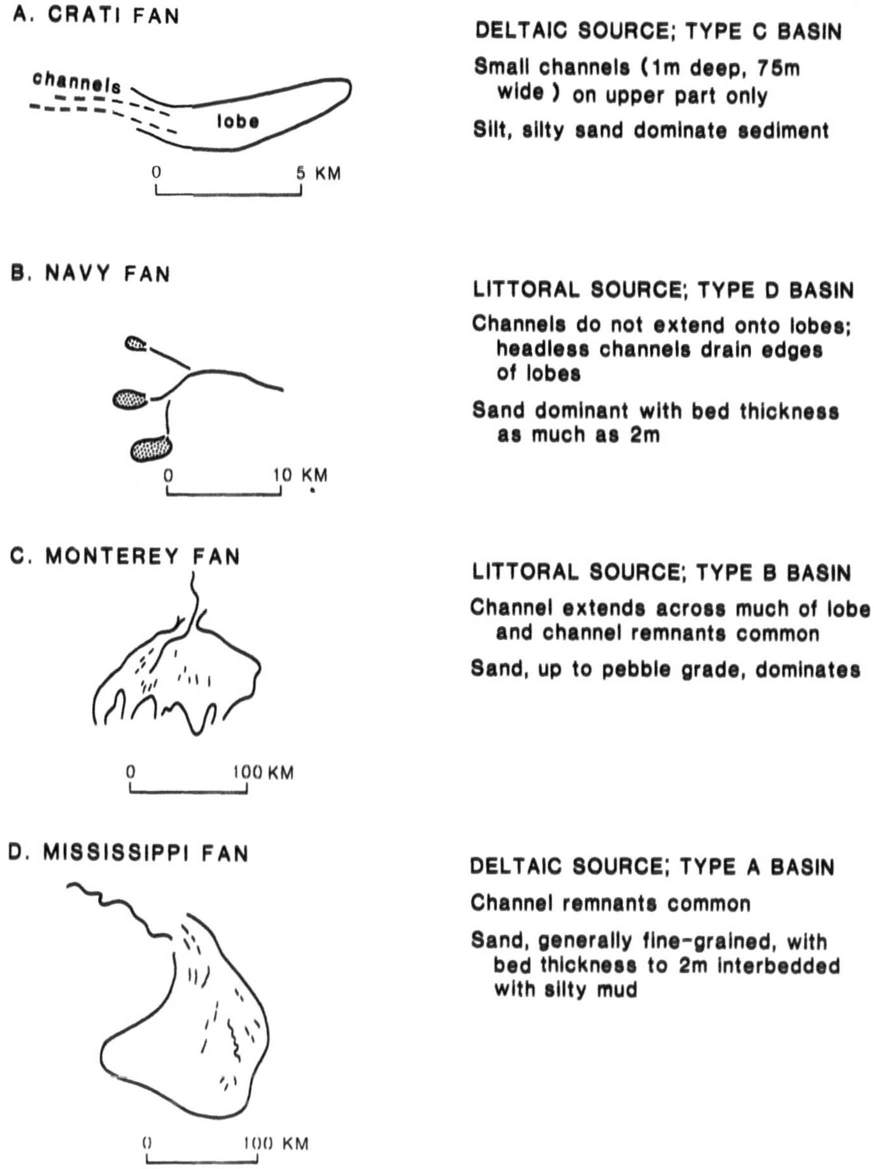

Fig. 12. Comparison of size and type of lobes suggested for modern fans (adapted from Barnes and Normark, 1985).

Salvatore Sandstone in the Bobbio tectonic window of the northern Appenines (Mutti and Ghibaudo, 1972). This depositional element was then incorporated into the facies model of Mutti and Ricci Lucchi (1972). Sandstone lobes were originally defined as non-channelized bodies that ranged from 3 to 90 m in thickness and were composed of thick-bedded sandstone alternating with thinner-bedded and finer-grained interlobe facies. The association of lobe and interlobe deposits was interpreted to be diagnostic of an outer-fan depositional environment. Each lobe was considered to be a progradational feature analogous to delta-front channel-mouth bars. Subsequent surface and subsurface studies emphasized the tabular geometry and great lateral extent of sandstone lobes as well as the lack of channelized deposits commonly associated with inner and middle fan environments (Casnedi, 1981; Casnedi *et al.*, 1978; Mutti *et al.*, 1978; Pickering, 1981; Ricci Lucchi and Pignone, 1978; Ricci Lucchi and Valmori, 1980; Van Vliet, 1978).

Work in Greece (Piper *et al.*, 1978), California

(Link and Nilsen, 1980), and northern Italy (Mutti, 1979) led to the recognition of a second type of non-channelized sandstone lobe. This type is of generally smaller areal extent and is characterized by the ubiquitous association, both vertically and laterally, with channelized or extensively scoured, coarse-grained facies. This sandstone-lobe type was thought to be equivalent to the smooth, outer part of a suprafan deposit as described for modern turbidite fans (Normark, 1970a).

The growing evidence for two types of sandstone lobes also showed that they are associated with basically different turbidite systems, which were termed sand-rich, poorly efficient and mixed-sediment, highly efficient types (Mutti, 1979; Nilsen, 1980). In sand-rich systems the sand lobes are channel-attached features, while in the mixed-sediment systems the sand lobes are primarily channel-detached and were thought to be sheet-like bodies deposited in the flatter, distal fan regions.

Walker (1980) and, more recently, Shanmugam and Moiola (1985) questioned the distinction between the two lobe types. Detailed correlation patterns within a number of ancient turbidite systems, however, clearly supports such a distinction for both subsurface (Casnedi, 1981) and surface (e.g. Cazzola *et al.*, 1981, 1985; Cazzola and Rigazio, 1983; Mutti *et al.*, 1978; Remacha, 1983; Ricci Lucchi and Valmori, 1980) examples. Therefore, we accept the distinction between the two types of sandstone lobes as part of our conceptual framework for comparing turbidite systems.

Regardless of type, as discussed in this paper, lobe deposits are considered as a depositional element composed of non-channelized sandstone bodies that occur downstream of channel-fill sediment (and/or the fill of major submarine erosional features) and of the channel–lobe transition (see below).

Lobe deposits are easy to recognize in the field if exposures are large enough to allow the recognition of the non-channelized geometry of the sandstone packages. In most cases, however, the size of the exposures does not permit direct observation of lobe geometry, which must therefore be inferred from facies characteristics and vertical sequence analysis. Detailed correlation patterns based on carefully measured sections still provide the most reliable evidence for lobe geometry. In subsurface studies, seismic reflection profiles and electric-log correlation patterns can provide similar information. In some cases, lobe identification can be difficult, and it is possible to confuse a lobe sequence with that found in a depositional channel (e.g. Mutti, in preparation; Walker, 1980).

The main characteristics of ancient sandstone lobe deposits include the following:

1. Lobes are packages bounded by even, parallel surfaces that consist primarily of thick and relatively coarse-grained sandstone beds. Lobes are commonly between 3 and 15 m thick and, except for shallow scours, the beds are generally parallel-sided even at the scale of large exposures.
2. Individual sandstone lobes occur as isolated bodies within mudstone sequences or, more commonly, form vertical successions of several hundred meters where the sandstone bodies alternate with variable thicknesses of mudstone interbedded with finer-grained and thinner-bedded sandstone. The down-current (basinward) extent of individual sandstone lobes varies from a few kilometers (Cazzola and Rigazio, 1983) to several tens of kilometers (Casnedi *et al.*, 1978; Remacha, 1983; Ricci Lucchi and Valmori, 1980).
3. Sandstone lobes can be basin-wide features that exhibit abrupt onlap terminations against adjacent slope facies (e.g. Cazzola *et al.*, 1981, 1985) or that are broadly convex-upward bodies wedging gradually into thinner-bedded lobe-fringe deposits in a down-current direction (e.g. Cazzola and Rigazio, 1983; Estrada Aliberas, 1982; Ricci Lucchi and Valmori, 1980).
4. The internal structures of the thick sandstone beds that comprise individual lobe units reflect the grain-size distribution of sediment from the source area as well as basin size and configuration. Within any given system, however, the lobe sediments generally display better grading and better development of vertical sequences of internal sedimentary structures (Bouma sequences) than other sandy facies within the turbidite system. Figure 13 illustrates an idealized example of the down-current changes in sandstone bed thickness and internal structure within typical lobe facies. Complete, classical Bouma sequences are found in lobes where sufficient fines and sand are present (Facies C_1 and C_2 of Mutti and Ricci Lucchi, 1972, 1975). A lesser proportion of fine-grained material in the turbidity currents results in sandstone lobes in which coarse-grained, graded, and internally unstratified divisions of the Bouma sequence become dominant, and water-escape structures are common (Facies B_1 of Mutti and Ricci Lucchi, 1972, 1975).
5. Sandstone lobe packages commonly display superposed, small-scale thickening-upward sequences, each composed of a limited number (generally less than ten) of sandstone beds. These small-scale sequences were termed 'compensation cycles' by Mutti and Sonnino (1981) and are thought

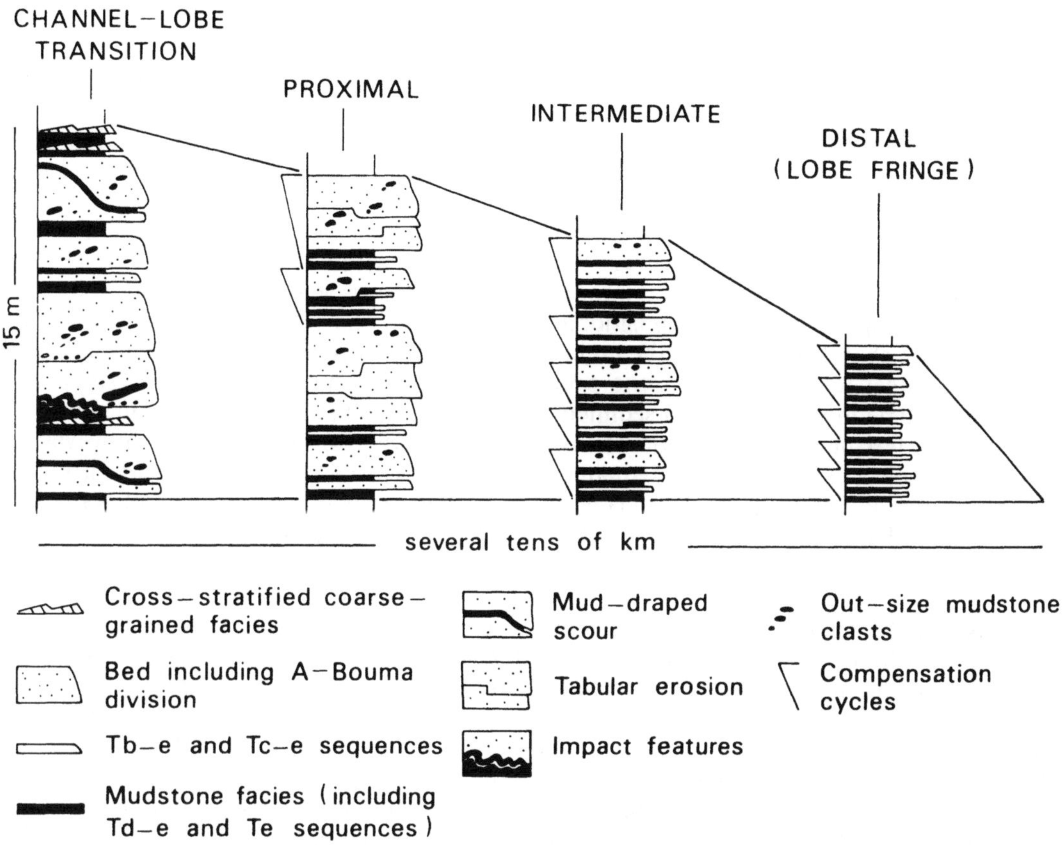

Fig. 13. Main down-current facies changes within an idealized individual sandstone lobe package based primarily on observations from the Eocene Hecho Group turbidite complex of the south-central Pyrenees.

to result from the progressive smoothing of subtle depositional relief produced during the upbuilding of each lobe.

6. Scours are common in many typical lobe deposits but they are neither as deep nor as abundant as those observed in the channel–lobe transition area. Scours in sandstone lobe packages are generally cut into mudstone beds and have sharp, step-like margins and flat floors (Fig. 8). As a result, many individual sandstone beds within lobe packages are amalgamated and contain an abundance of rip-up mud clasts that develop during the same depositional event (Mutti and Nilsen, 1981). The origin of these planar, or sheet-like, erosional features is apparently not related to the intense turbulence developed in the channel–lobe transition zone (see below), which is generally found a considerable distance up-current. The planar scours are apparently diagnostic of sandstone lobe deposits, and we follow Marschalko (1970) and Van Vliet (1978) in feeling that these features form by impact of the denser, lower part of a turbidity current on a relatively soft, muddy sea floor prior to deposition of the sand. Chaotic impact features that can occur with thin-bedded mud and sand units within a lobe sequence may be locally common but should not be misidentified as slump features. In other words, such chaotic features do not indicate slope conditions, as these sediments are deposited on virtually flat surfaces.

In any given turbidite system the lobe deposits generally represent the maximum extent for sand transported to the basin. Within some systems, especially those in tectonically active, confined settings and with abundant sediment supply, exceptionally large (mega-) turbidity currents can transport much of the sand component across the lobe area and deposit thick and laterally extensive beds in adjacent, ponded basin-plain areas. Documented examples include those from the Eocene Hecho Group (Mutti and Johns, 1978) and the Miocene Maranoso-

Arenacea in the Appenines (Ricci Lucchi and Valmori, 1980).

Channel–lobe transition

The relation between channels and lobes in both modern and ancient turbidite systems has received relatively little attention in previous work. We believe, conversely, that this relation is very important to an understanding of turbidite depositional patterns. Therefore, we shall briefly review the major character of the channel–lobe transition zone and attempt to highlight some morphologic and sedimentary features that we feel may be diagnostic of this zone. Because this subject has not been adequately discussed in the existing literature, this section is necessarily more detailed than those reviewing the other basic elements that we propose be used to map turbidite systems.

In a broad sense, the channel–lobe transition is the region that, within any turbidite system, separates well-defined channels or channel-fill deposits from well-defined lobes or lobe facies. As such, this region and its associated sediments can exhibit some of the characteristics of both the channels and lobes, with which they intervene. In our opinion, however, the characteristic features of this region are primarily related to changes that occur when turbidity currents undergo a hydraulic jump or other rapid changes in flow conditions. The hydraulic jump occurs where the flows pass from relatively high-gradient and channelized conditions into flatter and smoother, and locally unchannelized, regions of the fan. Analyses to date on hydraulic jumps in turbidity current flow (Komar, 1973; Menard, 1964; Middleton, 1970; Ravenne and Beghin, 1983) suggest that the jump is accompanied by: (1) dissipation of energy through increased internal turbulence; (2) enlargement and dilution of the flow; and (3) a marked increase in erosion of the sea floor as a result of the turbulence.

The location of and sedimentological effects produced by the flow transformation at the site of the hydraulic jump depends upon the composition and size of the turbidity current itself. The following analyses assume that the channel–lobe transition zone is in depositional 'equilibrium' with the medium-to-large turbidity currents, which probably undergo the hydraulic jump within this zone. For turbidity currents of the same general characteristics (grain-size distribution, grain composition, source area, mechanism of generation, etc.), the smaller volume flows would tend to undergo the hydraulic jump farther upstream within the canyon/channel system; likewise, the very large flows might not undergo the jump until reaching the basin plain area. Thus, the position of the hydraulic jump within the turbidite system is not only a function of change in gradient but is also affected by the relation between flow volume with distance along the longitudinal profile. The smaller flows, which are more commonly contained within the canyon/channel system, are probably more sensitive to changes in gradient for initiation of the hydraulic jump. Previous work on hydraulic jumps has generally emphasized that the jump would occur at the break in slope between the canyon (or delta slope) and the turbidite system (see above references). Recent work, however, has determined that a relatively recent turbidity current that deposited sediment across much of Navy Fan did not undergo the jump until near the termination of the levees of the main channel (Bowen *et al.*, 1984).

We discuss two general cases for features developed within the channel–lobe transition zone: (1) for turbidity currents that are composed dominantly of sand and coarser sediment, and (2) for those currents that transport sand together with a substantial muddy component (Fig. 14). When the relatively sandy turbidity currents undergo the hydraulic jump, the rapid dilution resulting from increased turbulence reduces the competence of the flow and leads to rapid deposition of the coarser material as well as the formation of large-scale scour features. The lack of any substantial muddy component in the flow results in little deposition of fines and an insufficient muddy component to maintain a driving force for the current now moving over a gentler slope. Thus, most of the sediment is rapidly deposited as a wedge that downlaps downfan from the channel mouth and that includes the region of intense scouring (Fig. 14). This wedge-shaped deposit is the 'suprafan' type of lobe (Normark, 1970a), and the flow regime has been termed 'poorly efficient' because most of the sand is deposited close to the feeding channel (Mutti, 1979).

When turbidity currents that have a substantial mud component undergo the hydraulic jump, the turbulent mixing provides a substantial flow thickness with dispersal of the fines, and a flow surge will keep most of the sediment in motion (Ravenne and Beghin, 1983). Scouring of the sea floor is generally less intense than for the sand-rich turbidity currents. Some deposition of the coarsest sediment does occur but, in general, most of the sediment bypasses the zone of the hydraulic jump and is carried farther downfan to lobes that are essentially detached from the channel (Fig. 14). This system has been termed 'highly efficient' (Mutti, 1979).

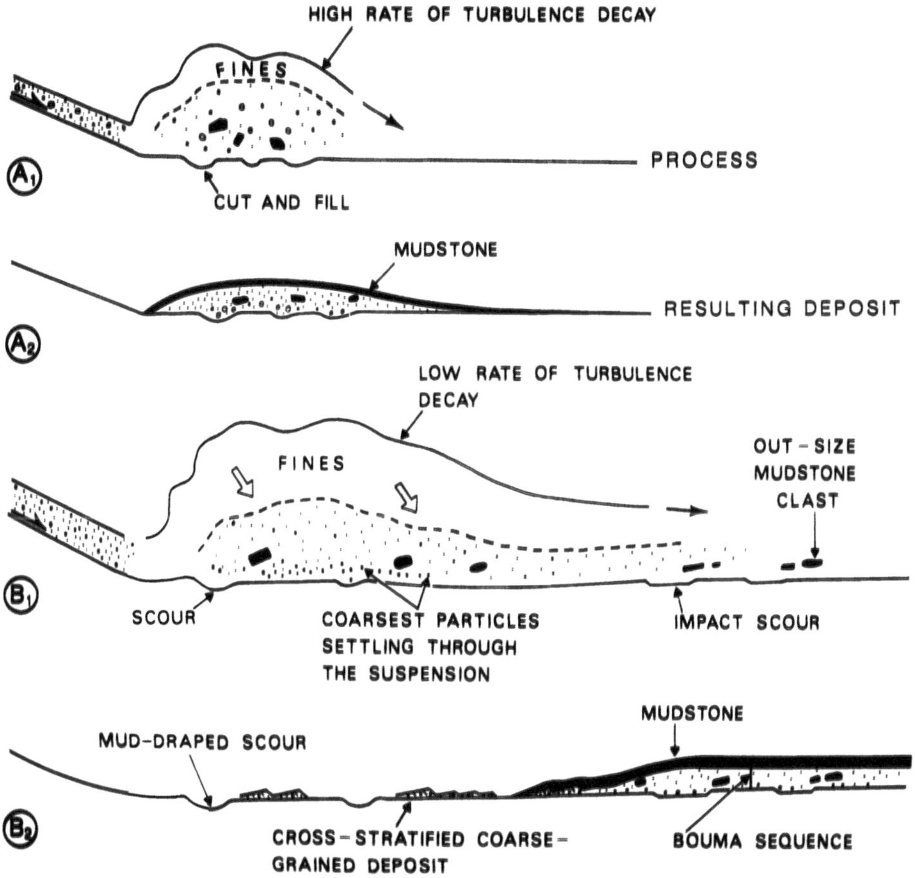

Fig. 14. Processes and resulting erosional and depositional features related to hydraulic jumps in turbidity currents. A: Relatively small-volume, sand-laden flows with minor amounts of fines. B: Large-volume, sand-laden flows that contain a large proportion of fines. See text for discussion, and Fig. 23, p. 409, in Mutti (1979).

Ancient channel–lobe transition characteristics

In ancient turbidite systems, specific facies and facies associations related to the channel–lobe transition have been briefly described only by Cazzola *et al.* (1981) for the Tertiary Piedmont Basin in northwestern Italy and by Mutti *et al.* (1985) for the Eocene Hecho Group in northern Spain.

The channel–lobe transition deposits of the Tertiary Piedmont Basin consist of thick-bedded, and commonly conglomeratic, sandstone facies that are typically amalgamated because of extensive scouring. These sediments comprise volumetrically important (and locally dominant) sections of relatively small and coarse-grained turbidite systems that formed through the resedimentation of alluvial and fan-delta marginal facies. In a downcurrent direction, commonly over distances of a few tens of meters up to several kilometers, the channel–lobe transition facies pass into well-defined, plane-parallel bedded sandstone lobes with more abundant and laterally persistent mudstone partings.

The channel–lobe transition deposits are characterized by an abundance of shallow but relatively narrow cut-and-fill features as well as by broad channels filled with a limited number of beds. Outsize mudstone clasts may be locally abundant. More rarely, these sediments include medium-scale megaripple-shaped sandstone units with or without internal cross-stratification (see below). Disrupted and contorted chaotic units composed of originally thin-bedded sandstone and mudstone may locally develop as impact features at the base of thick, and laterally discontinuous, sandstone beds. We interpret these facies to be produced by rapid deposition from sand-laden turbidity currents during and immediately after undergoing a hydraulic jump (Fig. 14).

In the Hecho Group of the south-central Pyrenees in northern Spain, the channel–lobe transition deposits are characterized by an abundance of impact and cut-and-fill features as well as by outsize mudstone clasts. These features occur within the proximal sectors of sandstone lobes and are therefore interpreted to be the result of intense turbulence and sea-floor erosion that develop within the hydraulic jump zone of turbidity currents. In the Hecho Group sequences, however, sandstone lobe facies extend several tens of kilometers downstream from the jump, and the channel–lobe transition features appear to characterize a volumetrically minor part of the overall non-channelized sandstone bodies, or lobes (see Mutti *et al.*, 1985). Our interpretation is that the hydraulic jump zone was *not* related to a substantial reduction in the capacity of the flows to carry sand farther basinward.

In addition, the Hecho Group channel–lobe transition deposits display an abundance of sedimentological features that directly indicate sediment bypassing (i.e. limited or no deposition during passage of the flow) by turbidity currents both within the transition zone and locally within the proximal sector of the lobe region. These features include (1) mud-draped scours, and (2) coarse-grained, cross-stratified facies produced by tractional processes.

Mud-draped scours (Fig. 8) clearly indicate that the large-scale turbulence developed during a hydraulic jump led to substantial scouring of the sea floor but that little or no sediment was deposited after the scouring. This implies that turbulence was sufficient to maintain even the coarser particles within the flow after undergoing the jump; coarse sand and granule-size sediment are found in the typical lobe facies found farther basinward that are devoid of scour surfaces. We interpret the mud-draped scours to be genetically similar to the scours observed on modern fans, although the exposed examples within the Hecho Group are on a considerably smaller scale.

Coarse-grained sandstone and granule conglomerate in markedly lenticular and internally cross-stratified beds are probably the most common type of facies developed at the channel–lobe transition in the Hecho Group as well as in many other deposits classified as highly efficient (Mutti, 1979), or Type I (Mutti, 1985) deposits (i.e. systems characterized by development of extensive non-channelized sandstone-lobe facies). These characteristic coarse-sediment units can occur (1) within the proximal sectors of non-channelized sandstone-lobe facies, (2) as discrete, isolated bodies enclosed within mudstone and lying between (at the same stratigraphic level) channel-fill and lobe deposits, and (3) at channel terminations.

These facies are characterized by fairly well-sorted coarse-grained sediment that is internally stratified and, therefore, shows evidence of tractive transport as bed load at the base of the flow, rather than being structureless or graded as in classical turbidite beds. Depending upon the velocity and transport capacity of the flow, the coarse-grained sediment is deposited in ripple, megaripple, and, more rarely, planar bed forms. Some of these features are illustrated in Fig. 15.

Most of the bed forms probably were not in equilibrium with steady flow conditions, as indicated by irregular wavelengths in ripple and megaripple units and by marked lateral discontinuity within these beds. Ripple-shaped beds with internal small-scale cross-stratification were referred to as Facies E in the Mutti and Ricci Lucchi (1975) scheme. Megaripple-shaped beds with internal medium-scale cross-stratification, locally underlain by plane-bed horizontal stratification, were referred to as Facies B_2 in the same scheme.

The importance of these coarse-grained facies within the depositional pattern of the Hecho Group turbidite systems was emphasized in previous papers (Mutti, 1977, 1979) that also interpreted the facies as evidence for sediment bypassing at channel mouths after hydraulic readjustment of the turbidity currents. These coarse-grained and internally stratified turbidite beds represent the coarsest fraction of the sediment load of the turbidity currents before the hydraulic jump. This coarsest fraction settles through the flow within the hydraulic jump zone as flow competence decreases. The relatively good sorting observed in these beds suggests the grains settled through a relatively dilute flow consistent with the expected flow dilution and enlargement during the jump (Middleton, 1970, p. 259). The resulting internal stratification forms under tractive conditions as the flow continues moving basinward. These sediments are therefore considered to be essentially a residual facies of the coarsest grains left behind as the bulk of the sediment load is carried farther downslope (Fig. 14).

Both mud-draped scours and coarser-grained, cross-stratified turbidite beds, as well as the length of the unchannelized sandstone lobes developed downcurrent, suggest that the rate of decay of the turbulence within a flow is relatively low for systems like the Hecho Group compared to the sand-rich systems described earlier. The narrow width of the basin for the Hecho Group also helps maintain the flow as the entire basin becomes the effective channel. Under such conditions, turbidity current reflection can occur (Hiscott and Pickering, 1984) providing further evidence for the slow decay of flow turbulence.

The main characteristics of the channel–lobe tran-

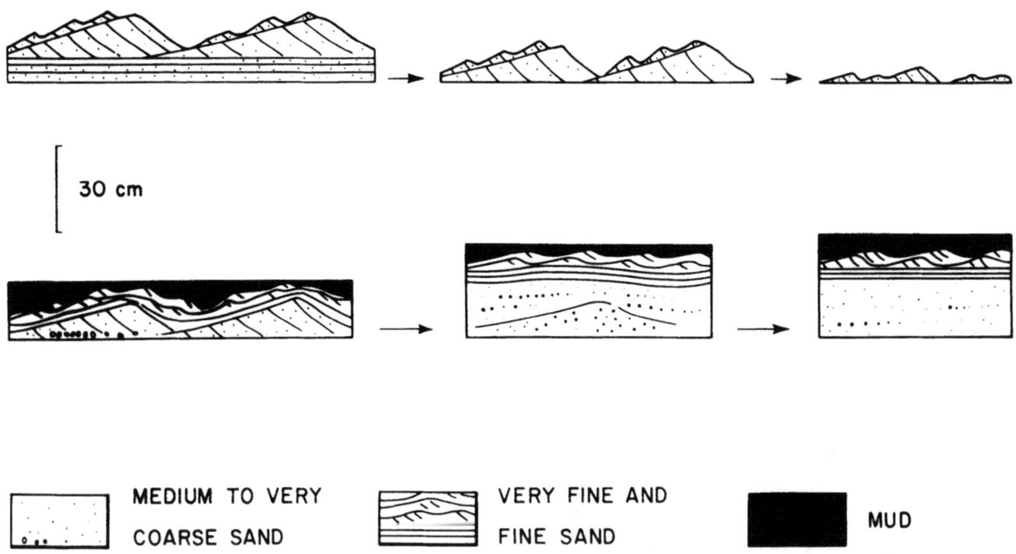

Fig. 15. Major facies types characterized by coarse-grained and internally stratified sandstone in channel–lobe transition zone (partly included in Facies B_2 and E of Mutti and Ricci Lucchi classifications, 1972, 1975). The upper part shows the downcurrent evolution of beds that lack current-laminated fines in their upper parts. The lower part of the figure depicts more complex beds within which a basal, coarse-grained and internally stratified division grades downcurrent into a faintly stratified and, finally, to a structureless division. This coarse-grained division is overlain by thoroughly current-laminated divisions of fine to very fine sandstone and by an upper mudstone division. All these beds are here collectively referred to as Facies E. (Mutti, 1984.)

sition deposits are compared to those of channel-fill and lobe sequences in the diagram of Fig. 7.

Modern channel–lobe transition features

Although the general zone of the channel–lobe transition can be identified from morphologic criteria on modern fans, few data are available to characterize this zone. For channels wider than about one kilometer, the loss of levee relief can be recognized on surface-ship reflection profiles, and the termination of the levees probably marks the upstream limit of the transition zone. Morphologic features that are of the scale of sedimentological structures and bedforms observed from channel–lobe-transition sediments in ancient turbidite systems can only be mapped on modern fans through the use of deep-towed, high-resolution reflection and side-scan sonar systems. Thus, there are only a few modern deposits for which we can suggest either the character of the morphology within the transition zone or its basinward extent. Available deep-tow records suggest that the downslope extent is probably limited to the sectors with mappable surface relief (i.e. one to two meters in height and several tens of meters across) as the only character to distinguish the transition zone from the apparently smooth lobe areas.

If the limitation of the channel–lobe transition suggested above is basically correct, then the following comments cover the available observations from modern turbidite systems.

1. The size of the transition zone appears to be roughly proportional to the size of the feeding channel where the channel reflects active growth or maintenance of channel size. This relation is tenuous because of the lack of data but is consistent with the hydraulic scenario previously discussed.

2. The basinward extent of the transition zone will be most easily recognized on morphological grounds. The inability to obtain core samples from coarse-grained deposits will severely limit opportunities to map such zones on sedimentological criteria. The reliance on morphologic data, however, means that the extent of the zone could be overestimated where a very large turbidity current has produced scouring in proximal lobe areas.

3. The only detailed studies using deep-tow side-scan data that include the channel–lobe transition on modern sand-rich turbidite systems are the inner

suprafan sectors on San Lucas and Navy fans (Normark, 1970a; Normark *et al.* 1979). The fine-scale morphology (mesotopography) of Navy Fan is better understood because the deep-tow vehicle was generally maintained much closer to the sea floor than for the earlier San Lucas Fan study, so that there is far greater detail and resolution on the side-scan sonar records. Both fans show extensive channel-like and scour relief in the region immediately in front of the channel mouth. Most channel features are isolated (San Lucas) or are cut off by low levees leaving only one channel tributary as a continuous distributary feature (Navy Fan).

The extent of mesotopographic relief on Navy Fan (Fig. 16) is greatest within an area that extends three to four kilometers from the mouth of the main channel. Large scours, 20–400 m across, described earlier are the most striking bedforms. The larger scours are isolated but many of the smaller ones occur nested in groups (like flutes), especially across channel margins and at the levee terminations. Between the channel segments, numerous small, irregular shadows and patches of high acoustic backscatter on the side-scan records suggest even smaller bedforms, other erosional features, and high variability in grain size or surface relief (of the order of centimeters) across this area. Small-scale relief features (ripples, flutes, grooves, etc.) and textural (grain-size) variations cannot be distinguished from backscatter data alone on currently available deep-water, side-scan sonar records. Nevertheless, such features are not observed with side-scan anywhere else on the fan surface and are, therefore, probably unique to the channel–lobe transition zone. Much higher resolution side-scan sonar records (e.g. 50 m swath width) or large-scale bottom photographs will be necessary to actually define the relief features in this zone.

4. For large modern fans, limited surveying with the SeaMARC I deep-tow system appears to provide the only other glimpse of channel–lobe transition features (O'Connell, 1986). The area of termination of the main leveed channel is not well constrained with existing data, however, and channel segments of varying relief and extent cover a large area (10 000 km^2) that has been termed a depositional lobe (Normark and Meyer, 1986). The area with channel segments shows extreme variability both in surficial backscatter amplitude on SeaMARC I profiles and in the shallow structure observed on surface-ship and deep-tow, high-resolution profiles. The acoustic records also show large semicircular depressions, long, narrow grooves (20 m × 2 km), and breaches of small levees on channel segments. Bottom photographs, although very limited in number and areal distribution, show apparent mud-clast lag (?) deposits in areas outside the channels (O'Connell, 1986). Many of the complexities in the acoustic records seen on this part of the Mississippi Fan could be indicative of a channel–lobe transition zone, although O'Connell (1986) suggests that a major (mega)turbidity current/mass-flow event may have been the primary cause.

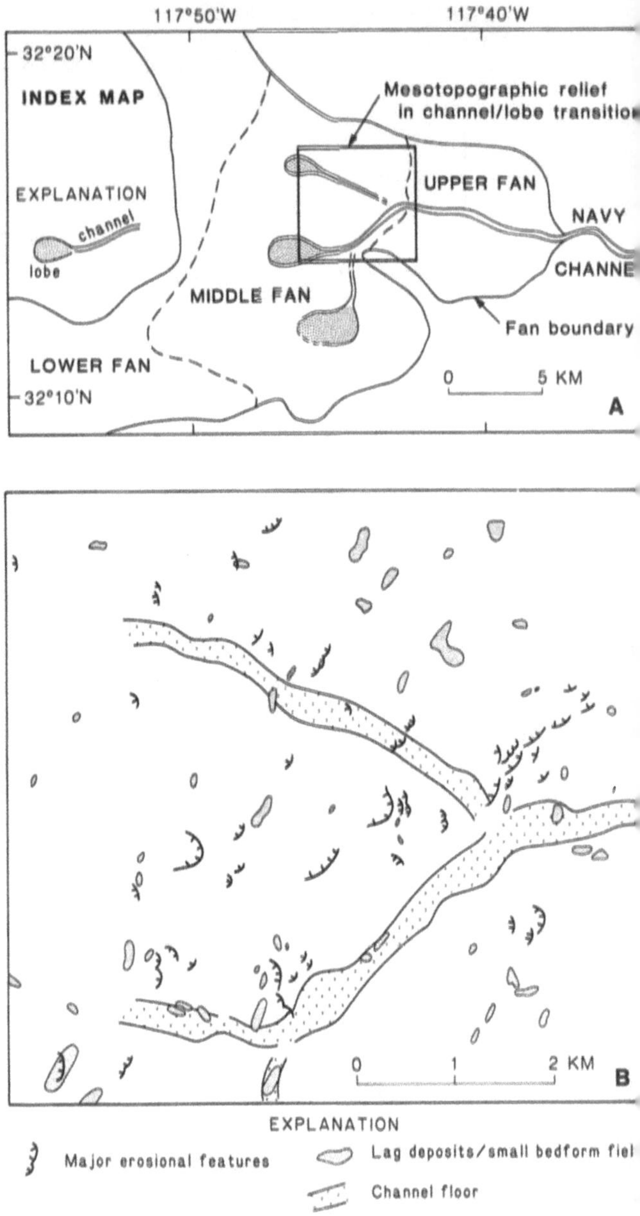

Fig. 16. Mesotopographic relief in the channel–lobe transition zone on Navy Fan (from Normark *et al.*, 1979).

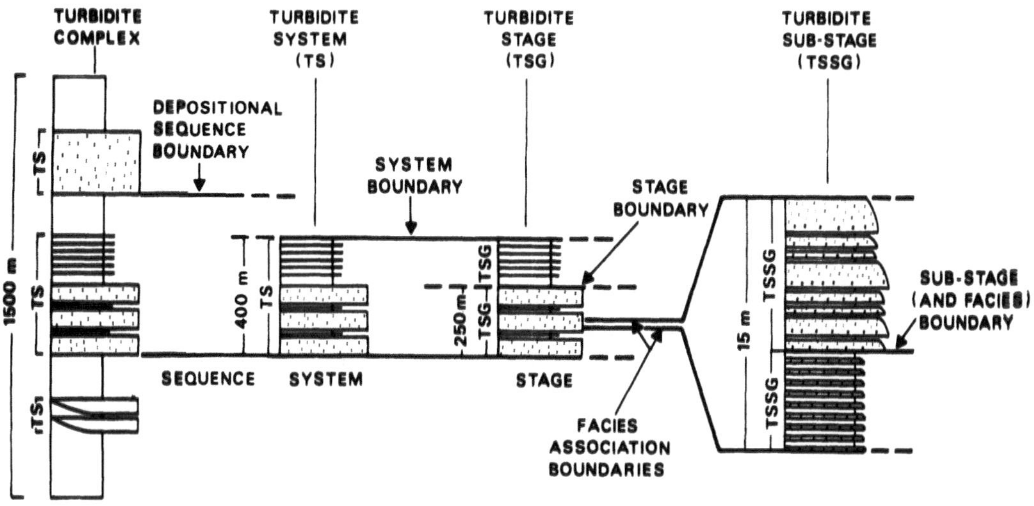

Fig. 17. Conceptual classification framework for turbidites. The classifcation is based on the stratigraphic rank of each unit and, therefore, on a physical scale that, in many cases, also reflects a temporal scale (re: Fig. 2).

Generalizations about the morphology and sediment distribution indicative of the channel–lobe transition are speculative, and this element of modern turbidite systems requires extensive deep-tow investigations to improve our understanding. Neither the transition zone nor the lobes themselves have been adequately studied in modern basins because (1) the relief is too small to resolve by surface-ship techniques alone, (2) the sediments are relatively coarse grained and not easily sampled, and (3) the area involved is generally too large to map completely with standard deep-tow techniques.

SYSTEMS, STAGES, AND SUBSTAGES

The definition of elements in the previous section was to suggest what we consider to be recognizable and mappable features common to most turbidite systems (*sensu lato*). Basically, our elements are depositional and erosional features that develop at the scale of stages and substages during the growth of submarine turbidite systems (Fig. 17). Recognition and mapping of these elements and the framing of these features within well-established correlation patterns will allow (1) definition of the type of system, (2) recognition of the different stages of growth of the system under consideration, and (3) permit significant comparisons with other, similarly defined systems, both modern and ancient.

The main types of systems and their component stages and substages are briefly reviewed below with special emphasis on the significance of the different elements described in the previous section.

Systems

Both modern and ancient turbidite systems differ considerably in terms of size, overall geometry, types of facies and facies associations, and geometry and distribution of sand bodies. These systems essentially form a poorly understood continuum controlled primarily by long-term sea-level variation, local and regional tectonic setting, and amount and type of sediment available for resedimentation processes. Most ancient systems differ from each other mainly in terms of where the bulk of the sandstone facies occurs within the system (Mutti, 1979; 1985). On this basis, the scheme of Fig. 18 depicts the three end members that can be used as a tentative reference in describing the depositional pattern of most ancient systems (see also Slaczka, 1983).

Type I deposits, which are relatively well described in previous literature (Mutti, 1979; Nilsen, 1980) and are characterized by a 'highly efficient' sediment dispersal pattern, are dominantly composed of unchanneled sandstone lobes. These lobes are correlative with erosional channels and particularly with large-scale erosional features that are probably extensive slump scars in shelf-edge deposits from which the bulk of the sand for the lobes is derived.

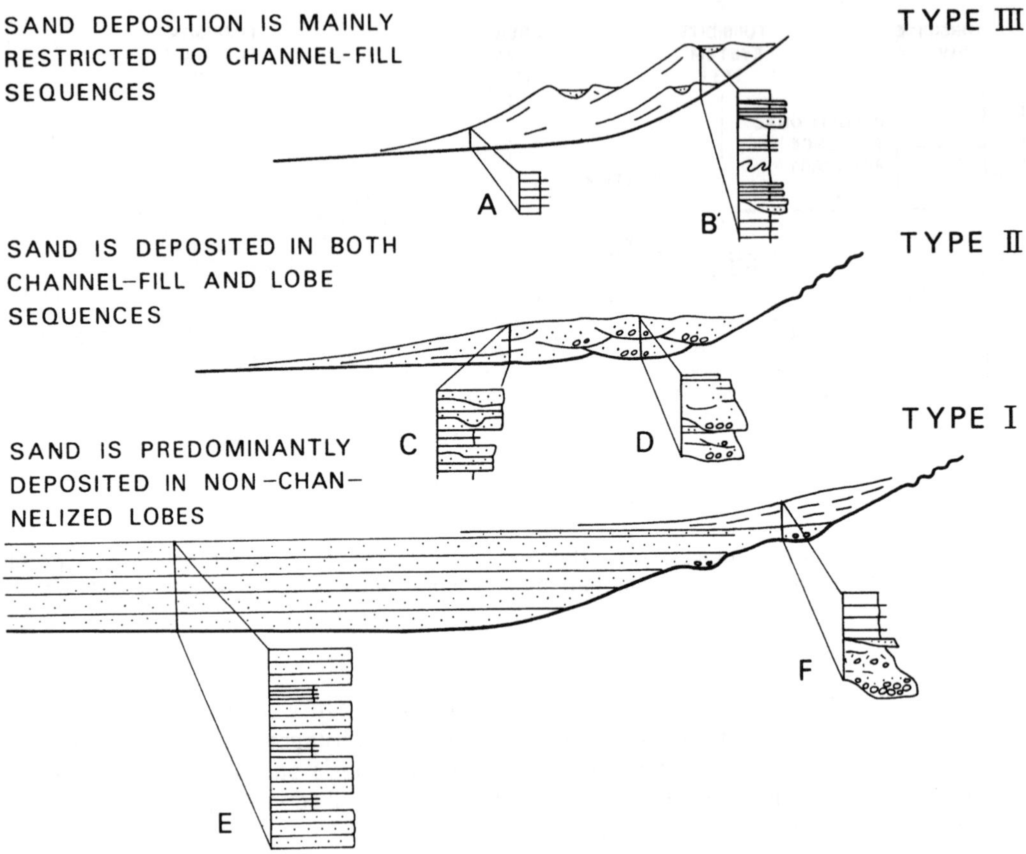

Fig. 18. Schematic representation of three main types of turbidite deposits as characterized by relative amount of and position within the deposit of the sandy component (see text for details).

Type I deposits are apparently characteristic of elongate foreland basins (Type C basins) where tectonic activity produces and maintains narrow basins that enhance the distance of transport of the sand (e.g. Mutti, 1979, 1985; Mutti and Ricci Lucchi, 1981; Ricci Lucchi, 1985; Ricci Lucchi and Ori, 1985), but trenches as well as other types of elongate basins may be equally suitable for the development of similar depositional patterns (e.g. Nilsen, 1985, Surlyk and Hurst, 1984). Provided that large amounts of fines are involved in the resedimentation processes to enhance the mobility of turbidity currents, Type I deposits can probably form also in relatively unconfined basins. Modern, well-documented examples of Type I deposits are unfortunately unavailable at present.

Type II deposits are typically smaller and have coarser-grained sediment than do Type I. Type II deposits are composed of depositional and mixed channel-fill types that grade downslope into sandstone lobes. These lobes are considerably less extensive than in Type I deposits. For this reason, the sediment dispersal pattern of such systems has been referred to as 'poorly efficient' (Mutti, 1979). Type II deposits are commonly developed in Type D basins, including rifted (e.g. Cazzola *et al.*, 1981; Stow, 1985; Surlyk, 1978) and borderland (e.g. Link, 1981; Link and Nilsen, 1980; McLean and Howell, 1985; Nilsen, 1980) basins, as well as in continental arc settings (e.g. Busby-Spera, 1985). Possible modern analogs of Type II systems include some fans of the California Continental Borderland in transform-controlled basins that have been mentioned in a previous section (p. 4).

Type III deposits are basically composed of fine-grained and thin-bedded sediments that probably record channel–levee complexes and various types of slope drapes. The Delgada Fan levee–valley complex may be a modern Type III deposit. Within these deposits, which in general comprise overbank wedges as described in a previous section, sandstone bodies are almost entirely restricted to the infill of depositional channels. This type is conspicuous for the lack of associated sandstone lobes.

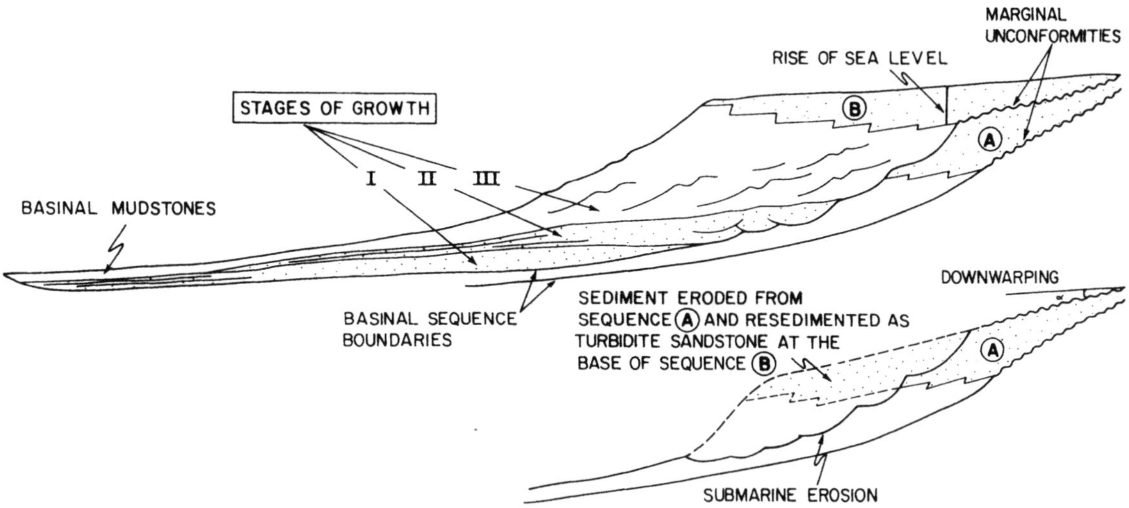

Fig. 19. Conceptual diagram relating relative changes of sea level and different stages of growth of turbidite systems within depositional sequences developed in elongate flysch basins (from Mutti, 1985).

Stages

Some turbidite systems apparently express only one stage of growth; this can occur, for example, where tectonism within the depositional basin is rapid enough to prevent the development of a vertical succession of different types of facies associations (e.g. Navy Fan: Normark and Piper, 1972; Smith and Normark, 1976), or where there are only minor changes with time in the type and rate of sediment supply (San Lucas Fan: Normark, 1970a). Most ancient Type II deposits also display an apparent lack of vertical organization related to different periods of growth, probably because many of these systems were deposited during relatively short periods of time through very high-frequency and essentially similar flows. Stages of growth in Type III deposits are basically unknown in detail, although recent detailed work from the Indus Fan (McHargue and Webb, 1986) shows that inner fan channel–levee complexes and associated sand bodies probably underwent a very complex history of stage development through time.

Conversely, other turbidite systems clearly develop as composite features displaying multiple stages, of which one type may predominate. These stages are expressed by the lateral and vertical succession of facies associations that record a specific period of growth of the system. In the Hecho Group, Type I deposits are characteristically composed of three stages of growth (Mutti, 1985; Mutti *et al.*, 1985) that are diagrammatically shown in Fig. 19. Stage I is volumetrically predominant and consists almost entirely of sandstone lobes that are correlative with large-scale submarine erosional features and erosional channels. Stage II is primarily expressed by channel-fill sequences, and Stage III deposits consist of an overbank wedge that essentially expresses basinward progradation of channel–levee complexes and/or thin-bedded, predominantly muddy slope turbidites. A very similar vertical sequence of turbidite stages is depicted by the lateral and vertical succession of seismic facies within the inner portion of individual turbidite systems, or fan lobes of the modern Mississippi Fan (Feeley *et al.*, 1985). In both the Eocene Hecho Group and the modern Mississippi Fan, the vertical succession of these stages has been similarly interpreted as the product of the gradual reduction in the volume of mass flows associated with progressive relative rise of sea level and/or slope retreat with time (Feeley *et al.*, 1985; Mutti, 1985; see also Bouma *et al.*, 1985b).

We think that stages, which mimic systems on a smaller scale, are certainly common to most modern and ancient turbidite systems, and should be particularly well developed in all those systems which develop as sufficiently long-lived features to undergo variations in the type of sediment supplied and in the volume of flows produced by tectonic activity and/or short-term relative sea-level variations.

Future work to understand stage development has great potential for construction of depositional models and significant comparison of modern fans and ancient turbidite systems. Stage development may also represent a powerful tool for the distinction between systems where deposition is controlled by relative or global sea-level variations and those controlled by tectonic activity and/or shelf-edge subsidence. Theoretically, in basins where turbidite deposition is primarily controlled by sea-level variations, extensive sea-level lowering should result in a pattern of growth where larger volume flows succeed smaller; conversely, in basins where tectonic activity with shelf-edge subsidence are the dominant factors in controlling turbidite deposition, the resulting large-scale shelf-edge failures should produce stage developments that would be the reverse. Progressive slope retreat would result here in an overall decrease in the volume of flows and give way to stage successions as observed in the Hecho Group and the Mississippi Fan.

Substages

Major or long-term (10^4–10^5 years) tectonic activity and relative sea-level variations primarily affect the vertical evolution of stages and overall development of the turbidite systems. Little is known about the physical and temporal factors that control substage development and succession. These short-term variations in sediment supply are recorded as substages, i.e. individual facies with thicknesses in the order of meters or a few tens of meters that are laterally and vertically organized into facies associations. The basic vertical cyclic organization which is observed in turbidite successions is expressed at this physical and temporal scale.

Part of the above cyclicity is certainly related to autocyclic mechanisms such as lateral shifting of channel axes with time or compensation of smooth depositional relief in the lobe regions. This type of facies cyclicity is generally easy to detect, particularly in sufficiently large exposures or through detailed correlation patterns. Most commonly, however, the vertical and sequential arrangement observed in turbidite facies can only be explained in terms of variations of sediment supply. Individual facies that record such variations, as well as the lateral and vertical relations among these facies, define our substages.

The two basic different types of substage development are tentatively depicted in the diagrams of Fig. 20. The upper model emphasizes decrease in sediment supply with time, and is apparently the most common pattern observed in large, type C basin, ancient sediments; the lower model emphasizes the reverse, i.e. an increase in sediment supply. The two models bear obvious differences in correlating channel-fill and lobe sediments genetically related to the same substage development. As noted earlier, some systems only display one substage character throughout their development, and other sequences involving only two types of substage deposits are probable. In modern systems, for example, some deposits lose most sources for sand during a relatively short period of sea-level rise (10^4 years $\pm$), and there are insufficient deposits to reflect a substage development between times of abundant sandy turbidity currents (low sea level) and a little or no sand for turbidity currents (high sea level).

Regardless of the type of system and stage to which they belong (and, therefore, regardless of their areal extent, facies characteristics, and volume of sediment involved), facies associations and their component substages represent the building blocks of ancient turbidite systems as do their equivalent morphologic elements for modern systems. Figure 21 shows the relation between elements, facies/facies associations, and stage/substage distinctions that represent successive levels for understanding a turbidite system (*sensu stricto*). Effective comparison of modern and ancient turbidite systems can rely on using the descriptive elements alone, but the development of comprehensive models for submarine fan growth can be greatly enhanced when facies associations and stage/substage distinctions can be determined for modern systems as well.

CONCLUDING OBSERVATIONS

In undertaking to write this paper (beginning in 1983), we both realized how difficult it was—and is!—to understand and compare modern and ancient turbidites. Although we (the authors of this paper) basically know why our own fan models are dissimilar, the further we pushed to bring our ideas together, the more problems we encountered. One of the problems is simply our different backgrounds and experiences in working on turbidite systems. In our attempts to understand each other's work as clearly as possible and to use cautiously such newly gained insights in this paper, our text probably includes needless repetition of some concepts. The redundancy, in part, reflects our effort to educate each other, but we hope that the readers will benefit more from repetition than they would from omission. Our experience thus emphasizes the need for closer cooperation between marine and land-based geologists and sedimentologists to provide a more common language and background for the study of turbidite systems.

In part because of the length problems, we deliberately did not discuss in this paper two important

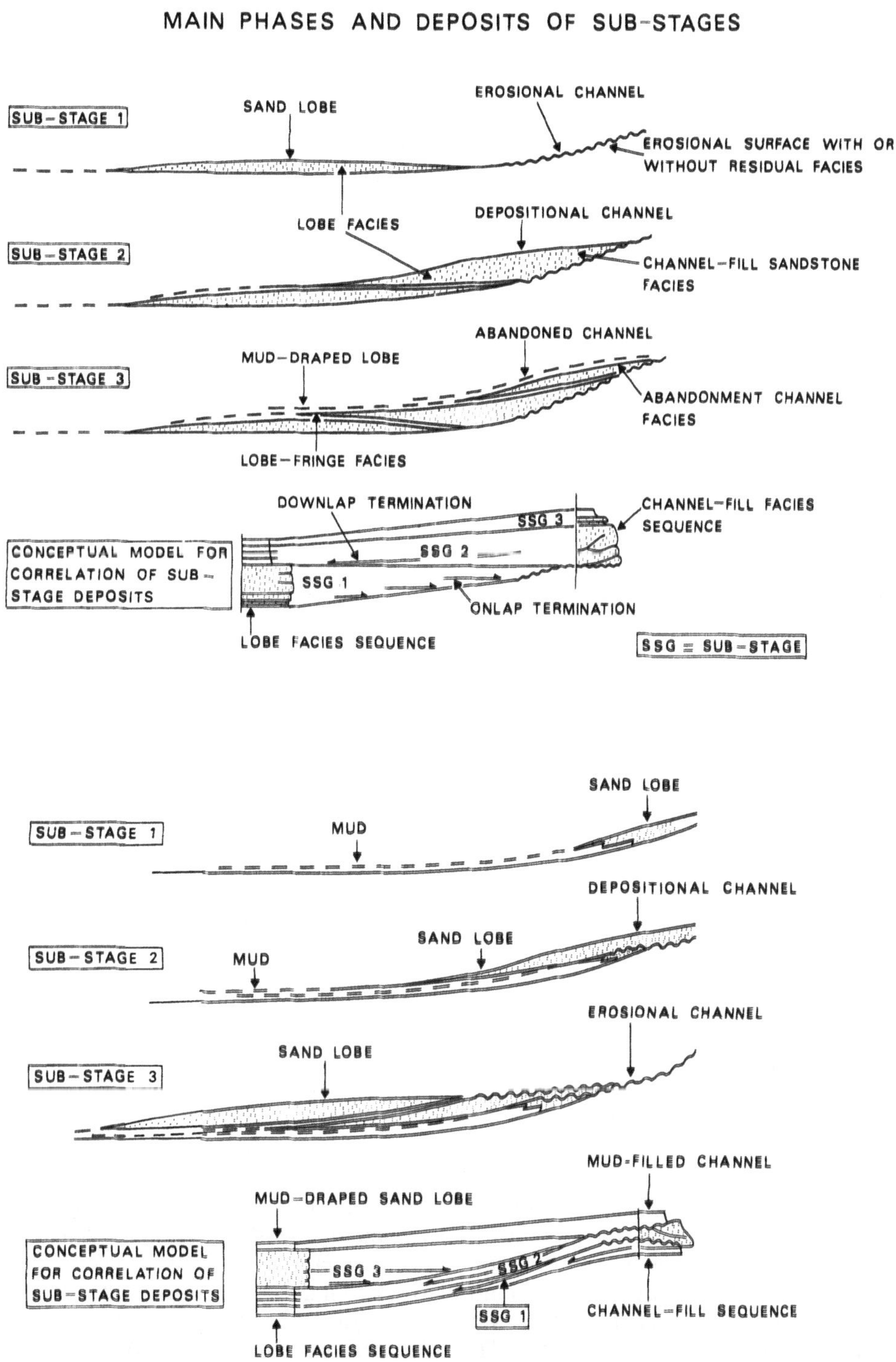

Fig. 20. Conceptual diagrams showing vertical (temporal) and lateral relations of substages within a facies association consisting of an upcurrent channel-fill sequence and a downcurrent lobe sequence. The upper diagram depicts a substage evolution from relatively large-volume to small-volume flows. The lower diagram depicts an apparently less common evolution from relatively small-volume to large-volume flows. The vertical stacking of facies associations of this type comprises the composite, or multistory, fill of an individual channel and its corresponding lobe package. This high-frequency cyclicity displayed by most turbidite deposits remains one of the major and more interesting problems for future work. Although allocyclic mechanisms are emphasized in this paper, normal autocyclic mechanisms (e.g. lateral shifting of channels and lobe axes with time but *within* the boundary of the same facies association) are also important.

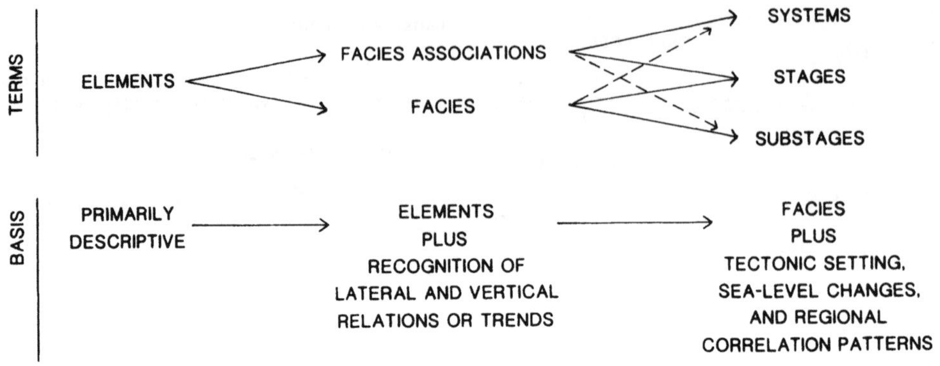

Fig. 21. Summary of terminology for mapping turbidite systems emphasizing successive levels of knowledge required.

points: (1) turbidite systems modified by bottom currents; and (2) the significance of fining- and coarsening-upward facies sequences. We shall briefly comment upon both points in this concluding section.

Submarine fans that exhibit substantial modification by bottom currents are common, if not the rule, along those modern continental margins where large-scale oceanic circulation is involved. Contour currents and their resulting deposits and erosional features have been amply discussed in papers by Bouma and Hollister (1973) and Lovell and Stow (1979) (see also Shunmugam and Moiola, 1982). Examples of ancient submarine systems that have been winnowed by such currents include the Eocene deposits of the Brazilian offshore (Mutti *et al.*, 1980) and the Paleocene and Eocene of the North Sea (Enjolras *et al.*, 1986; Heritier *et al.*, 1979). It appears, however, that these winnowed turbidite systems are still basically unknown and cannot, as such, be incorporated in the general discussion of this paper. We strongly feel that most modern submarine fans on passive margins are, in one way or another, affected by bottom currents; conversely, most of the ancient turbidite systems that have been described formed in narrow inter- and/or intra-plate basins and therefore escaped such modifications. Consequently, the comparison between modern and ancient turbidites is even more difficult than we have indicated. The turbidite-bearing, foreland basins of thrust belts such as those in the northern Apennines and south-central Pyrenees probably have little in common with the modern submarine fans of Atlantic-type margins, both tectonically and depositionally.

Thickening- (and coarsening-) and thinning- (and fining-) upward facies sequences have long been essentially synonymous with lobe and channel-fill deposits. These concepts, introduced in the turbidite literature by Mutti and Ricci Lucchi (1972) and Mutti and Ghibaudo (1972) have probably helped sedimentologists in discriminating the more obvious channel-fill and lobe sediments in reconnaissance type of work in ancient turbidite basins. More recent work has questioned the validity of the concepts and the way such sequences can be objectively recognized (e.g. Walker, 1975, 1980; Hiscott, 1981). We fully agree with this criticism. Channel-fill and lobe sequences obviously should be recognized on the basis of their external geometry. However, in both surface and subsurface work, the common lack of direct control on the geometry has forced the use of vertical sequence models to discriminate between channel and lobe deposits. Such models carry numerous pitfalls that commonly lead to misinterpretation of relatively small exposures or of e-log patterns. These pitfalls, as well as the vertical sequence analysis of turbidite facies, are discussed elsewhere (Mutti, in preparation). Implicitly, we have dealt with the problem of vertical facies sequences in previous sections of this paper (see Figs 19 and 20) and emphasized that such sequences are essentially restricted to two basic types that, in turn, are controlled by the decrease or increase with time of the volume of the flows (see also Mutti, 1985). Much work remains to be done in this field.

Neither of our original fan models was meant to be comprehensive—the Mutti–Ricci Lucchi model was a *preliminary* description of facies that coincidentally happened to merge in a coherent model for deposition. The Normark (suprafan) model was based on comparison of three fan systems described in the original work (Normark, 1970a, b), but only one system exhibited an unmodified depositional (growth) pattern. We may suggest new or modified models, but not at this time. We shall concentrate on finding several well-studied fans to compare first.

In suggesting methods for comparison of turbidite systems, we have tried to avoid terms derived from interpreted parameters as much as possible. For example, the COMFAN compilations (Barnes and Normark, 1985) show that there is no consistent usage of inner, middle, outer or upper, middle, lower divisions of modern-fan surfaces, and this makes comparison with facies-derived inner-middle-outer divisions very misleading. Therefore, we have avoided these terms in our framework approach.

Turbidite facies and facies associations are still valid as criteria to guide basin analysis, but mere recognition of them is not sufficient for accurate comparisons. Application of the suggested framework (Fig. 21) to systems described in existing literature will be difficult in many cases. Primary problems are a failure to recognize scale differences between channels in ancient examples and channels from modern. In *most* cases, the largest channels described in outcrop are below resolution if found in the modern environment. Lobes in ancient sequences have been identified from one measured section only—this is not enough to define a lobe, as coarsening-and-thickening-upward sequences can develop in other depositional environments just as thinning-upward sequences can develop outside a channel.

More generally, it is apparent that very few documented examples are available from the literature in which turbidite systems and their component stages and substages have been adequately recognized and mapped in both modern and ancient settings. The reasons for this are certainly the limitations that we have discussed and emphasized in many sections of this paper. These limitations do exist and in most cases prevent better stratigraphic and depositional reconstructions of both modern and ancient turbidite systems. However, we think that such limitations could be at least partly overcome if more attention were paid to the stratigraphic framework (both temporal and spatial) of these systems. We do not know whether the approach we have suggested in this paper can become a routine one for turbidite basin analysis. We can only say that, based on our experience, it is probably worth trying.

We are also aware that determining stratigraphic frameworks within which we can analyze our observations generally requires much time-consuming and often boring work. This work can be enhanced by a closer collaboration with other specialists, particularly biostratigraphers and sedimentary petrologists. We hope this comprehensive approach will be followed by those just beginning their careers, and for this reason we dedicate this paper to the memory of Tarquin Teale, who, among these young people, was certainly one who, in addition to his great, natural talent, fell so in love with geology that he did not regret spending years of his young life in the mountains of Calabria to obtain data upon which to base his interpretations. *Ciao*, Tarquin.

ACKNOWLEDGEMENTS

The ideas and problem review presented in this paper involve re-evaluation of more than 15 years of field work by each of the authors as well as results of recent and continuing research. The list of colleagues, students and technical personnel who have helped in one way or another in the germination and development of the concepts is so extensive that any attempt to document fully the contributions would, we are afraid, only highlight inadvertent omissions. Thus, instead we would like to note those who have been particularly helpful in the last few years, including C. Cazzola, F. Fonnesu, M. Sgavetti, D. J. W. Piper, C. E. Gutmacher, and G. R. Hess.

The forum for review of modern and ancient turbidite systems provided by the COMFAN meeting in 1982 was obviously timely.

E. Mutti acknowledges financial support provided by the Italian Ministry of Public Education (MPI). W. R. Normark especially acknowledges the financial support of travel for teaching and field work in Italy and Spain provided by the University of Parma, the University of Bologna, and CNR (Italy).

The manuscript was improved by critical reviews from H. E. Clifton and D. G. Howell. Most drafting was kindly done by Edwige Masini and Phyllis Swenson.

References

Bachman, S. B. and Graham, S. A. (1985). La Jolla Fan, Pacific Ocean, *in* A. H. Bouma, W. R. Normark and N. E. Barnes (Eds), *Submarine Fans and Related Turbidite Systems*, Springer-Verlag, New York, pp. 65–70.

Barnes, N. E. and Normark, W. R. (1985). Diagnostic parameters for comparing modern submarine fans and ancient turbidite systems, *in* A. H. Bouma, W. R. Normark and N. E. Barnes (Eds), *Submarine Fans and Related Turbidite Systems*, Springer-Verlag, New York, pp. 13–14.

Belderson, R. H., Kenyon, N. H., Stride, A. H. and Pelton, C. D. (1984). A 'braided' distributary system on the Orinoco deep-sea fan, *Marine Geology*, **56**, 195–206.
Bouma, A. H. and Hollister, C. D. (1973). Deep ocean basin sedimentation, *in Turbidites and deep water sedimentation*, SEPM Pacific Section Short Course, Anaheim 1973, pp. 79–118.
Bouma, A. H., Normark, W. R. and Barnes, N. E. (1985a). COMFAN: needs and initial results, *in* A. H. Bouma, W. R. Normark and N. E. Barnes (Eds), *Submarine Fans and Related Turbidite Systems*, Springer-Verlag, New York, pp. 7–11.
Bouma, A. H., Stelting, C. E. and Coleman, J. M. (1985b). Mississippi Fan, Gulf of Mexico, *in* A. H. Bouma, W. R. Normark and N. E. Barnes (Eds), *Submarine Fans and Related Turbidite Systems*, Springer-Verlag, New York, pp. 143–150.
Bouma, A. H., Coleman, J. M. and DSDP Leg 96 Shipboard Scientists (1985c). Mississippi Fan: Leg 96 program and principal results, *in* A. H. Bouma, W. R. Normark and N. E. Barnes (Eds), *Submarine Fans and Related Turbidite Systems*, Springer-Verlag, New York, pp. 247–252.
Bowen, A. J., Normark, W. R. and Piper, D. J. W. (1984). Modelling of turbidity currents on Navy Submarine Fan, California Continental Borderland, *Sedimentology*, **31**, 169–185.
Busby-Spera, C. (1985). A sand-rich submarine fan in the lower Mesozoic Mineral King Caldera Complex, Sierra Nevada, California, *Journal of Sedimentary Petrology*, **55**, 376–391.
Carter, R. M. (1975). A discussion and classification of subaqueous mass transport with particular application to grain-flow, slurry-flow, and fluxoturbidites, *Earth Science Review*, **11**, 145–177.
Carter, R. M. and Lindquist, J. K. (1975). Sealers Bay submarine fan complex, Oligocene, southern New Zealand, *Sedimentology*, **22**, 465–483.
Carter, R. M. and Norris, R. J. (1977). Redeposited conglomerates in Miocene flysch sequence at Blackmount, Western Southland, New Zealand, *Sedimentary Geology*, **18**, 289–319.
Casnedi, R. (1981). The hydrocarbon-bearing submarine fan system of Cellino (central Italy): Abstracts Volume, 2nd European Regional Meeting of International Association of Sedimentologists, Bologna, pp. 22–25.
Casnedi, R., Moruzzi, G. and Mutti, E. (1978). Correlazioni elettriche dei lobi deposizionali torbiditici nel Pliocene del sottosuolo abruzzese, *Memorie della Societa Geologica Italiana*, **18**, 23–30.
Cazzola, C. and Rigazio, G. (1983). Caratteri sedimentologici dei corpi torbiditici di valla e Mioglia. Formazione di Rocchetta (Oligocene–Miocene) del Bacino Terziario Piemontese, *Giornale di Geologia*, **65**, 87–100.
Cazzola, C., Fonnesu, F., Mutti, E., Rampone, G., Sonnino, M. and Vigna, B. (1981). Geometry and facies of small, fault controlled deep-sea fan systems in a transgressive depositional setting (Tertiary Piedmont Basin, north-western Italy), *in* F. Ricci Lucchi (Ed.), *Excursion Guidebook*, 2nd European Regional Meeting of International Association of Sedimentologists, Bologna, pp. 5–56.
Cazzola, C., Mutti, E. and Vigna, B. (1985). Cengio turbidite system, Italy, *in* A. H. Bouma, W. R. Normark and N. E. Barnes (Eds), *Submarine Fans and Related Turbidite Systems*, Springer-Verlag, New York, pp. 179–183.
Clifton, H. E. (1981). Submarine canyon deposits, Point Lobos, California, *in* V. Frizzell (Ed.), *Upper Cretaceous and Paleocene Turbidites, Central California Coast*, Pacific Section Society of Economic Paleontologists and Mineralogists Field Trip Guidebook, Field Trip No. 6, San Francisco Annual Meeting, pp. 79–92.
Clifton, H. E. (1984). Sedimentation units in stratified deep-water conglomerate, Paleocene submarine canyon fill, Point Lobos, California, *in* E. H. Koster and R. J. Steele (Eds), *Sedimentology of Gravels and Conglomerates*, Canadian Society of Petroleum Geologists, Mem. 10, pp. 429–441.
Coleman, J. M., Prior, D. B. and Lindsay, J. F. (1983). Deltaic Influences on Shelf Edge Instability Processes, *SEPM Special Publication* **33**, pp. 121–127.
Cremer, M., Orsolini, P. and Ravenne, C. (1985). Cap-Ferret Fan, Atlantic Ocean, *in* A. H. Bouma, W. R. Normark and N. E. Barnes (Eds), *Submarine Fans and Related Turbidite Systems*, Springer-Verlag, New York, pp. 113–120.
Damuth, J. E. and Flood, R. D. (1985). Amazon Fan, Atlantic Ocean, *in* A. H. Bouma, W. R. Normark and N. E. Barnes (Eds), *Submarine Fans and Related Turbidite Systems*, Springer-Verlag, New York, pp. 97–106.
Droz, L. and Bellaiche, G. (1985). Rhone deep-sea fan: morphostructure and growth pattern, *American Association of Petroleum Geologists Bulletin*, **69**, 460–479.
Dupuy, G. L. P., Oswaldt, G. and Sens, J. (1963), *Champ de Cazaux, Géologie et Production*, Sixth World Petroleum Congress, Proceedings, Section 2, 199–213.
Emmel, F. J. and Curray, J. R. (1985). Bengal Fan, Indian Ocean, *in* A. H. Bouma, W. R. Normark and N. E. Barnes (Eds), *Submarine Fans and Related Turbidite Systems*, Springer-Verlag, New York, pp. 107–112.
Enjolras, J. M., Gouadain, J., Mutti, E. and Pizon, J. (1986). New turbiditic model for Lower Tertiary Sands in the south Viking Graben, *in* A. M. Spencer *et al.* (Eds), *Habitat of Hydrocarbons on the Norwegian Continental Shelf*, Norwegian Petroleum Society, Graham and Trotman, London, pp. 171–178.
Enos, P. (1969). Anatomy of a flysch, *Journal of Sedimentary Petrology*, **39**, 680–723.
Estrada Aliberas, M. R. (1982). *Lobulos deposicionales de la parte superior del Grupo de Hecho entre el anticlinal de Boltana y el Rio Aragon* (*Huesca*), Tesis, Univ. Aut. Barcelona.

Feeley, M. H., Buffler, R. T. and Bryant, W. R. (1985). Depositional units and growth pattern of the Mississippi Fan, *in* A. H. Bouma, W. R. Normark and N. E. Barnes (Eds), *Submarine Fans and Related Turbidite Systems*, Springer-Verlag, New York, pp. 253–257.

Haner, B. E. (1971). Morphology and sediments of Redondo submarine fan, *Geological Society of America Bulletin*, **82**, 2413–2432.

Hein, F. J. (1982). Depositional mechanisms of deep-sea coarse clastic sediments, Cap Enragé Formation, Quebec, *Canadian Journal of Earth Sciences*, **19**, 267–287.

Hein, F. J. and Walker, R. G. (1982). The Cambro–Ordovician Cap Enragé Formation, Quebec, Canada: conglomeratic deposits of a braided submarine channel with terraces, *Sedimentology*, **29**, 309–329.

Heritier, F. E., Lossel, P. and Wathne, E. (1979). Frigg Field—large submarine-fan trap in lower Eocene rocks of North Sea Viking Graben, *American Association of Petroleum Geologists Bulletin*, **63**, 1999–2020.

Hess, G. R. and Normark, W. R. (1976). Holocene sedimentation history of the major fan valleys of Monterey Fan, *Marine Geology*, **22**, 233–251.

Hirayama, J. and Nakajima, T. (1977). Analytical study of turbidites, Otadai Formation, Boso Peninsula, Japan, *Sedimentology*, **24**, 747–779.

Hiscott, R. N. (1981). Deep-sea fan deposits in the Macigno Formation (middle–upper Oligocene) of the Gordana Valley, northern Apennines, Italy—Discussion, *Journal of Sedimentary Petrology*, **51**, 1015–1021.

Hiscott, R. N. and Pickering, K. T. (1984). Reflected turbidity currents on an Ordovician basin floor, Canadian Appalachians, *Nature*, **311**, 143–145.

Howell, D. G. and Normark, W. R. (1982). Sedimentology of submarine fans, *in* P. A. Scholle and D. Spearing (Eds), *Sandstone Depositional Environments*, American Association of Petroleum Geologists Memoir 31, pp. 365–404.

Johnson, B. A. and Walker, R. G. (1979). Paleocurrents and depositional environments of deep-water conglomerates in the Cambro-Ordovician Cap Enragé Formation, Quebec Appalachians, *Canadian Journal of Earth Sciences*, **16**, 1375–1387.

Kelling, G. and Woollands, M. A. (1969). The stratigraphy and sedimentation of the Llandoverian rocks of the Rhayader District, *in* A. Wood (Ed.), *The Pre-Cambrian and Lower Palaeozoic Rocks of Wales*, University of Wales Press, pp. 255–282.

Kolla, V. and Coumes, F. (1985). Indus Fan, Indian Ocean, *in* A. H. Bouma, W. R. Normark and N. E. Barnes (Eds), *Submarine Fans and Related Turbidite Systems*, Springer-Verlag, New York, pp. 129–136.

Komar, P. D. (1973). Continuity of turbidity current flow and systematic variations in deep-sea channel morphology, *Bulletin of the Geological Society of America*, **84**, pp. 3329–3338.

Link, M. H. (1981). Sand-rich turbidite facies of the Upper Cretaceous Chatsworth Formation, Simi Hill, California, *in* M. H. Link, R. L. Squires and I. P. Colburn (Eds), *Simi Hills Cretaceous Turbidites, Southern California*, Pacific Section, Society of Economic Paleontologists, Fall Field Trip Guidebook, pp. 63–71.

Link, M. H. and Nilsen, T. H. (1980). The Rocks Sandstone, an Eocene sand-rich deep-sea fan deposit, northern Santa Lucia Range, California, *Journal of Sedimentary Petrology*, **50**, 583–602.

Lovell, J. P. B. and Stow, D. A. V. (1981). Identification of ancient sandy contourites, *Geology*, **9**, 347–349.

Lowe, D. R. (1972). Implications of three submarine mass-movement deposits, Cretaceous, Sacramento Valley, California, *Journal of Sedimentary Petrology*, **42**, 89–101.

Lowe, D. R. (1979). Sediment gravity flows: their classification and some problems of application to natural flows and deposits, *in* L. J. Doyle and O. H. Pilkey (Eds), *Geology of Continental Slopes*, SEPM Special Publication No. 27, pp. 75–82.

Lowe, D. R. (1982). Sediment gravity flows: two depositional models with special reference to the deposits of high-density turbidity currents, *Journal of Sedimentary Petrology* **52**, 279–297.

Marschalko, R. (1970). The origin of disturbed structures in Carpathian turbidites, *Sedimentary Geology*, **4**, 5–18.

McHargue, T. R. and Webb, J. E. (1986). Internal geometry, seismic facies, and petroleum potential of canyons and inner fan channels of the Indus Submarine Fan, *American Association of Petroleum Geologists Bulletin*, **70**, 161–180.

McLean, H. and Howell, D. G. (1985). Blanca Turbidite System, California, *in* A. H. Bouma, W. R. Normark and N. E. Barnes (Eds), *Submarine fans and related turbidite systems*, Springer-Verlag, New York, pp. 167–172.

Menard, H. W. (1964). *Marine Geology of the Pacific*, McGraw-Hill, 271 p.

Middleton, G. V. (1970). Experimental studies related to problems of flysch sedimentation, *in* J. Lajoie (Ed.), *Flysch Sedimentology in North America*, Geological Association of Canada Special Paper 7, pp. 253–272.

Middleton, G. V. and Hampton, M. A. (1973). Sediment gravity flows: mechanics of flow and deposition, *in Turbidites and Deep-Water Sedimentation*, SEPM Pacific Section Short Course, Anaheim 1973, pp. 1–38.

Mitchum, R. M., Jr (1985). Seismic stratigraphic expression of submarine fans, *in* O. R. Berg and D. G. Woolverton (Eds), *Seismic Stratigraphy II*, AAPG Memoir No. 39, pp. 117–136.

Mutti, E. (1969). Sedimentologia delle Arenarie di Messanagros (Oligocene-Aquitaniano) nell' isola di Rodi, *Memorie della Società Geologica Italiana*, **8**, 1027–1070.

Mutti, E. (1977). Distinctive thin-bedded turbidite facies and related depositional environments in the Eocene Hecho

Group (south-central Pyrenees, Spain), *Sedimentology*, **24**, 107–131.
Mutti, E. (1979). *Turbidites et cones sous-marins profonds*, *in* Peter Homewood (Ed.), *Sédimentation Détritique* (*Fluviatile, Littorale et Marine*), Institut de Géologie, Université de Fribourg, Suisse, pp. 353–419.
Mutti, E. (1985). Turbidite systems and their relations to depositional sequences, *in* G. G. Zuffa (Ed.), *Provenance of Arenites*, NATO-ASI Series, Reidel Publishing Company, pp. 65–93.
Mutti, E. and Ghibaudo, G. (1972). Un esempio di torbiditi di conoide sottomarina esterna—Le Arenarie di San Salvatore (Formazione di Bobbio, Miocene) nell' Appennino di Piacenza, *Memorie dell'Accademia della Scienze di Torino*, Classe di Scienze Fisiche, Matematiche e Naturali, Serie 4, no. 16, 40 p.
Mutti, E. and Johns, D. R. (1978). The role of sedimentary by-passing in the genesis of fan fringe and basin plain turbidites in the Hecho Group system (south-central Pyrenees), *Memorie della Società Geologica Italiana*, **18**, 15–22.
Mutti, E. and Nilsen, T. H. (1981). Significance of intraformational rip-up clasts in deep-sea fan deposits, Abstracts Volume, 2nd European Regional Meeting of International Association of Sedimentologists, Bologna, pp. 117–119.
Mutti, E. and Ricci Lucchi, F. (1972). Le torbiditi dell' Appennino settentrionale: introduzione all' analisi di facies, *Memorie della Società Geologica Italiana*, **11**, 161–199.
Mutti, E. and Ricci Lucchi, F. (1975). Turbidite facies and facies associations, *in* E. Mutti *et al.* (Eds), *Examples of Turbidite Facies and Associations from Selected Formations of the Northern Apennines*, Field Trip Guidebook A-11, 9th International Association of Sedimentologists Congress, Nice.
Mutti, E. and Ricci Lucchi, F. (1981). Introduction to the excursions on siliciclastic turbidities, Abstracts of the 2nd European Regional Meeting International Association of Sedimentologists, Bologna, pp. 1–3.
Mutti, E., Nilsen, T. H. and Ricci Lucchi, F. (1978). Outer fan depositional lobes of the Laga Formation (Upper Miocene and Lower Pliocene), east-central Italy, *in* D. J. Stanley and G. Kelling (Eds), *Sedimentation in Submarine Canyons, Fans, and Trenches*, Dowden, Hutchinson and Ross, Stroudsburg, PA, pp. 210–223.
Mutti, E., Barros, M., Possato, S. and Rumenos, L. (1980). Deep-sea fan turbidite sediments winnowed by bottom currents in the Eocene of the Campos Basin, Brazilian offshore (abs.), Bochum, 1st European Meeting, International Association of Sedimentologists, p. 114.
Mutti, E., Remacha, E., Sgavetti, M., Rosell, J., Valloni, R. and Zamorano, M. (1985). Stratigraphy and facies characteristics of the Eocene Hecho Group turbidite systems, south-central Pyrenees, *in* M. D. Mila and J. Rosell (Eds), *Excursion Guidebook of the 6th European Regional Meeting of International Association of Sedimentologists, Lerida*, pp. 521–576.
Nardin, T. R., Edwards, B. D. and Gorsline, D. S. (1979). Santa Cruz Basin, California Borderland: dominance of slope processes in basin sedimentation, *in* L. J. Doyle and O. H. Pilkey (Eds), *Geology of Continental Slopes*, SEPM Special Publication No. 27, pp. 209–221.
Nelson, C. H. (1985). Astoria Fan, Pacific Ocean, *in* A. H. Bouma, W. R. Normark and N. E. Barnes (Eds), *Submarine Fans and Related Turbidite Systems*, Springer-Verlag, New York, pp. 45–50.
Nelson, C. H. and Nilsen, T. H. (1974). Depositional trends of modern and ancient deep-sea fans, *in* A. H. Dott and R. H. Shaver (Eds), *Modern and Ancient Geosynclinal Sedimentation*, SEPM Special Publication No. 19, pp. 69–91.
Nilsen, T. H. (1980). Modern and ancient submarine fans: discussion of papers by R. G. Walker and W. R. Normark, *American Association of Petroleum Geologists Bulletin*, **64**, 1094–1101.
Nilsen, T. H. (1985). Chugach Turbidite System, Alaska, *in* A. H. Bouma, W. R. Normark and N. E. Barnes (Eds), *Submarine Fans and Related Turbidite Systems*, Springer-Verlag, New York, pp. 185–192.
Nilsen, T. H. and Abbate, E. (1985). Gottero turbidite system, Italy, *in* A. H. Bouma, W. R. Normark and N. E. Barnes (Eds), *Submarine Fans and Related Turbidite Systems*, Springer-Verlag, New York, pp. 199–204.
Normark, W. R. (1970a). Growth patterns of deep-sea fans, *American Association of Petroleum Geologists Bulletin*, **54**, 2170–2195.
Normark, W. R. (1970b), Channel piracy on Monterey deep-sea fan, *Deep-Sea Research*, **17**, 837–846.
Normark, W. R. (1978). Fan valleys, channels, and depositional lobes on modern submarine fans—characters for recognition of sandy turbidite environments, *American Association of Petroleum Geologists Bulletin*, **62**, 912–931.
Normark, W. R. (1985). Local morphologic controls and effects of basin geometry on flow processes in deep marine basins, *in* G. G. Zuffa (Ed.), *Provenance of Arenites*, NATO-ASI Series, D. Reidel Publishing Company, pp. 47–63.
Normark, W. R. and Gutmacher, C. E. (1985). Delgada Fan, Pacific Ocean, *in* A. H. Bouma, W. R. Normark and N. E. Barnes (Eds), *Submarine Fans and Related Turbidite Systems*, Springer-Verlag, New York, pp. 59–64.
Normark, W. R. and Piper, D. J. W. (1972). Sediments and growth pattern of Navy deep-sea fan, San Clemente Basin, California Borderland, *Journal of Geology*, **80**, 198–223.
Normark, W. R., Barnes, N. E. and Bouma, A. H. (1985). Comments and new directions for deep-sea fan research, *in* A. H. Bouma, W. R. Normark and N. E. Barnes (Eds), *Submarine Fans and Related Turbidite Systems*, Springer-Verlag, New York, pp. 341–343.
Normark, W. R., Hess, G. R. and Spiess, F. N. (1978). Mapping of small-scale (outcrop-size) sedimentological features on modern submarine fans, *Offshore Technology Conference, Houston, Texas*, pp. 593–598.
Normark, W. R., Meyer, A. W. and Leg 96 Scientific Staff (1986). Summary of drilling results for the Mississippi Fan and

considerations for application to other turbidite systems, *in* A. H. Bouma, J. M. Coleman *et al.*, *Initial Reports of the Deep Sea Drilling Project*, **96**, Washington, DC, US Government Printing Office, pp. 425–436.
Normark, W. R., Piper, D. J. W. and Hess, G. R. (1979). Distributary channels, sand lobes, and mesotopography of Navy submarine fan, California Borderland, with applications to ancient fan sediments, *Sedimentology*, **26**, 749–774.
Normark, W. R., Hess, G. R., Stow, D. A. V. and Bowen, A. J. (1980). Sediment waves on the Monterey fan levees: a preliminary physical interpretation, *Marine Geology*, **37**, 1–18.
O'Connell, S. (1986). Anatomy of modern submarine depositional and distributary systems, Thesis, Columbia University.
Pickering, K. T. (1981). Two types of outer fan lobe sequence, from the late Precambrian Kongsfjord Formation Submarine Fan, Finnmark, northern Norway, *Journal of Sedimentary Petrology*, **51**, 1277–1286.
Pickering, K. T. (1982a). Middle-fan deposits from the late Precambrian Kongsfjord Formation Submarine Fan, north-east Finnmark, northern Norway, *Sedimentary Geology*, **33**, 79–110.
Pickering, K. T. (1982b). The shape of deep-water siliciclastic systems: a discussion, *GeoMarine Letters*, **2**, 41–46.
Piper, D. J. W. (1970). A Silurian deep sea fan deposit in western Ireland and its bearing on the nature of turbidity currents, *Journal of Geology*, **78**, 509–522.
Piper, D. J. W. and Normark, W. R. (1983). Turbidite depositional patterns and flow characteristics, Navy Submarine Fan, California Borderland, *Sedimentology*, **30**, 681–694.
Piper, D. J. W., Panagos, A. G. and Pe, G. G. (1978). Conglomeratic Miocene flysch, western Greece, *Journal of Sedimentary Petrology*, **48**, 117–126.
Piper, D. J. W., Stow, D. A. V. and Normark, W. R. (1985). Laurentian Fan, Atlantic Ocean, *in* A. H. Bouma, W. R. Normark and N. E. Barnes (Eds), *Submarine Fans and Related Turbidite Systems*, Springer-Verlag, New York, pp. 137–142.
Prior, D. B. and Bornhold, B. D. (1986). Sediment transport on subaqueous fan delta slopes, Britannia Beach, British Columbia, *GeoMarine Letters*, **5**, 217–224.
Ravenne, C. and Beghin, P. (1983). Apport des expériences en canal à l'interprétation sédimentologique des dépots de cones détritiques sous-marins: *I.F.P.*, **38**, 280–297.
Remacha, E. (1983). 'Sand tongues' de le Unidad de Broto (Grupo de Hecho), entre el anticlinal de Boltana y el Rio Osia (Prov. de Huesca), Tesis Doctoral Publ. Univ. Autonoma Barcelona, Bellaterra, 163 p.
Ricci Lucchi, F. (1969). Channelized deposits in the Middle Miocene flysch of Romagna (Italy): *Giornale di Geologia* **36**, 203–282.
Ricci Lucchi, F. (1975). Depositional cycles in two turbidite formations of northern Apennines, *Journal of Sedimentary Petrology*, **45**, 1–43.
Ricci Lucchi, F. and Ori, G. G. (1985). Syn-orogenic deposits of migrating basin system in the north-west Adriatic Foreland, *in* P. Allen, P. Homewood and G. Williams (Eds), *Foreland Basins: Excursion Guidebook*, pp. 137–176.
Ricci Lucchi, F. and Pignone, R. (1978). Ricostruzione geometrica parziale di un lobo de conoide sottomarina: *Memorie della Società Geologica Italiana*, **18**, 125–133.
Ricci Lucchi, F. and Valmori, E. (1980). Basin-wide turbidites in a Miocene, over-supplied deep-sea plain: a geometrical analysis, *Sedimentology*, **27**, 241–270.
Shanmugam, G. and Moiola, R. J. (1982). Eustatic control of turbidites and winnowed turbidites, *Geology*, **10**, 231–235.
Shanmugam, G. and Moiola, R. J. (1985). Submarine fan models: problems and solutions, *in* A. H. Bouma, W. R. Normark and N. E. Barnes (Eds), *Submarine Fans and Related Turbidite Systems*, Springer-Verlag, New York, pp. 29–34.
Siemers, C. T., Tillman, R. W. and Williamson, C. R. (1981). Deep-water sediments—a core workshop, SEPM Core Workshop, No. 2, San Francisco, 416 p.
Simo, A. and Puigdefabregas, C. (1985). Transition from shelf to basin on an active slope, Upper Cretaceous, Tremp area, southern Pyrenees, *in* M. D. Mila and J. Rosell (Eds), *Excursion Guidebook of the 6th European Regional Meeting of International Association of Sedimentologists, Lerida*, pp. 63–108.
Slaczka, A. (1983). Some deep-sea fans from the Polish Flysch Carpathians, *Anuarul Institutului de Geologie si Geofizica*, **62**, 341–346.
Smith, D. L. and Normark, W. R. (1976). Deformation and patterns of sedimentation, South San Clemente Basin, California Borderland, *Marine Geology*, **22**, 175–188.
Stelting, C. E. and DSDP Leg 96 Shipboard Scientists (1985). Migratory characteristics of a mid-fan meander belt, Mississippi Fan, *in* A. H. Bouma, W. R. Normark and N. E. Barnes (Eds), *Submarine Fans and Related Turbidite Systems*, Springer-Verlag, New York, pp. 283–290.
Stevenson, A. J., Scholl, D. W. and Vallier, T. L. (1983). Tectonic and geologic implications of the Zodiac Fan, Aleutian Abyssal Plain, north-east Pacific, *Geological Society of America Bulletin*, **94**, 259–273.
Stow, D. A. V. (1985). Brae oilfield turbidite system, North Sea, *in* A. H. Bouma, W. R. Normark and N. E. Barnes (Eds), *Submarine Fans and Related Turbidite Systems*, Springer-Verlag, New York, pp. 231–236.
Stow, D. A. V. and Lovell, J. P. B. (1979). Contourites: their recognition in modern and ancient sediments, *Earth Science Review*, **14**, 251–291.

Stow, D. A. V. *et al.* (1985). Mississippi Fan sedimentary facies, composition, and texture, *in* A. H. Bouma, W. R. Normark and N. E. Barnes (Eds), *Submarine Fans and Related Turbidite Systems*, Springer-Verlag, New York, pp. 259–266.
Surlyk, F. (1978). Submarine fan sedimentation along fault scarps on tilted fault blocks (Jurassic–Cretaceous boundary, east Greenland), *Gronlands Geologiske Undersogelse Bulletin*, **128**, 108 p.
Tillman, R. W. and Ali, S. A. (1982). Deep-water canyons, fans, and facies; models for stratigraphic trap exploration, AAPG, Reprint Series 26, 596 p.
Tokuhashi, S. (1979). Three-dimensional analysis of a large sandy-flysch body, Mio-Pliocene Kiyosumi Formation, Boso Peninsula, Japan, *Memoirs of Faculty of Science, Kyoto University, Series of Geology and Mineralogy*, **46**, 1–60.
Vail, P. R., Mitchum, R. M. and Thompson, S. (1977). Seismic stratigraphy and global changes of sea level. Part 4: Global cycles of relative changes of sea level, *in* C. E. Payton (Ed), *Seismic Stratigraphy, Application to Hydrocarbon Exploration*, AAPG Memoir 26, pp. 83–97.
Van Vliet, A. (1978). Early Tertiary deepwater fans of Guipuzcoa, northern Spain, *in* D. J. Stanley and G. Kelling (Eds), *Sedimentation in Submarine Canyons, Fans and Trenches*, Dowden, Hutchinson and Ross, Stroudsbourg, PA, pp. 190–209.
Walker, R. G. (1966). Deep channels in turbidite-bearing formations, *American Association of Petroleum Geologists Bulletin*, **50**, 1899–1917.
Walker, R. G. (1975). Nested submarine-fan channels in the Capistrano Formation, San Clemente, California, *Geological Society of America Bulletin*, **86**, 915–924.
Walker, R. G. (1978). Deep-water sandstone facies and ancient submarine fans: model for stratigraphic traps, *American Association of Petroleum Geologists Bulletin*, **62**, 932–966.
Walker, R. G. (1980). Modern and ancient submarine fans: *American Association of Petroleum Geologists Bulletin*, **64**, 1101–1108.
Walker, R. G. (1984a). Mudstones and thin-bedded turbidites associated with the Upper Cretaceous Wheeler Gorge conglomerates, California: a possible channel–levee complex, *Journal of Sedimentary Petrology*, **55**, 279–290.
Walker, R. G. (1984b). Turbidites and associated coarse clastic deposits, *in* R. G. Walker (Ed.), *Facies Models*, Geoscience Canada Reprint Series 1, pp. 171–188.
Walker, R. G. (1985). Cardium formation at Ricinus Field, Alberta: a channel cut and filled by turbiditic currents in Cretaceous western interior seaway, *American Association of Petroleum Geologists Bulletin*, **69**, 1963–1981.
Walker, R. G. and Mutti, E. (1973). Turbidite facies and facies associations, *in Turbidites and Deep Water Sedimentation*, SEPM Pacific Section Short Course, Anaheim 1973, pp. 119–158.

Chapter 2

Unravelling Hinterland and Offshore Palaeogeography from Deep-water Arenites

Gian Gaspare Zuffa, Dipartimento di Scienze Geologiche, Università degli Studi di Bologna, Via Zamboni, 67, 40127 Bologna, Italy

ABSTRACT

Integration of basin analysis with arenite petrography provides a powerful tool in obtaining palaeogeographic reconstructions. However, several factors need to be considered if ancient geological settings are to be correctly interpreted.

The following concepts are outlined:

1. A classification of the main types of grains in arenite frameworks according to temporal (coeval versus non-coeval), spatial (intrabasinal versus extrabasinal) and compositional criteria.
2. A calibrated procedure for detecting coeval versus non-coeval carbonate grains. This technique is applied, and the palaeogeographic implications are illustrated, in a case study: the Paleogene arenites of the Ager valley, Pyrenees.
3. A classification scheme for volcanic grains, involving five classes dependent on palaeovolcanic and neovolcanic sources. A case history, volcanic Quaternary sands deposited in the Nankai Trough off south-western Japan, illustrates the application of the scheme.
4. Compositional and textural criteria which manifest the presence of multicycle as opposed to first-cycle components in the siliciclastic sand framework. The proposed criteria area applied in a case history: modern sands carried by rivers to the Northern Adriatic Sea.
5. The effect of grain size on detrital mode data obtained by optical analyses. A slight modification of the Dickinson point-counting method (1970) as originally proposed by Gazzi (1966) enables loss of information to be avoided and the dependence of composition on grain size to be minimized.
6. Factors which complicate the connection between source area and depositional basin in palaeogeographic reconstructions. Examples are presented for the influence of sea-level variations, littoral drift currents, dominant winds, continental glaciations, cannibalism and tectonic setting.

INTRODUCTION

The idea that the sand particles of an arenite can be considered a microscopic representation of the geological mosaic of the source area is very old. However, it is now clear that reliable palaeogeographic reconstructions of source areas cannot be obtained in a straightforward fashion from the constituent minerals and rock fragments in arenites. Because of the different mechanical and chemical stability of minerals, weathering of the rock exposed in the source area can generate detritus which may not be wholly representative, in either type or proportion, of the rocks exposed in the hinterland (e.g. Basu, 1985). Sand grains can be selected during transport towards their final depositional site according to

Marine Clastic Sedimentology © J. K. Leggett & G. G. Zuffa (Graham and Trotman, 1987) pp. 39–61

their mechanical durability, grain size, density and shape (e.g. Cameron and Blatt, 1971; Davis and Etheridge, 1975; Dietz, 1973; Keunen, 1959; Mack, 1978). Furthermore, the mineralogical and chemical make-up of sands is modified to varying degrees by diagenetic processes, and this can result both in losses of and enrichment in certain minerals (e.g. McBride, 1985).

These processes can make correct reconstructions difficult. In addition, in those cases where the influence of such factors appears to have been negligible we can easily deceive ourselves into thinking that a relatively straightforward provenance determination is correct. But this might not be so: in practice, every basin is fed by source areas in which old sandstone formations are present and multicycle siliciclastic particles can be mistaken for first-cycle grains (i.e. derived from igneous or metamorphic rocks), thus causing errors in reconstruction. Unfortunately, we currently lack the proper tools to evaluate these (Blatt, 1967).

Furthermore, it is commonly assumed that all analysts adopt a standard routine procedure for compositional analysis of arenite framework. This is not the case. Techniques for point-counting analysis differ from one author to another. The dependence of rock composition on grain size, and misinterpretation of carbonate and volcanic particles (which can be either coeval with or older than the deposit), can lead to errors and/or serious complications in reconstructing hinterland and offshore palaeogeography (Zuffa, 1980, 1985).

And, last but not least, it is necessary to consider the location both in space and in time of the source–area/depositional-basin complex. This requires that basin analysis be carefully carried out, and that sedimentary petrologists carefully integrate their data with the observations of field sedimentologists.

This paper concentrates on three of the numerous factors which complicate palaeogeographic reconstructions, and outlines methods to combat the problems involved:

1. the influence of methodology on compositional results: (a) how to deal with coeval versus non-coeval carbonate and volcanic particles, and (b) how to minimize the dependence of arenite composition on grain size without losing information;

2. the influence of recycling old sandstone formations;

3. the influence of the connection between source area and depositional basin in complicating palaeogeographic reconstructions.

Figure 1 shows a tentative scheme which can be used in reconstructing an ancient source-area/deep-water basin system. The flow diagram shows the relationships between source areas and basins in terms of their products, of the methods used, of the results obtained and of those factors which complicate or limit interpretation of hinterland and offshore palaeogeography within a particular geotectonic setting. Particular emphasis is given to the role played by the almost contemporaneous shallow-water deposits which, if present, act both as intermediate sedimentation site and as a source of coeval material available for the ultimate depositional basin (Ricci Lucchi, 1985).

INFLUENCE OF METHODOLOGY ON ANALYTICAL RESULTS

Types of grains

When considering arenite frameworks, it must be remembered that the composition and texture of each grain is essentially independent of surrounding grains. For this reason sedimentary petrology is a more difficult field of study than igneous and metamorphic petrology (Blatt, 1967). The problem in obtaining a quantitative optical description of an arenite framework lies in selecting an approach which is not only descriptive (mainly objective) but which can also take the genetic aspects into account. Every meaningful mineralogical, textural and palaeontological characteristic must be considered in selecting the petrographic classes to be used in obtaining palaeogeographic reconstructions both of the source area and depositional basin.

In Table 1, the main types of sand grains which make up the framework of marine arenites are listed in a scheme that takes into account temporal, spatial and compositional criteria. In order to understand the limits of the classification criteria adopted in this table, some definitions and assumptions must be introduced:

(i) The classification is mainly applicable to arenites from deep marine basins where the objective is to obtain a palaeogeographic reconstruction of the source/basin system.

(ii) Shallow-water areas in marine basins can occur either as ultimate depositional sites or as interim storage sites for sediments prior to later redeposition in a deeper depositional site.

(iii) Sand particles produced by chemical and biochemical processes while a particular stratigraphic

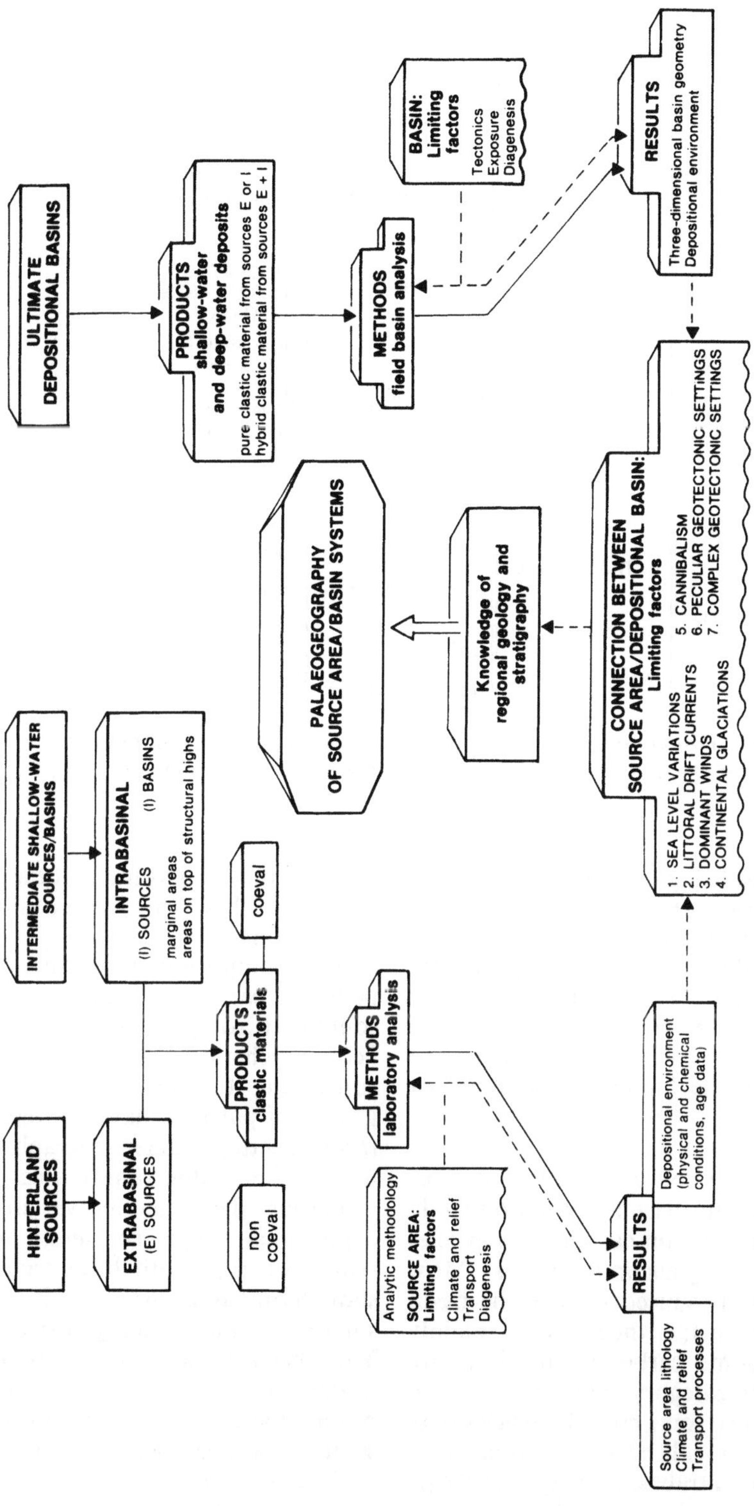

Fig. 1. Unravelling the palaeogeography of ancient source-area/deep-water basin systems: flow diagram.

TABLE 1. Classification of sand particles in palaeogeographic reconstruction of source area and depositional basin.

CRITERIA	TEMPORAL	SPATIAL	COMPOSITIONAL	SAND GRAINS	IDENTIFICATION CODE
TYPES OF GRAINS (The classification mainly applies to deep-marine depositional basins)	COEVAL (1)	EXTRABASINAL	CARBONATE	ooids bioclasts caliches travertine	CE_c
			NONCARBONATE	VOLCANICS (V_{2a}, V_{2b}, V_3) (4) vegetation particles	NCE_c
		INTRABASINAL	CARBONATE	OOIDS BIOCLASTS INTRACLASTS PELOIDS	CI_c
			NONCARBONATE	glauconite (2) iron-oxides gypsum phosphates rip-up clasts (3) volcanics (V_4) (4)	NCI_c
	NONCOEVAL (1)	EXTRABASINAL	CARBONATE	LIMESTONES DOLOSTONES	CE_{nc}
			NONCARBONATE	SILICICLASTS	NCE_{nc}
		INTRABASINAL	CARBONATE	limestones dolostones	CI_{nc}
			NONCARBONATE	siliciclasts	NCI_{nc}

The terms refer to the time interval of deposition of the sedimentary sequence under consideration. Those in capital letters indicate major components; those in small letters indicate minor components. (V_{2a}, V_{2b}, V_3, V_4 are explained in Fig. 5) (after Zuffa, 1985).

sequence (*sensu lato*) was being deposited, are regarded as coeval (penecontemporaneous) with that sequence. These particles provide information on the chemical–physical conditions of the depositional basin, whereas sand particles produced by the weathering of rocks are non-coeval with respect to the depositional sequence which contains them, and provide information on the hinterland source area (Fig. 2).

A large majority of the coeval particles are produced within the basin (intrabasinal) and are mainly carbonate, although some non-carbonate particles are also to be found (e.g., glauconitic grains). Most non-coeval particles are produced outside the basin (extrabasinal, terrigenous); they are mainly siliciclastic, although some carbonate particles are also to be found in this category. This classification has some limitations, since some coeval particles may be extrabasinal (e.g. soil caliches) and some non-coeval particles may be intrabasinal (e.g. siliciclastic grains from crystalline intrabasinal uplifted blocks). However, these types of grains do not comprise much of the total volume of the arenite record. Consequently they can be considered as geological oddities only.

The compositional distinction used in Table 1 between carbonate and non-carbonate grains is somewhat rough, and other more detailed classification schemes could be suggested. However, data concerning the age and genesis of these peculiar groups of particles have proved useful in practice in reconstructing palaeogeography from marine arenites in the Mediterranean area (Zuffa, 1980, 1985).

During the time interval in which a particular depositional sequence (*sensu lato*) forms we must remember that both local tectonics and long- and short-term variations in sea level can expose intrabasinal areas and submerge extrabasinal areas (Fig. 2). Thus between what might be called a 'permanent hinterland source area' and a 'final depositional basin' there lies an intermediate domain which switches between being an intrabasinal and an extrabasinal area. Evidence for the existence of such a domain is difficult to obtain from the sand particles themselves. If particles which originated within the

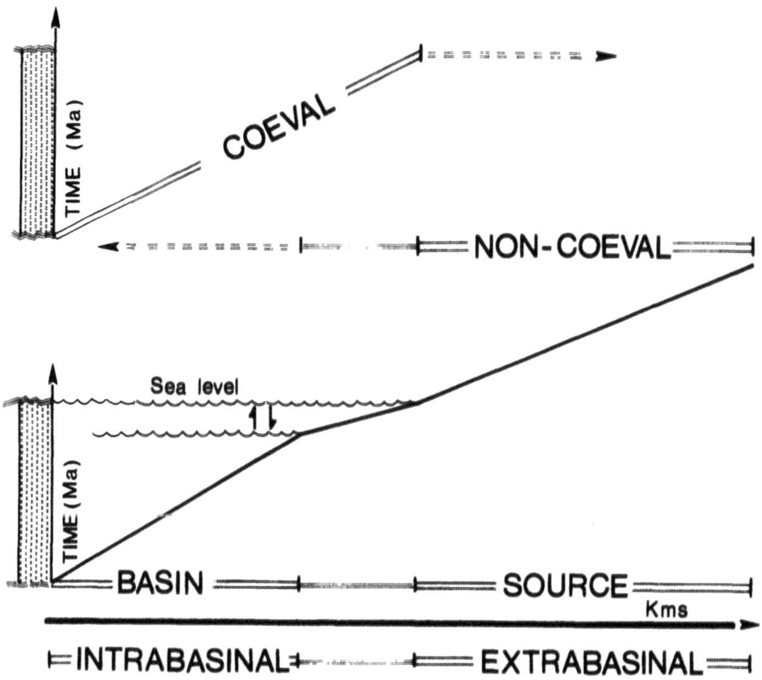

Fig. 2. Schematic illustration of temporal and spatial characteristics of sand particles in relation to basin and source area for a particular sedimentary sequence (*sensu lato*).

basin become subaereally exposed due to a lowering of sea level, they will occupy an extrabasinal position, but they will continue to provide data concerning the conditions of the depositional basin. On the other hand, if extrabasinal areas are flooded by a rise in sea level, they will become part of the intrabasinal domain. The arenitic sedimentary cover will still contain data relevant to the source area, even though new coeval grains may later mix with non-coeval grains. The status of temporal (coeval and non-coeval) and spatial (intrabasinal and extrabasinal) concepts does not hold perfectly, then, for coastal domains, but it can be considered a fairly good approximation if our interest is primarily focused on the hinterland areas and deep-sea deposition sites.

Although the compositional attributes of particles are obtained by objective criteria, their temporal and spatial characteristics cannot always be deduced with absolute confidence. However, criteria for distinguishing between extra- and intrabasinal carbonate particles and between the various types of volcanic grains have proved useful when applied by the author and co-workers in several studies of modern and ancient arenite sequences (see Zuffa, 1985: Tables 1 and 2 and Figs 2 and 3).

The types of grains listed in capital letters in Table 1 are the ordinary constituents of the marine arenite geological record. The types of grains listed in small letters are considered to be geological oddities only; they may be relevant only in very unusual settings.

Coeval extrabasinal carbonate grains

These include lacustrine ooids and bioclasts, soil caliche, and travertine originating in extrabasinal areas during deposition of the sequence considered. They provide data concerning the physical and chemical environment existing in the hinterland. Although they can readily be classified according to their palaeontological and textural characteristics, they are very rare in marine arenites.

Coeval extrabasinal non-carbonate grains

This class comprises volcanic and vegetal grains. The former originate in two different ways: from penecontemporaneous volcanism located in the source area (V_{2a} and V_{2b}: see later), or from active volcanoes located outside the geotectonic province considered (V_3). How these particles can be distinguished, and what implications they have in palaeogeography, are discussed in a later section.

Coeval intrabasinal carbonate grains

These are particles as classified petrographically by Folk (1962a). The class 'peloids' has been adopted to include all intrabasinal carbonate particles with cryptocrystalline features (McKee and Gutschick, 1969).

Coeval intrabasinal non-carbonate grains

These comprise non-carbonate particles of intrabasinal origin such as glauconite, iron-oxides, gypsum, phosphate, etc.

Rip-up fine-grained particles generated by erosion processes within the basin are also included; the textural features which enable these grains to be recognized have been discussed in Zuffa (1985, p. 172). Grains derived from penecontemporaneous volcanism located within the basin itself also form part of this group (V_4, see later).

Concerning intrabasinal coatings on extrabasinal grains, the following approach is recommended. Suppose a sample contains quartz sands coated by concentrically laminated intrabasinal phosphates (see, for example, Swett and Crowder, 1982). During point counting analysis we would apportion 'quartz' or 'phosphate' depending on what is beneath the cross hair. Given a statistically large enough sample, this would yield data both on the lithology of the source area and on the environmental conditions of the basin during deposition.

Non-coeval extrabasinal carbonate

In certain geotectonic settings such as foreland basins in collisional orogens, extrabasinal carbonate particles can constitute an important or even exclusive component of both ancient and modern arenitic sequences (Zuffa, 1980, 1984). Unreliable reconstructions of palaeobasin and palaeosource areas can result from misinterpretation of arenites containing a mixture of carbonate grains which are coeval or non-coeval with the depositional sequence (hybrid arenites). This problem is discussed in the next section.

Non-coeval extrabasinal non-carbonate grains

This class comprises the common siliciclasts of sandstone frameworks, such as quartz, feldspars, silicate rock fragments, etc. It also includes particles derived from the erosion of old volcanic sequences located in the source area (V_1, see later).

Non-coeval intrabasinal carbonate grains

These are particles which enter deeper environments as a result of mass-flow from structural highs within the depositional basin. This occurred, for example, in the Tethyan domain of the Southern Alps during Jurassic and Cretaceous times, when the drowning of carbonate platform areas resulted in a complex horst and graben morphology (Bernoulli and Jenkyns, 1974), and when later compressive tectonic phases triggered gravity-flows which introduced non-coeval particles from deep within the stratigraphy of the platform into intrabasinal settings (E. Garzanti, personal communication).

Non-coeval intrabasinal non-carbonate grains

These particles originate in the same geological setting as non-coeval intrabasinal carbonate grains where the stratigraphy on structural highs contains siliciclastic sediments. These last two categories are not very common in the geological record of arenites, and field data generally provide conclusive evidence concerning the original sources of these particles (e.g. Santantonio and Teale, this volume; Teale and Young, this volume).

Although the classification proposed may appear somewhat speculative, I feel that it provides a useful template for organizing data in the study of arenites for palaeogeographic reconstructions. In Table 1, those groups of grains denoted by small letters are geological oddities and cannot always be recognized on an optical basis alone. Even some of those denoted by capital letters, i.e. the most common constituents of arenite, cause difficulties, since their spatial and/or temporal attributes are not always easy to decipher. But among those grains which make up the bulk of marine arenites, the main problems are linked with the distinction between coeval and non-coeval carbonate and volcanic grains. Accordingly, the following two sections will concentrate on the difficulties present both at the analytical and interpretative stages, and will outline methods to overcome the problems involved.

Carbonate grains

Analytical technique

The importance of distinguishing in hybrid arenites between terrigenous limestone/dolostone rock fragments and intrabasinal carbonate allochems for the purposes of palaeogeographic reconstruction has been emphasized by Zuffa (1980). Results from recent symposia, and much data from modern marine sands, have highlighted the fact that hybrid arenites are more abundant than is generally believed (Doyle and Roberts, 1983; McIlreath and Ginsburg, 1982; Mack, 1984; Mount, 1984, 1985; Walker *et al.*, 1983;

TABLE 2. Criteria for distinguishing between extrabasinal (non-coeval) and intrabasinal (coeval) carbonate particles of hybrid arenites.

MAIN TYPES OF CARBONATE SAND PARTICLES		LIMECLASTS: INTRABASINAL-COEVAL	LIMECLASTS: EXTRABASINAL-NONCOEVAL
COMPOSITION	MINERALOGY	Calcite (Mg-calcite, aragonite, and rarely aphanitic dolomite for modern sediments)	Calcite, dolomite and ankerite
COMPOSITION	FAUNA	Not older than the host sediment Bioclasts may be abundant	Fossils may be present in carbonate rock fragments which are older than the host formation (1) – The occurrence of reworked old single bioclasts in the host formation would be unusual
TEXTURE	GRAIN SIZE	May be different, typically coarser than terrigenous particles	Generally same as other terrigenous siliciclasts
TEXTURE	SPHERICITY AND ROUNDNESS	**Shallow water: winnowing absent** a) Peloids (McKee and Gutschik, 1969) and some fossil molds tend to have high values of sphericity b) Soft intraclasts occur with irregular and/or sinuous contours, also they exhibit an elongated and/or platy shape – Skeletal debris is generally angular **Shallow water: winnowing present** All allochemical components are well-rounded and tend to have high values of sphericity	Generally rounded and with high values of sphericity
TEXTURE	INTERNAL FEATURES	Common fabric of allochemical components (see Folk, 1974) Absence of particles with internal veins Absence of recrystallized particles	Particles with characteristics (i.e., color, oxidized contours) which can not be associated with the common characteristics of intrabasinal allochems Particles with internal veins are present Recrystallized particles (e.g., dolostones, metacarbonates) may be present (2)
INDIRECT EVIDENCE		IF BIOCLASTS OF THE SAME AGE AS THE DEPOSIT ARE PRESENT, THEN OTHER INTRABASINAL CARBONATE PARTICLES ARE LIKELY TO BE PRESENT	IF DOLOMITE AND/OR ANGULAR CHERT ROCK FRAGMENTS ARE PRESENT, THEN OTHER EXTRABASINAL CARBONATE PARTICLES ARE LIKELY TO BE PRESENT (3)
INFORMATION FROM FACIES STUDIES		TOTAL EVIDENCE OF THE DEPOSITIONAL ENVIRONMENT	TOTAL EVIDENCE OF THE DEPOSITIONAL ENVIRONMENT
GEOLOGY		GEOTECTONIC FRAMEWORK	GEOTECTONIC FRAMEWORK

When the distinction cannot be made with confidence, particles are assigned to a neutral class (limeclasts).

Wolf, 1965; Zuffa, 1984). However, the detection of terrigenous versus intrabasinal carbonate particles is still considered 'a difficult and often interpretative process' (Mount, 1985). At a recent NATO-ASI meeting held in Cetraro (Southern Italy) on 'provenance of arenites', some criteria for the distinction of carbonate grains during optical analysis were proposed (De Rosa and Zuffa, 1979; Gandolfi *et al.*, 1983; Zuffa, 1980; Zuffa and De Rosa, 1978). Particularly useful has been a study of arenites from the Hecho Group (Eocene, Pyrenees) (Mutti *et al.*, 1985), which are rich (up to more than 50% of the framework) in carbonate clasts, either coeval or non-coeval with respect to the time interval of deposition of the turbidite sequence. A revised list of criteria used is reported in Table 2, and Fig. 3 shows a flow diagram which illustrates the use of a binary process for distinguishing between intrabasinal (coeval) and extrabasinal (non-coeval) carbonate grains.

The thin section (stained for carbonate) is first checked in order to test for the presence of dolostone lithoclasts (or recrystallized and/or foliated carbonate rock fragments); in clastic depositional systems these can be considered not to be coeval with the deposit. If, as in most cases, other carbonate particles are present, we shall look for fossiliferous clasts which can be used in determining age. If on this basis non-coeval carbonate clasts are detected, we can

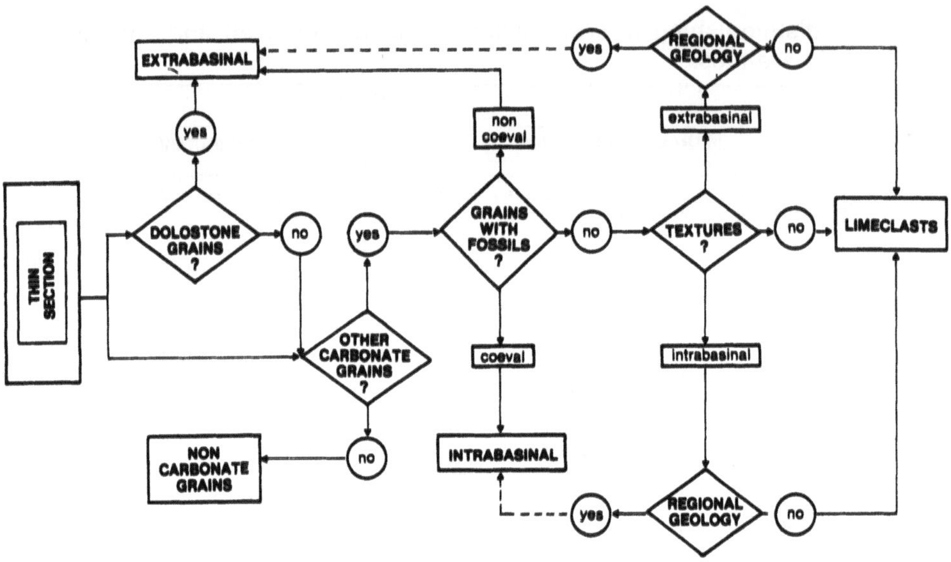

Fig. 3. Operative flow diagram for detection of coeval (mainly intrabasinal) versus non-coeval (mainly extrabasinal) carbonate grains.

assign the population of grains similar to the dated particles to the extrabasinal terrigenous group. In my experience, the tintinnid fauna of Late Jurassic–Early Cretaceous age is the most common and useful indicator of non-coeval, i.e. terrigenous, clasts in the Tertiary deposits of the Pyrenees and Apennines. When only terrigenous clasts are present in the carbonate component, determination of affinities is generally a trivial task (Fontana and Mantovani, in press). If the age of single bioclasts (which are normally abundant in the intrabasinal component) or fossil-bearing intraclasts enables coeval grains to be recognized, we can similarly assign to the intrabasinal population all grains similar to or compatible with the dated particles.

Once we know that coeval grains are also present we can further separate the two components by looking at the textural characteristics of the two populations; these are texturally different and allow more objective determinations of individual grains.

In some cases fossiliferous clasts may be absent, and the sample may consist only of micritic structureless particles. The age of these particles may sometimes be determined by extracting small amounts of micritic material from some grains using an uncovered slide on the stage of an optical microscope, and examining the nannoflora.

If all attempts are unsuccessful, we then go on to investigate the fabric and texture of grains. Table 1 shows what can be noted about grain size, sphericity, roundness and internal texture of particles. Particles may display a terrigenous character (oxidized rims, veins, size in the range of the siliciclastic population, high roundness and sphericity values, well lithified appearance, grains not squeezed among the more rigid siliciclasts, pressure solution contacts, etc.). We may also know from the regional geology and geotectonic setting that old carbonate sequences were likely to crop out in the source area and that their lithological characteristics are similar to those of the grains observed. For example, in collision-orogen foreland basins a supply of carbonate detritus from subaerial areas is commonplace. If all the information is consistent we may assign carbonate particles to the extrabasinal group with reasonable confidence.

If particles display evidence of intrabasinal character (soft grains with irregular or sinuous contours, elongated and/or flat-shaped particles showing a different mode—generally coarser with respect to the terrigenous quartz–feldspar population, peculiar diagenetic features such as extensive replacements or glauconitization) then we can weigh these observations against all the available information from regional geology and geotectonic setting. We may know that wide platform areas with carbonate 'factories' were likely to have been active during deposition, thus allowing turbidite resedimentation of both extrabasinal and intrabasinal detritus to the deep marine basin. The geotectonic setting could have been favourable for shelf carbonate production, as in the case for a normal passive margin sequence.

If all the data are consistent we may assign carbonate particles to this group.

When only one textural category of grains is present we can usually be sufficiently confident to attribute the grains in total either to the intrabasinal or extrabasinal classes. But, if both components are present, quantitative data cannot normally be completely relied on: only a semi-quantitative indication of the carbonate component can be given for the arenitic sequence studied.

Finally, if all determinations are unsuccessful we must assign the carbonate grains to the class '*limeclasts*'. This includes all the fine-grained particles which cannot be recognized with certainty as terrigenous or intrabasinal (Blatt *et al.*, 1972; Wolf, 1965; Zuffa, 1980). Even if we end up with most grains apportioned to the limeclast class, we have discovered more about our arenite than those petrographers who—in thinking that carbonate distinction is a difficult task—have chosen from the beginning to assign all the carbonate particles to a single class of carbonate clasts.

Note also that fluorescence and cathodoluminescence microscopy techniques (cf. Dravis and Yurewicz, 1985; Matter and Ramseyer, 1985) could well enable the nature of intrabasinal versus terrigenous carbonate particles to be better detected.

Case history

The application of the criteria described and the implications for palaeogeographic reconstructions can be shown by a case study of the marine Paleogene arenites of the Ager Valley in the Pyrenees (De Rosa and Zuffa, 1979; Zuffa, 1985). The sequence is about three hundred metres thick; sedimentation in the basin starts with carbonates and evolves by means of a transitional facies to a proximal littoral facies where the fluvial influence is felt. Figure 4 shows compositional plots of these arenites using three different approaches:

1. the carbonate particles are all considered to be intrabasinal and are consequently ignored (Fig. 4a), in which case we end up with mature litharenites;
2. the carbonate particles are all considered to be terrigenous rock fragments (Fig. 4b), in which case we end up with compositionally immature litharenites;
3. the carbonate particles are split into extrabasinal and intrabasinal according to their features (Fig. 4c).

The last approach enables a separation of the different grain populations which demonstrates how the sequence evolves from tidal bar–platform deposits (lower right corner of Fig. 4c) into fluvially dominated littoral deposits. It is clear that, even assuming some analytical bias, the present approach enables the evolution of the Ager Valley Paleogene sequence to be elucidated on a petrographic basis alone. This result was achievable because early carbonate particles were demonstrably intrabasinal on textural grounds, whereas the later carbonate particles had clearly been derived by the erosion of old carbonate complexes incorporated in an uplifted orogen.

Volcanic grains

Analytical technique

Grains derived by the erosion of ancient volcanics, either on continental areas or incorporated in orogenic belts, provide information on the old geological history of the source area and are here referred to as palaeovolcanic (non-coeval) grains (V_1 of Fig. 5).

Grains generated by active volcanism, either in the main source area or by submarine activity within the basin itself provide information on the geological setting of both source area and basin during deposition and are here referred to as neovolcanic (coeval) grains (V_{2a}, V_{2b} and V_4 of Fig. 5).

A third category of neovolcanic particles (V_3), which can be transported by wind or ejected by catastrophic explosions from geotectonic domains not directly related to the subaerial source area depositional basin system, are also considered in the scheme of Fig. 5.

Although distinguishing between the above categories is perhaps one of the most intricate tasks in optical analysis, it can shed new light on palaeotectonic reconstructions even when only qualitative information is obtainable. Using criteria proposed by the author (Zuffa, 1985; Table 2), quantitative data on these categories have been recently obtained in a study of Quaternary sands from the Nankai Trough (De Rosa *et al.*, 1986). These data are slightly modified here, and displayed in the form of a flow diagram.

In the flow chart in Fig. 5, the first easy distinction comes from field data. If volcanic grains are concentrated in single distinct, almost pure, pyroclastic beds, they can be safely identified as grains of the coeval groups V_4, V_3 and V_{2b}. Thin-section analyses involving compositional and textural determinations, integrated with specific information from the regional geology, can further help apportion particles to the various groups. Layers of type V_4 are graded and made of altered, poorly sorted, and morphologically

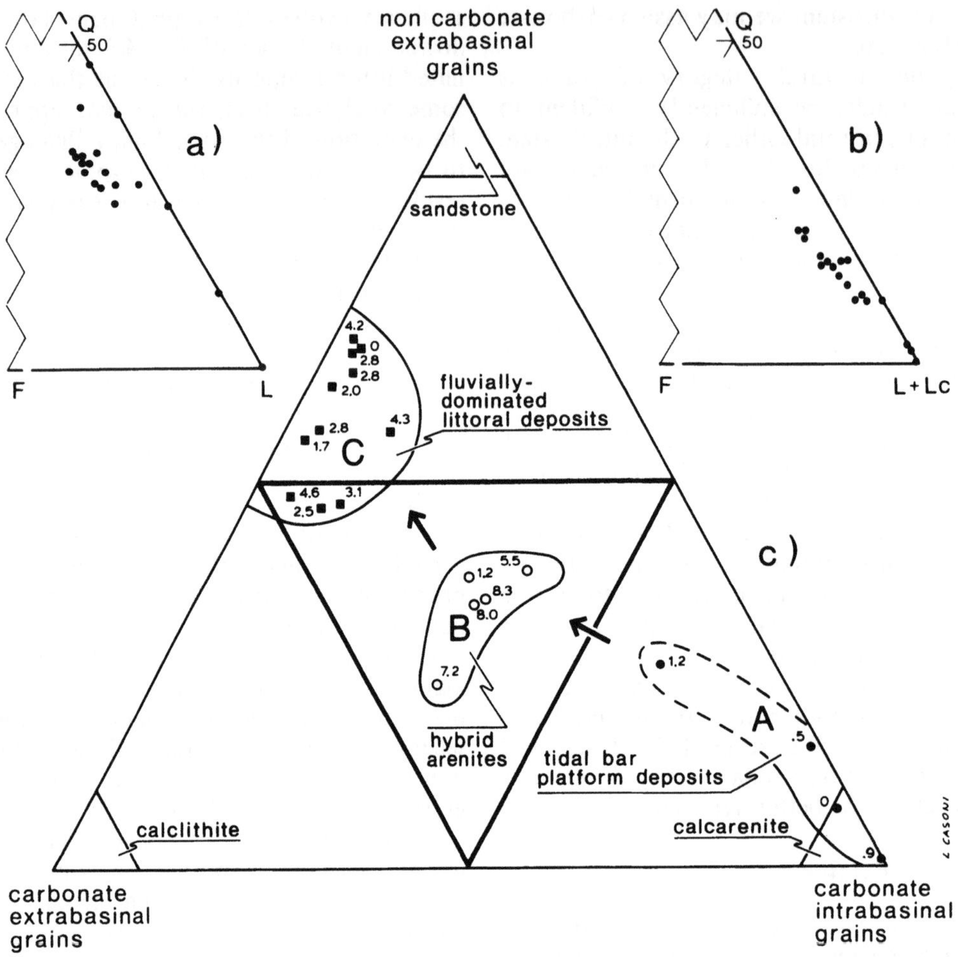

Fig. 4. Composition of Paleogene arenites of the Ager Valley (Pyrenees, Spain: data from De Rosa and Zuffa, 1979): an example of applying the criteria for distinction between coeval and non-coeval carbonate grains. (a) The L pole includes silicate rock fragments only; (b) the L + Lc pole also includes all the carbonate grains present in the rocks; (c) the diagram gives a first-level classification (Zuffa, 1980) of all the sand grains according to their intrabasinal or extrabasinal nature. The numbers beside the representative points indicate the non-carbonate intrabasinal grain content (NCI: glauconitic grains). NCE: non-carbonate extrabasinal; CE: carbonate extrabasinal; CI: carbonate intrabasinal. See text for further explanations.

immature grains which can be derived by direct ejection from submarine or even subaerial vents, and/or resedimentation from intrabasinal volcanic highs. The age of these grains is normally coeval with the depositional sequence; however, in some cases, non-coeval particles can be generated through gravity-flow mechanisms from faults which dig deeply into the body of a submarine volcanic high.

Layers of type V_3, because of eolian differentiation due to their long aerial transport, are made up of fine-grained, well-sorted vitric to vitrolithic grains of peculiar composition (e.g., De Rosa *et al.*, 1986: Table 1, sample 1).

Layers of type V_{2b}, which are derived by direct ejection of pyroclastic material from explosive volcanoes located in the main source area, can be very common in sedimentary sequences. They are usually graded, and the grain size population is likely to be distinct from that of terrigenous detritus produced through ordinary surface processes in the subaerial hinterland. The grains will be morphologically immature, generally unaltered and enriched in crystals and vitric particles which should be compositionally consistent with group V_{2a}.

The coeval particles of type (V_{2a}) have a similar grain size and morphology to that of other non-volcanic clasts. Alteration can be slight or moderate and the composition is usually different from palaeo-

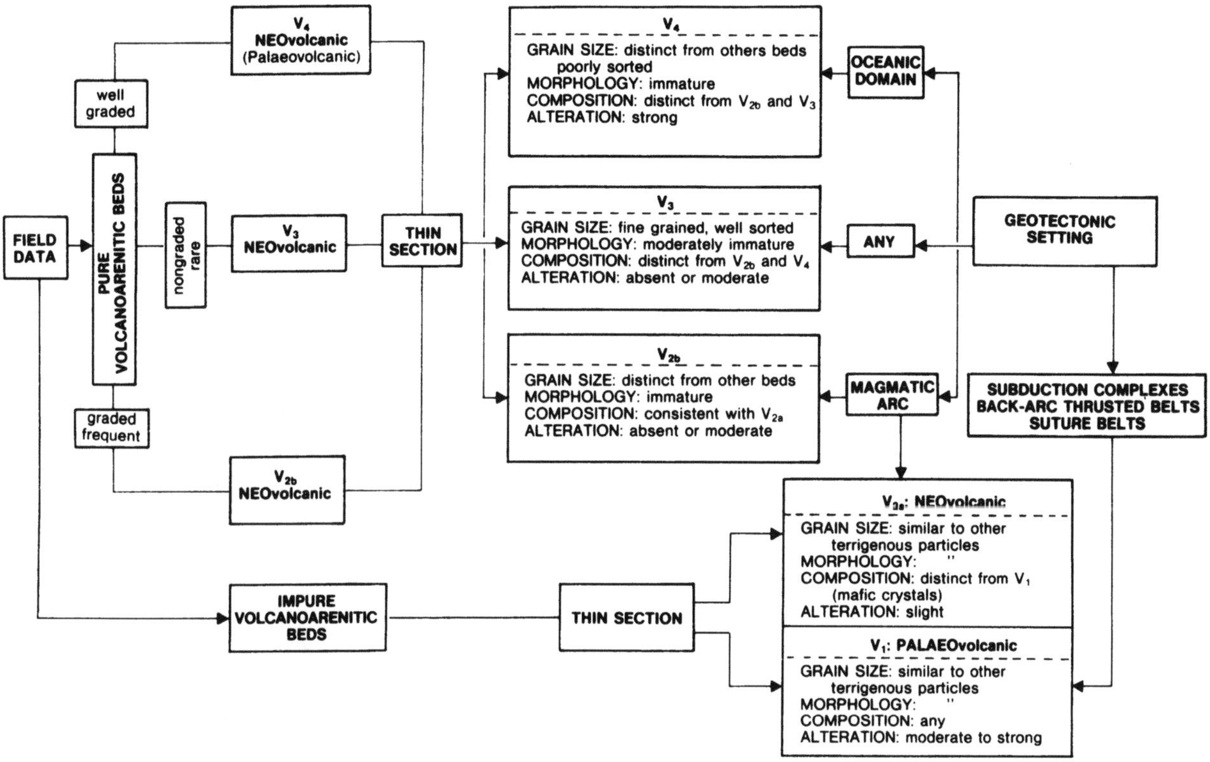

Fig. 5. Flow diagram for distinguishing types of volcanic grains in deep-sea arenites (where diagenetic alteration is absent or slight). V_1: derived from erosion of old volcanic suites; V_2: derived from active volcanism located in the source area; V_3: derived from active volcanoes not located in the main source area; V_4: derived from active volcanoes located within depositional basin.

volcanic particles of group V_1. The presence of coeval particles of type V_{2a} commonly implies the occurrence of type V_{2b} beds derived from the same volcanic source, but emplaced through mechanisms of pyroclastic fall, flow and surges and not through surface erosion and transportation processes.

The non-coeval particles (V_1) have a similar grain size and morphology to the other types of grains since they have common predepositional processes. Grain alteration can be moderate or strong according to their composition and type of eruption (Heiken, 1972).

Impure volcanoarenitic beds have a composite framework made up by volcanics and other types of grains. The volcaniclastic component may be entirely coeval, non-coeval or even a mixture of both. When mixtures of both coeval and non-coeval particles are found in the same bed it is difficult to distinguish them during optical analysis when sequences have been subject to strong diagenesis.

This technique for classifying volcanic grains has so far been applied in the study of sands from modern settings only (De Rosa *et al.*, 1986; Gazzi *et al.*, 1973) where the geological setting is known and diagenesis absent. For ancient settings it has not yet been tested as a working procedure.

Case history

The likely utility of the technique in palaeotectonic reconstructions is illustrated by the results of a study of Quaternary sands deposited in the Nankai Trough of South-west Japan (Fig. 6, modified from De Rosa *et al.*, 1986).

Twenty-three samples of Quaternary sands from DSDP Leg 87 Sites 582 (trench axis) and 583 (lowermost terrace of uplifted trench sediments in the accretionary prism) show a 70–80% volcanic component in the terrigenous grain population. This component comprises 30–40% neovolcanic grains, among which mafic and intermediate types are present in roughly equal proportions, and 60–70% palaeovolcanic grains, which are predominantly of acidic composition. No volcanic terranes occur in the hinterland of the Shikoku portion of the Nankai Trough, and the

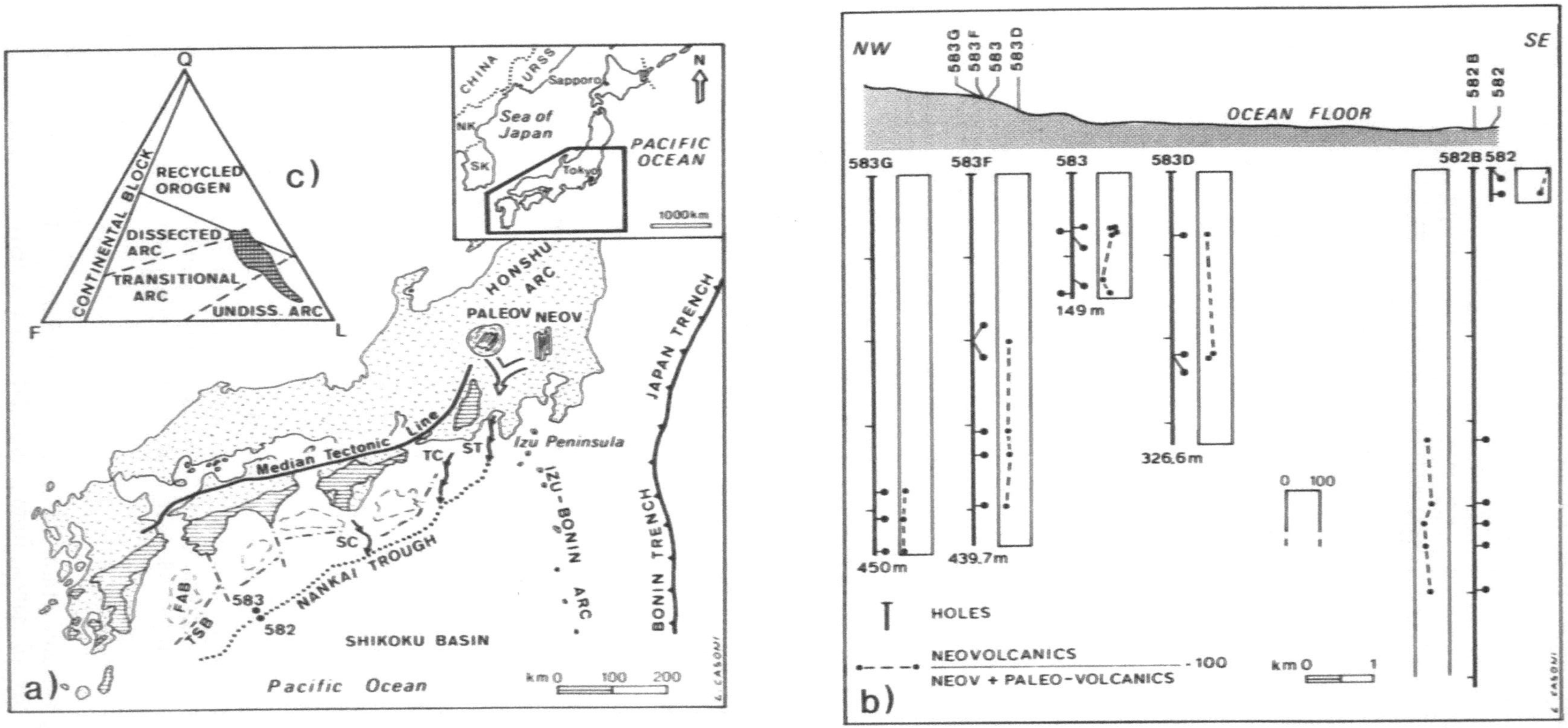

Fig. 6. (a) Regional setting of the Nankai Trough (South-west Japan) and location of DSDP sites 582 and 583. The dark pattern onland is the Shimanto terrane. All other inland terranes are dotted. The only three submarine canyons to penetrate all the way from the shelf to the trench-floor are shown by arrows. FAB: fore-arc basins; TSB: trench slope break; ST: Suruga Trough; TC: Tenryu Canyon; SC: Shionomisaki Canyon. Terrigenous volcanic material comprising both palaeovolcanic (PALEOV) and neovolcanic (NEOV) (Quaternary) detritus (70–80% of the sand fraction) is derived from the Tokai drainage basin and then emplaced, via the Suruga Trough, to the Nankai Trough. Fore-arc basins (FAB) trap the majority of the detritus reworked from the onland accretionary prism; (b) Log diagram showing the amount of neovolcanic versus total volcanic contributions (after De Rosa *et al.*, 1986). (c) QFL plot of the Quaternary sand modes of the Nankai Trough. Q: quartz; F: feldspars; L: lithics (including both palaeo- and neovolcanic rock fragments). The provenance categories after Dickinson *et al.* (1983) are shown.

first such rocks to the east (up the very slight depositional slope of the Nankai Trough axis) are not encountered for more than 500 km. These, occupying the Izu Peninsula and the majority of the Tokai drainage basin to the north, are Neogene and Recent volcanics which are of comparable variability to volcanic grains in the sands off Shikoku. This detritus also most likely derives from the Tokai drainage basin, where the easternmost outcrops of the above-mentioned terranes occur, because most sediments deriving from the Shikoku and the Kii regions are ponded in terraced fore-arc basins or in ponds on the lower trench slope. Only three major submarine canyons debouch into the floor of the Nankai Trough. The easternmost of these, the Suruga Trough, taps the volcanic Izu/Tokai hinterland, and is therefore the conduit for most sand fed to the trench off Shikoku.

Active westward subduction of the Pacific Plate under the Eurasian and Philippine Plates at various times during the Tertiary can thus be detected in the Quaternary sands of the Nankai Trough by means of neovolcanic and palaeovolcanic detritus. The trench fill records important features of the known geotectonic setting of the region. But would the same approach prove workable in the case of an ancient terrane, where the tectonic environment was unknown? This must to a large extent depend on the degree to which diagenetic processes have obscured the original features of palaeo- and neovolcanic grains. More work is necessary on refining the operational criteria used for distinguishing the types of volcanic particles.

The effect of grain size on detrital modes and the use of the QFL diagram

Compositional comparisons between arenites plotted in the widely used QFL diagrams are limited by two fundamental factors:

1. the criteria used in the literature for grouping compositional classes (rock fragments in particular) are both numerous (see, for example, Breyer, 1983; de Vries Klein, 1963, Table 3, p. 568) and frequently unexplained;
2. samples of the same unit counted by means of the traditional method (Ingersoll *et al.*, 1984) plot in different areas of a QFL diagram depending on their grain size.

The first factor has been tested (Zuffa, 1985, Figs 4 and 5, pp. 176–177) by plotting modal data from 15 samples, collected from the base to the top of a single turbidite bed, according to three different classifications. The representative points of the same bed cluster in three quite separate fields. This emphasizes that the use of the common glossary of names like arkose, litharenite, etc., can be misleading if not viewed in association with the classification criteria adopted by the original authors during point-counting analyses.

The second factor has been tested by Gazzi (1966) and discussed by Dickinson (1970). Both proposed a special counting technique which reduces the dependence of rock composition on grain size, improving clustering of data and allowing better discrimination of different compositional groups. Zuffa (1980) discussed this point-counting procedure and Ingersoll *et al.* (1984) recommended universal use of the technique (called the Gazzi-Dickinson method by these authors).

This recommendation has recently been the object of a lively discussion among sedimentary petrologists (Decker and Helmold, 1985; Ingersoll *et al.*, 1985a and b; Suttner and Basu, 1985).

The so-called 'G–D method', as discussed by Ingersoll *et al.* (1984) differs from the traditional point-counting technique in that coarse-grained rock fragments (single crystals more than 0.0625 mm in size) are assigned according to the mineral beneath the cross hair. Thus the 'L pole' includes only fine-grained rock fragments (single crystals less than 0.0625 mm in size) (see also Zuffa, 1980, Fig. 2, for further details). Suttner and Basu (1985) claim that this technique sacrifices the potential for detailed provenance interpretations, and that data collected in this way are useful in 'achieving palaeotectonic setting alone, and not reconstruction of palaeogeology'. In their opinion, the dependence of rock composition on grain size can be reduced simply by counting grains of a specific size (medium sand size) with a grid micrometer ocular, thus avoiding elimination of coarsely crystalline lithic grains. Decker and Helmold (1985) believe that the 'G-D method' fails to reduce effectively the influence of grain size on detrital modes, and that it misrepresents the detrital composition and mineralogical maturity of the sample.

In my opinion, such objections to the 'G–D method' can be countered. First, use of the technique does not necessarily involve ignoring coarse-grained crystalline rock fragments. In fact, in the original Gazzi (1966) paper, published in Italian, the classes Quartz, K-feldspar and Plagioclase are each split into a number of subclasses indicating whether the mineral is single-crystal or constitutes part of a specific type of rock fragment (Gazzi, 1966, Table 1, p.

88; see also Zuffa, 1985, Table 3, p. 181). Secondly, although the 'G–D method' does not eliminate grain-size bias altogether—because not only does the coarsely crystalline lithic component depend on grain size, but quartz and feldspar monocrystal contents are also heterogeneous between the various sand fractions (Huckenholtz, 1963: Figs 1 and 2, p. 916; Mack, 1978, Fig. 3, p. 593; Odom *et al.*, 1976, Fig. 3, p. 864)—it nonetheless offers an opportunity to reduce the grain-size effect appreciably. Thirdly, the suggestion that grain-size dependence can be reduced by restricting the grain-size counting interval to medium to coarse sand samples places impractical sampling constraints on the field sedimentologist. In several cases the coarse-grained proximal part of thick marine turbidite bodies lies buried under tectonic nappes and consequently only fine to medium grained samples are available for collection (e.g. Marnoso-Arenacea Formation, Gandolfi *et al.*, 1983; Laga Formation, Messinian to Lower Pliocene, author's unpublished data).

The central problem in this discussion, then, is that while it is imperative to arrive at a technique for quantitative optical descriptions of the arenite framework which allows comparisons to be made between data collected by different analysts, and one which minimizes the variation of composition with grain size, equally we must strive to avoid loss of information. My preferred procedure is as follows:

(a) to collect, whenever possible, samples of medium sand size;

(b) to perform point-counting using the Gazzi-Dickinson technique for minimizing the dependence of rock composition on grain size but, in doing this, to record both which crystal is underneath the cross hair and the type of rock fragment in which the crystal is located (e.g. quartz in medium-grade metamorphic or K-feldspar in intermediate volcanic rock fragment, or plagioclase in sandstone, etc: see, for example, the categories reported in Zuffa, 1985, Table 3, p. 181);

(c) to construct a QFL diagram. This will be the best plot for classification purposes—because comparisons using the procedure guarantee that similar source rocks will produce similar modal detrital composition regardless of grain size, and together with secondary parameter diagrams will also be suitable for tectonic setting interpretations (Dickinson, 1985);

(d) to make parallel examination of one hundred to two hundred rock fragments using thin sections of impregnated coarse sand fractions, and to use the resulting data to augment the QFL diagram.

Finally, we must remember that source area reconstructions should ideally not be performed with gross compositional data alone, since they need to be supported by other lines of evidence (light and heavy minerals, palaeocurrents, facies relations, etc.) (Pettijohn, 1975).

IDENTIFICATION OF FIRST-CYCLE MULTICYCLE DETRITUS

The occurrence of multicycle detritus in arenites has always been the weakest point in provenance determinations. Despite the fact that, since the early history of sedimentary petrology, authors have pointed out the gravity of this problem, no comprehensive procedure is available for estimation of the extent to which first-cycle grains are mixed with multicycle particles. Krynine (1942) estimated that some 20–25% of polycyclic particles are present in the average sandstone, and experimental studies carried out by several authors give contradictory results (Blatt, 1967, p. 1033). Several scientists have tried to determine the amount of recycling from indirect evidence such as the presence or absence of feldspar (i.e. Folk, 1962b) or the determination of large amounts of metamorphic rock fragments (Ferm, 1962). The thoughtful paper by Blatt (1967), which may be considered a milestone in the field, suggests that more integrated approaches should be used in studying arenites, so that a number of lines of evidence can be assembled to overcome the errors which arise from recycling.

Techniques for discrimination

Recent evaluations of the proportions of exposed igneous, metamorphic and sedimentary rocks in the Earth's crust by Blatt and Jones (1975) make a good starting point in considering this limiting factor in reading provenance from arenites. It is concluded by these authors 'that probably 80% of the grains in sandstone are derived from older sedimentary rocks'. Putting on one side the influence of such limiting factors as climate and relief, transport and diagenesis in reading provenance, what we wish to know are the rock types cropping out in the source area and, possibly, their relative areal extent.

The recycling of old sedimentary rocks poses different problems depending on whether the recycled rock is a mudrock, a carbonate rock or a sandstone. By simple recalculation of the data reported by Blatt (1982), exposures at the Earth's surface comprise 34% eruptives and metamorphics, 43% mudrocks,

13% sandstones, 7% carbonate rocks, and 3% other sedimentary types. Data from the study of sands sampled at the mouths of rivers in northern Italy which carry detritus from the Alps and Apennines to the Northern Adriatic Sea (Gazzi *et al.*, 1973) indicate that, unless mudrocks are well lithified, they have little chance of reaching the final site of deposition in the form of sand grains. In fact, the shale and silstone outcrops in several drainage systems exceed 50% of the surface area, whereas the grains of this lithology found in river sediments comprise only 2–3% of the sand framework. Limestones can pass into solution where the climate and morphology of the source area are favourable, or they can be transported to produce carbonate sands. Among sedimentary rocks, sandstones are the most suitable for generating sand grains as a result of erosion and weathering. Accordingly, going back to Blatt and Jones (1975) data, it seems reasonable to a petrologist reading provenance from arenites to assume that, all things being equal, anything up to 30% of siliciclastic arenites could be present in the source area.

The implication of this is quite important if we consider that sandstones mainly release quartz, feldspar and rock-fragment grains: the end numbers of the largely used classification triangles. It follows that the QFL diagram is by itself of little use in determining the ultimate source rocks without the integration of various petrographic data, like rock fragment analysis of coarse-grained impregnated thin sections, heavy minerals, etc., and basin analysis data completed by a knowledge of regional geology. However, some criteria to detect the presence of multicycle detritus as opposed to first-cycle, are tentatively classified here through the integration of data extracted from the literature and data tested by the author during the study of modern and ancient arenite formations in the Mediterranean area (Fig. 7).

The only direct evidence of quartzo-feldspathic multicycle detritus reported in Fig. 7 are obviously the sandstone lithoclasts both in conglomerates and in the arenite framework or the classic rounded overgrowths on detrital cores of quartz or other ultrastable grains (Sanderson, 1984). Some single minerals can be radiometrically dated, but this is not an easy and routine procedure. Finally, we may have geological constraints for the occurrence of sandstone in the source area. The practice of examining thin sections of impregnated coarse sand fractions allows the detection of sandstone fragments in the arenite framework and permits a semi-quantitative estimate of the importance of the recycling factor. Field examinations of pebbles in conglomerates, when present in the formation, are of great help but must be used with caution because conglomerates may reflect local and limited source areas.

The indirect evidence reported in Fig. 7 implies careful analysis of quartz and feldspar types, of the low-rank metamorphic varieties (Blatt, 1967), of morphology, and composition-related grain size (Folk, 1974). Research on modern sands (Gazzi *et al.*, 1973) suggests that the presence of significant quantities of calcite spar grains is a very useful indicator of recycling. In the work cited these grains were found to be invariably present in amounts of up to 16% when carbonate-cemented (pore-filled) sandstones crop out in the drainage area. A second and useful indicator of recycling is the presence of significant quantities of single, mostly unworn, Cenozoic bioclasts: chiefly planktonic foraminifera. They are likely to be derived from the shaley portion of the turbidites, and from other mudrocks of which we have little record in the sand fraction.

Case history

To illustrate the criteria presented here for the recognition of recycled material in ancient arenitic sequences, data from the study of Gazzi *et al.* (1973) of sands carried by modern rivers to the northern Adriatic Sea are summarized in Fig. 8. According to gross composition, rock fragments, heavy minerals, and X-ray analysis of carbonates, five petrographic provinces can be distinguished (A, B, C, D, E). Provinces A, D and E are entirely sedimentary, province C is largely igneous–metamorphic, whereas province B is characterized by volcanic rocks. In Fig. 8, bar diagrams show the principal rock fragment components. The compositional triangles show how crucial the detection of extrabasinal carbonate lithoclasts can be in thin-section analyses of ancient sequences. The five provinces cannot be detected from the QFL diagram, whereas they can be easily recognized in the QFL + Lc diagram if carbonate lithoclasts (Lc) are included in the rock fragment end member of the triangle. Province D is of the greatest interest to the recycling problem. This province is entirely sedimentary, with sandstones (turbidites of the Marnoso-Arenacea Formation, Miocene: Gandolfi *et al.*, 1983) cropping out in most of the drainage area along with mudstone and carbonate formations. All the quartz, feldspar and igneous–metamorphic rock fragment grains are thus second-cycle particles. Could we in such a case recognize the correct source area in a study of ancient sandstones? By using the criteria listed in Fig. 7 we may conclude that this province is chiefly sedimentary with a

CRITERIA		DATA	PRESENCE OF MULTICYCLE QFL DETRITUS CAN BE DEMONSTRATED OR INFERRED (DIRECT AND INDIRECT EVIDENCE) BY:
DIRECT EVIDENCE OF MULTICYCLE DETRITUS	COMPOSITIONAL	CONGLOMERATE	SANDSTONE LITHOCLASTS IN PEBBLE MICROCONGLOMERATE AND CONGLOMERATE
		SANDSTONE ROCK FRAGMENT	SANDSTONE LITHOCLASTS IN THE ARENITE FRAMEWORK
	TEXTURAL	Q: QUARTZ T: TOURMALINE	ROUNDED OR ABRADED OVERGROWTHS ON DETRITAL CORES OF QUARTZ OR OTHER ULTRASTABLE GRAINS
INDIRECT EVIDENCE OF MULTICYCLE DETRITUS	COMPOSITIONAL	SHALE SILTSTONE LIMESTONE DOLOSTONE AND CHERT	COMMON LITHOLOGIES ASSOCIATED WITH SANDSTONES IN TURBIDITES OR PRESENT AS DISTINCT FORMATIONS IN THE SEDIMENTARY SEQUENCE CROPPING OUT IN THE SOURCE AREA
		SINGLE BIOCLASTS	SINGLE UNWORN SAND-SIZED BIOCLASTS WHICH ARE OLDER THAN THE HOST FORMATION ARE MAINLY RELEASED BY MUDROCKS AND SHALES, THE LATTER COMMONLY ASSOCIATED TO ARENITIC SEQUENCES
		CARBONATE CRYSTAL SPARS	SIGNIFICANT AMOUNTS OF SINGLE CRYSTALS ARE LIKELY TO BE RELEASED BY EROSION OF CARBONATE CEMENTED (Pore filled) ARENITES
		ZIRCON TOURMALINE RUTILE	HIGH ZTR INDEX IN SHALLOW BURIED SEQUENCES
		Q = QUARTZ F = FELDSPARS L = ROCK FRAGMENTS - MONOCRYSTALLINE NON-ONDULATORY Q. - MONOCRYSTALLINE ONDULATORY Q. - POLYCRYSTALLINE Q. (BLATT 1967)	TRIANGULAR PLOTS SHOWING A WAY OF ESTIMATING THE IMPORTANCE OF MULTICYCLE SAND CLASTS IN AN ARENITE The circles within each triangle represent arbitrary initial compositions and the tracks show the probable change in composition from sedimentary processes (after BLATT, 1967)
	TEXTURAL	T: TOURMALINE O: ORTHOCLASE (FOLK 1974)	ABNORMAL RELATION BETWEEN ROUNDNESS AND MINERAL HARDNESS
		ORTHOCLASE	ABNORMAL RELATION BETWEEN SIZE AND ROUNDNESS
		ORTHOCLASE	POOR ROUNDNESS SORTING
DIRECT OR INDIRECT EVIDENCE OF MULTICYCLE DETRITUS	ABSOLUTE AGE	K-FELDSPAR GLAUCOPHANE GLAUCONITE ZIRCON, etc.	RADIOACTIVE DATING OF ABSOLUTE AGE OF SINGLE CRYSTALS
	KNOWLEDGE OF REGIONAL GEOLOGY AND STRATIGRAPHY		

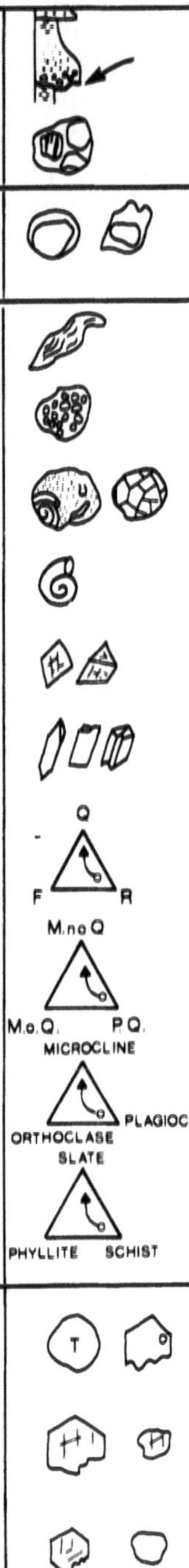

Fig. 7. Some criteria for recognition of a siliciclastic sand framework which is multicycle as opposed to first cycle.

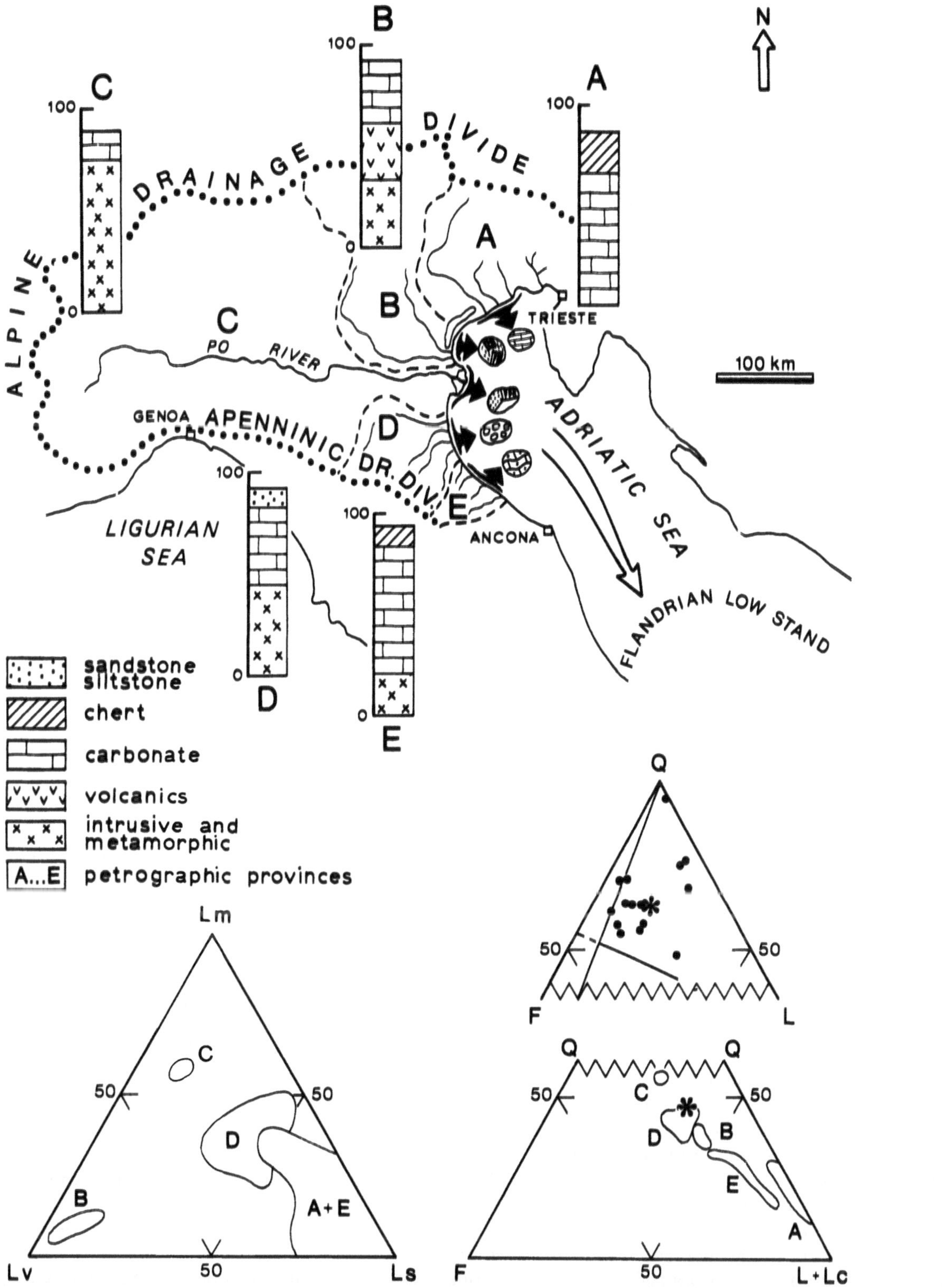

Fig. 8. Sands carried by modern rivers to the Northern Adriatic Sea. Q, quartz; F, feldspars; L, fine-grained rock fragments (including carbonate lithoclasts); Lm, Ls, Lv, metamorphic, sedimentary and volcanic rock fragments.

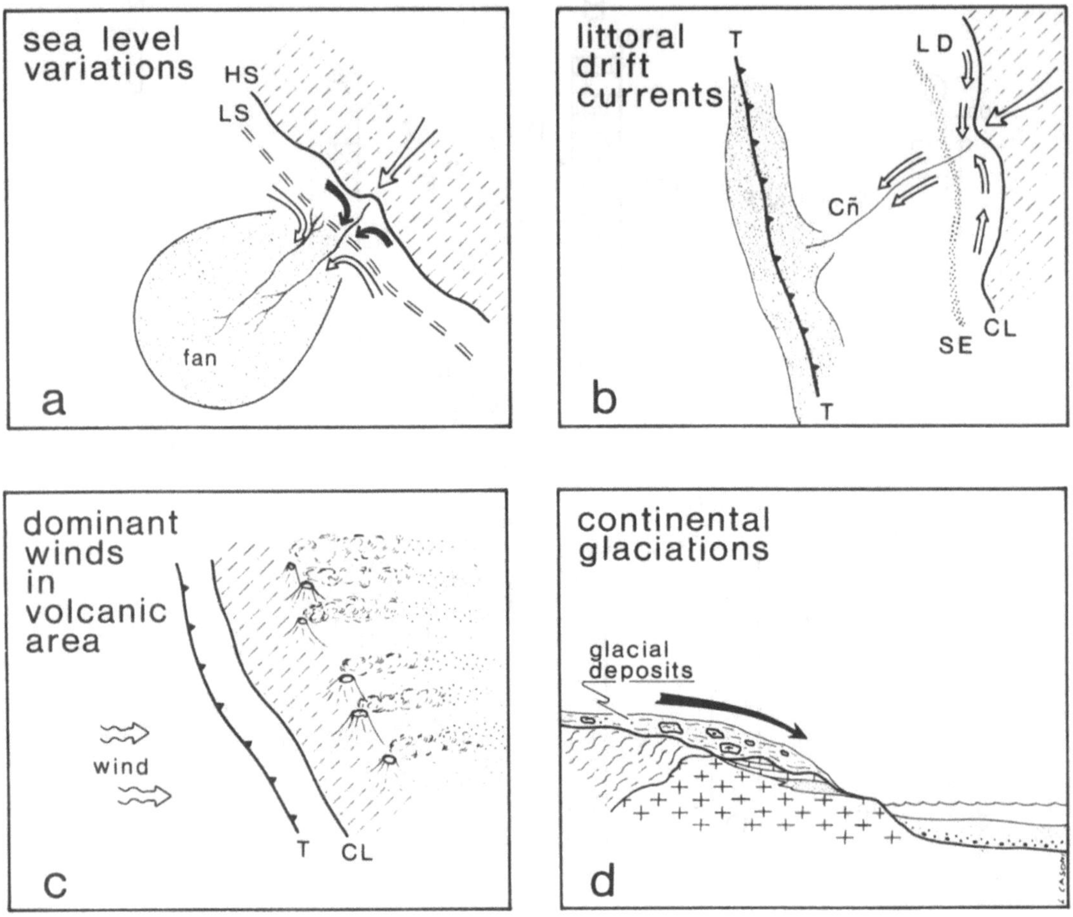

strong contribution from sandstone formations. Direct evidence is the sandstone lithoclast content determined by rock fragment analysis (11%). Indirect evidence is the calcite crystal spar (14%), the single bioclasts (5.3%) and the high K/P ratio (0.82).

A further and surprising observation is that the maturity index (Fig. 8) of this fully recycled province is lower than that of province C which is primarily a first-cycle province! The ZTR indices of provinces C and A are perfectly similar and very low, being 2.2% and 2.5% respectively. This suggests that the predepositional history of these sediments is not at all affected by climatic and transport processes in this particular morphotectonic setting and climatic zone.

The asterisks in the triangles of Fig. 8 show what the composition of the sand modes should be if we mix the detritus of the five provinces by assuming that contributions are proportional to the various drainage areas. The star of triangle QFL plots perfectly within the field of 'recycled orogenic provenance' proposed by Dickinson (1985). This result, which is in agreement with the regional tectonic setting, lends support to the validity of Dickinson's diagrams for the recognition of tectonic settings in ancient orogenic belts.

THE CONNECTION BETWEEN SOURCE AREA AND DEPOSITIONAL BASIN IN PALAEOGEOGRAPHIC RECONSTRUCTIONS

Connecting the source area and the ultimate depositional basin is generally a complex task; many factors (Fig. 1) can influence patterns of sediment distribution and lead to misinterpretations (Fig. 9).

Sea-level variations (Fig. 9a)

Oscillations in sea level may result in changes in the pathway of the sediment to a deep-sea fan, as in the Monterey Fan of California. Sediment from different sources might be mixed in the main depositional

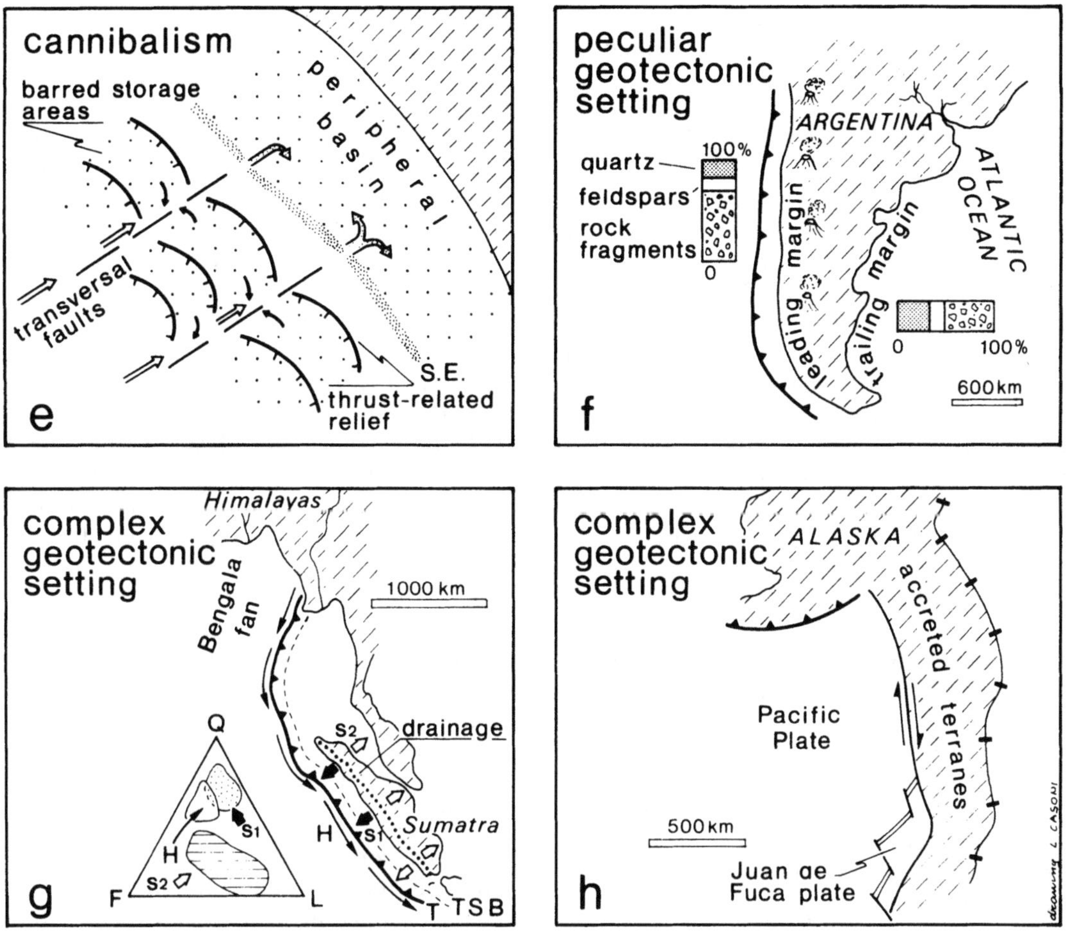

Fig. 9. Main limiting factors for the source-area/depositional-basin connection in palaeogeographic reconstructions. (a) Sketch from the North American western continental margin: HS, sea-level highstand; LS, sea-level lowstand; (plan view, not to scale). (b), (c) Examples from the Middle America Trench (plan view, not to scale): LD, littoral drift; CL, coast line; Cn, canyon; SE, shelf edge; T, trench line. (d) Speculative sketch section across glaciated hinterland. (e) Example from Northern Apennines (plan view, not to scale). Redrawn from Ricci Lucchi (1985). (f) South America (plan view). Redrawn from Potter (1984). (g) Sunda Arc (plan view): TSB, trench–slope break. Redrawn from Moore *et al.* (1982) and Velbel (1985). (h) North America western margin (plan view). Based on data from Jones *et al.* (1984) and Coney *et al.* (1980).

basin (Hess and Normark, 1976; Normark, 1985; Stow *et al.*, 1984), thus giving rise to a superposition of compositionally different sedimentary bodies. Integration between compositional data and basin analysis may clarify whether this process is occurring, and if so whether it can be ascribed to tectonics, or eustatic sea-level variations, or both.

Littoral drift currents (Fig. 9b)

Sediment moving along the coast in the littoral drift can be captured and resedimented via canyons to deep-water basins. An example of this situation, which is likely to be very common, can be found in the Middle America Trench, where quartzose littoral sands not related to the onland area are transported via canyons directly into the trench (Bachman and Leggett, 1982; Moore *et al.*, 1982). Textural analysis of sand grains (grain morphology), basin analysis and knowledge of the regional geology might enable this factor to be detected in an ancient setting.

Dominant winds (Fig. 9c)

In volcanic areas, products from explosive volcanism can be transported by dominant winds to the backarc area; some forearc basins can be mostly devoid of volcanic debris. This may create errors in palaeogeographic interpretation. Knowledge of the basin characteristics and regional geology might be the only way to avoid misinterpretations.

Continental glaciations (Fig. 9d)

Source area definition and basin connection can both be obscured by widespread continental glaciations (Potter, 1978). Glacial sediments may contribute conspicuous bodies of materials which have been transported in a very short geological time across the boundaries of tectonic provinces. In this case the ultimate source (glacial deposits) has no connection with the nature of the 'substratum' in the immediate hinterland of a basin. Provided sediments have not been affected by intense post-depositional processes, grain-size parameters and micromorphology analysis may help to identify the glacial nature of particles.

Cannibalism (Fig. 9e)

In a mobile orogenic belt, growing structures result in an alternation of morphological highs and lows. As a result, intrabasinal highs and/or exposed margins can release detritus both from previously deposited sediments and/or from uplifted blocks of basement (Gandolfi *et al.*, 1983; Ricci Lucchi, 1985; Zuffa, 1985). This condition creates problems in determining the palaeogeography through time from arenite frameworks alone.

Peculiar geotectonic settings (Fig. 9f)

Where narrow continental blocks are characterized by different types of continental margin, as in the case of South Argentina and Chile, the composition of sands released to both leading and trailing margins can prove very similar (Potter, 1984). In these cases errors may arise in reconstructing a particular geotectonic setting. Thus it is essential to incorporate information from the regional setting.

Complex geotectonic settings (Figs 9g and 9h)

The tectonic configuration of the Sunda Arc region constitutes an interesting case-history, showing how sands can be transported across the boundaries of tectonic provinces, not always reflecting the tectonic terranes of the active arc itself. In the north and central Sumatra trench-arc system, the petrology of Quaternary sands deposited in the trench, forearc and backarc basins differs markedly (Moore *et al.*, 1982; Velbel, 1985): the trench sands have a Himalayan provenance, the forearc sands are likely derived from the south-western flank of Sumatra where crystalline rocks are exposed, and the recent volcanoes lying north-east of the drainage divide supply appreciable quantities of detritus only to the backarc region. The triangle in the lower left of Fig. 9g, reported by Moore *et al.* (1982), shows the QFL composition of these sands.

When, as in the case of the western margin of north America (Fig. 9h), microplates of various tectonic domains are accreted by complex processes of subduction, collision, and strike-slip, 'the key relations between provenance and basins are certainly governed by plate tectonics' (Dickinson and Suczek, 1979), but it can be very difficult to unravel the palaeogeography of such a setting (Underwood and Bachman, 1986).

Of the factors described, the last three all deal with peculiar or complex geotectonic settings which were probably more common than we think in geological history. A good knowledge of the regional geology and stratigraphy thus becomes an essential facet of reconstructions of ancient source/basin systems.

Having considered this long list of complicating factors in the unravelling of hinterland palaeogeography, basin configuration and the link between the source and basin, aspiring sedimentary petrographers might be somewhat discouraged. But they should not be. As soon as the investigator has put all the available lines of evidence together, the petrology of arenites can prove to be one of the most rewarding tools available in understanding palaeogeography.

ACKNOWLEDGEMENTS

I thank D. Fontana, P. Forti, F. Ricci Lucchi, R. De Rosa, R. Valloni, G. G. Dri and students of the University of Bologna for comments on a previous version of the paper. I am very grateful to J. K. Leggett, W. Cavazza, E. Garzanti, R. Ingersoll, D. Scott and J. M. Arribas who kindly reviewed and improved the final version of the manuscript.

The financial support for this study was provided by the Italian Ministry of Education (MPI 40%, 1984 grant and MPI 60% 1985 grant).

References

Bachman, S. B. and Leggett, J. K. (1982). Petrology of Middle America trench and trench slope sands, Guerrero margin, Mexico, *in* J. C. Moore and J. S. Watkins (Eds), *Initial Reports of DSDP*, US Government Printing Office, Washington DC, **66**, 429–436.

Basu, A. (1985). Influence of climate and relief on compositions of sands released at source areas, *in* G. G. Zuffa, (Ed.), *Provenance of Arenites*, NATO-ASI Series, Dordrecht, D. Reidel, pp. 1–18.

Bernoulli, D. and Jenkyns, H. C. (1974). Alpine, Mediterranean, and central Atlantic Mesozoic facies in relation to the early evolution of the Tethys, *in* R. H. Dott and R. H. Shaver (Eds), *Modern and Ancient Geosynclinal Sedimentation, Soc. Econ. Paleont. Mineral. Spec. Publ.*, **19**, 129–160.

Blatt, H. (1967). Provenance determinations and recycling of sediments, *Journal of Sedimentary Petrology*, **37**, 1031–1044.

Blatt, H. (1982). *Sedimentary Petrology*, San Francisco, W. H. Freeman, 564 p.

Blatt, H. and Jones, R. L. (1975). Proportions of exposed igneous, metamorphic, and sedimentary rocks, *Geological Society of America Bulletin*, **86**, 1085–1088.

Blatt, H., Middleton, G. V. and Murray, R. C. (1972). Origin of sedimentary rocks, Englewood Cliffs, N.J., Prentice-Hall, 634 p.

Breyer, J. A. (1983). Sandstone petrology: a survey for the exploration and production geologist, *The Mountain Geologist*, **20**, 15–40.

Cameron, K. L. and Blatt, H. (1971). Durabilities of sand size schist and 'volcanic' rock fragments during fluvial transport, Elk Creek, Black Hills, South Dakota, *Journal of Sedimentary Petrology*, **41**, 565–576.

Coney, P. J., Jones, D. L. and Monger, J. W. H. (1980). Cordilleran suspect terranes, *Nature*, **288**, 329–333.

Davies, D. K. and Ethridge, F. G. (1975). Sandstone composition and depositional environment, *American Association of Petroleum Geologists Bulletin*, **59**, 239–264.

Decker, J. and Helmold, K. P. (1985). The effect of grain size on detrital modes: a test of the Gazzi-Dickinson point-counting method: discussion, *Journal of Sedimentary Petrology*, **55**, 618–620.

De Rosa, R. and Zuffa, G. G. (1979). Le areniti ibride della Valle di Ager (Eocene, Pirenei centro-meridionali), *Mineralogica et Petrographica Acta*, **23**, 1–12.

De Rosa, R., Zuffa, G. G., Taira, A. and Leggett, J. K. (1986). Petrography of trench sands from the Nankai Trough, SW Japan: implications for long-distance turbidite transportation, *Geological Magazine*, **123**, 477–486.

Dickinson, W. R. (1970). Interpreting detrital modes of graywacke and arkose, *Journal of Sedimentary Petrology*, **40**, 695–707.

Dickinson, W. R. (1985). Interpreting provenance relations from detrital modes of sandstones, *in* G. G. Zuffa, (Ed.), *Provenance of Arenites*, NATO-ASI Series, Dordrecht, D. Reidel, pp. 333–361.

Dickinson, W. R. and Suczek, C. A. (1979). Plate tectonics and sandstone compositions, *American Association of Petroleum Geologists Bulletin*, **63**, 2164–2182.

Dickinson, W. R., Beard, L. S., Brakenridge, G. R., Erjavec, J. L., Ferguson, R. C., Inman, K. F., Knepp, R. A., Lindberg, F. A. and Rybert, P. T. (1983). Provenance of North American Phanerozoic sandstones in relation to tectonic setting, *Geological Society of America Bulletin*, **94**, 222–235.

Dietz, V. (1973). Experiments on the influence of transport on shape and roundness of heavy minerals, *Contributions to Sedimentology*, **1**, 69–102.

Doyle, L. and Roberts, H. H. (conveners). (1983). AAPG/SEPM carbonate to clastic facies change, I., *American Association of Petroleum Geologists Bulletin*, **67**, 404.

Dravis, J. J. and Yurewicz, D. E. (1985). Enhanced carbonate petrography using fluorescence microscopy, *Journal of Sedimentary Petrology*, **55**, 795–804.

Ferm, J. C. (1962). Petrology of some Pennsylvanian sedimentary rocks, *Journal of Sedimentary Petrology*, **32**, 104–123.

Folk, R. L. (1962a). Spectral subdivision of limestone types, *in* W. E. Ham (Ed.), *Classification of Carbonate Rocks*, American Association of Petroleum Geologists Memoirs, pp. 62–84.

Folk, R. L. (1962b). Petrography and origin of the Silurian Rochester and McKenzie shales, Morgan County, West Virginia, *Journal of Sedimentary Petrology*, **32**, 539–578.

Folk, R. L. (1974). *Petrology of Sedimentary Rocks*, Austin, Texas, Hemphill's Bookstore, 182 p.

Fontana, D. and Mantovani, M. P. (1986). La frazione terrigena carbonatica nelle arenarie della Pietraforte (Cretaceo Superiore, Toscana meridionale), *Società Geologica Italiana Bulletin*, in press.

Gandolfi, G., Paganelli, L. and Zuffa, G. G. (1983). Petrology and dispersal pattern in the Marnoso-Arenacea Formation (Miocene, Northern Apennines), *Journal of Sedimentary Petrology*, **53**, 493–507.

Gazzi, P. (1966). Le arenarie del flysch sopracretaceo dell'Appennino modenese; correlazioni con il flysch di Monghidoro, *Mineralogica et Petrographica Acta*, **16**, 69–97.

Gazzi, P., Zuffa, G. G., Gandolfi, G. and Paganelli, L. (1973). Provenienza e dispersione litoranea delle sabbie delle spiagge adriatiche fra le foci dell'Isonzo e del Foglia: inquadramento regionale, *Società Geologica Italiana Memorie*, **12**, 1–37.

Heiken, G. (1972). Morphology and petrography of volcanic ashes, *Geological Society of America Bulletin*, **83**, 1961–1988.

Hess, G. R. and Normark, W. R. (1976). Holocene sedimentation history of the major fan valleys of Monterey Fan, *Marine Geology*, **22**, 233–251.

Huckenholtz, H. G. (1963). Mineral composition and texture in graywackes from the Harz Mountains (Germany) and in arkoses from the Auvergne (France), *Journal of Sedimentary Petrology*, **33**, 914–918.

Ingersoll, R. V., Bullard, T. F., Ford, R. L., Grimm, J. P., Pickle, J. D. and Sares, S. W. (1984). The effect of grain size on detrital modes: a test of the Gazzi–Dickinson point-counting method, *Journal of Sedimentary Petrology*, **54**, 103–116.

Ingersoll, R. V., Bullard, T. F., Ford, R. L., Grimm, J. P., Pickle, J. D. and Sares, S. W. (1985a). The effect of grain size on detrital modes: a test of the Gazzi–Dickinson point-counting method: reply to discussion of Lee J. Suttner and Abhijit Basu, *Journal of Sedimentary Petrology*, **55**, 617–618.

Ingersoll, R. V., Bullard, T. F., Ford, R. L., Grimm, J. P., Pickle, J. D. and Sares, S. W. (1985b). The effect of grain size on detrital modes: a test of the Gazzi–Dickinson point-counting method: reply to discussion of John Decker and Kenneth P. Helmold, *Journal of Sedimentary Petrology*, **55**, 620–621.

Jones, D. L., Cox, A., Coney, P. and Beck, M. (1984). La crescita del Nord America, *Le Scienze* (*Quaderni*), **13**, 28–42.

Klein, G. de V. (1963). Analysis and review of sandstone classifications in the North American geological literature, 1940–1960, *Geological Society of America Bulletin*, **74**, 555–576.

Krynine, P. D. (1942). Provenance versus mineral stability as a controlling factor in the composition of sediments (abs.), *Geological Society of America Bulletin*, **53**, 1850–1851.

Kuenen, Ph. H. (1959). Experimental abrasion. 3. Fluviatile action on sand, *American Journal of Science*, **257**, 172–190.

Mack, G. H. (1978). The survivability of labile light-mineral grains in fluvial, aeolian and littoral marine environments: the Permian Cutler and Cedar Mesa Formations, Moab, Utah, *Sedimentology*, **25**, 587–604.

Mack, G. H. (1984). Exceptions to the relationship between plate tectonics and sandstone compositions, *Journal of Sedimentary Petrology*, **54**, 212–220.

Matter, A. and Ramseyer, K. (1985). Cathodoluminescence microscopy as a tool for provenance studies of sandstones: *in* G. G. Zuffa (Ed.), *Provenance of Arenites*, NATO-ASI, Dordrecht, D. Reidel, pp. 191–211.

McBride, E. F. (1985). Diagenetic processes that affect provenance determinations in sandstone, *in* G. G. Zuffa (Ed.), *Provenance of Arenites*, NATO-ASI Series, Dordrecht, D. Reidel, pp. 95–113.

McIlreath, I. and Ginsburg, R. N. (conveners), (1982). Symposium 27: mixed deposition of carbonate and siliciclastic sediments, International Association of Sedimentologists Abstract Papers, Eleventh International Congress of Sedimentology, pp. 109–113.

McKee, E. D. and Gutschick, R. C. (1969). History of Redwall limestone of northern Arizona, *Geological Society of America Memoirs*, **114**, 726 p.

Mount, J. F. (1984). Mixing of siliciclastic and carbonate sediments in shallow shelf environments, *Geology*, **12**, 432–435.

Mount, J. F. (1985). Mixed siliciclastic and carbonate sediments: a proposed first-order textural and compositional classification, *Sedimentology*, **32**, 435–442.

Moore, G. F., Curray, J. R. and Emmel, F. J. (1982). Sedimentation in the Sunda Trench and forearc region, *in* J. K. Leggett (Ed.), *Trench Forearc Geology*, Geological Society of London, London, Blackwell Scientific Publications, pp. 245–258.

Mutti, E., Sgavetti, M., Valloni, R. and Zuffa, G. G. (1985). Turbidite systems and depositional sequence in the inner Eocene basin of the South-Central Pyrenees, Meeting on Foreland Basins, Fribourgh, Sept. 1985, abs., p. 86.

Normark, W. R. (1985). Local morphologic controls and effects of basin geometry on flow processes in deep marine basins, *in* G. G. Zuffa (Ed.), *Provenance of Arenites*, NATO-ASI, Dordrecht, D. Reidel, pp. 47–63.

Odom, I. E., Doe, T. W. and Dott, R. H. Jr. (1976). Nature of feldspar–grain size relations in some quartz-rich sandstones, *Journal of Sedimentary Petrology*, **46**, 862–870.

Pettijohn, F. J. (1975). *Sedimentary Rocks*, New York, Harper and Brothers, 628 p.

Potter, P. E. (1978). Petrology and chemistry of modern big river sands, *Journal of Geology*, **86**, 423–449.

Potter, P. E. (1984). South American beach sand and plate tectonics, *Nature*, **311**, 645–648.

Ricci Lucchi, F. (1985). Influence of transport processes and basin geometry on sand composition, *in* G. G. Zuffa (Ed.), *Provenance of Arenites*, NATO-ASI Series, Dordrecht, D. Reidel, pp. 19–45.

Sanderson, I. D. (1984). Recognition and significance of inherited quartz overgrowths in quartz arenites, *Journal of Sedimentary Petrology*, **54**, 473–486.

Stow, D. A. V., Howell, D. G. and Nelson, C. H. (1984). Sedimentary, tectonic, and sea-level controls on submarine fan and slope–apron turbidite systems, *Geo-Marine Letters*, **3**, 57–64.

Suttner, L. J. and Basu, A. (1985). The effect of grain size on detrital modes: a test of the Gazzi–Dickinson point-counting method: discussion, *Journal of Sedimentary Petrology*, **55**, 616–617.

Swett, K. and Crowder, R. K. (1982). Primary phosphatic oolites from the lower Cambrian of Spitsbergen, *Journal of Sedimentary Petrology*, **52**, 587–593.

Velbel, M. A. (1985). Mineralogically mature sandstones in accretionary prisms, *Journal of Sedimentary Petrology*, **55**, 685–690.

Underwood, M. B. and Bachman, S. B. (1986). Sandstone petrofacies of the Yazer complex and the Franciscan Coastal belt Paleogene of northern California, *Geological Society of America Bulletin*, **97**, 809–817.

Walker, K. R., Shanmugan, G. and Ruppel, S. C. (1983). A model for carbonate to terrigenous clastic sequences, *Geological Society of America Bulletin*, **94**, 700–712.

Wolf, K. H. (1965). Gradational sedimentary products to calcareous algae, *Sedimentology*, **5**, 1–37.

Zuffa, G. G. (1980). Hybrid arenites: their composition and classifications, *Journal of Sedimentary Petrology*, **50**, 21–29.

Zuffa, G. G. (1984). A model for carbonate to terrigenous clastic sequences: Discussion and reply, *Geological Society of America Bulletin*, **95**, 753.

Zuffa, G. G. (1985). Optical analyses of arenites: influence of methodology on compositional results, *in* G. G. Zuffa (Ed.), *Provenance of Arenites*, NATO- ASI Series, Dordrecht, D. Reidel, pp. 165–189.

Zuffa, G. G. and De Rosa, R. (1978). Petrologia delle successioni torbiditiche eoceniche della Sila nord-orientale (Calabria), *Società Geologica Italiana Memorie*, **18**, 31–55.

Chapter 3

An Example of the Use of Detrital Episodes in Elucidating Complex Basin Histories: the Caloveto and Longobucco Groups of N.E. Calabria, S. Italy

Massimo Santantonio, Servizio Geologico d'Italia, Largo S. Susanna 13, 00187, Rome, Italy; fieldwork carried out whilst at Dipartimento di Scienze della Terra, Università della Calabria
C. Tarquin Teale, Geology Department, Imperial College of Science and Technology, Prince Consort Road, London SW7 2BP, UK

ABSTRACT

Important detrital episodes at various stratigraphic levels in the Caloveto and Longobucco Groups provide evidence for discrete phases during the evolution of the basins in which the strata were deposited. The main episodes appear to be related to: (1) rifting and transgression over a crystalline basement and birth of a carbonate platform (Middle Lias); (2) progressive collapsing and drowning of the platform (Late Lias); (3) further extension of the basin during deposition of radiolarites and coeval with oceanic spreading documented in other Calabrian basins (Late Jurassic); (4) local clastic accumulations related to small peri-insular environments; (5) lowstands of the sea-level (Middle–Late Lias boundary) causing turbiditic deposition in basins (Longobucco) and emergence-related stratigraphic gaps in platform areas (Caloveto); (6) orogenic compression causing closure of the basin and deposition of the Upper Cretaceous–Eocene conglomeratic–turbiditic Paludi Formation, which documents denudation of all older (already deformed) formations. Speculations based on these events, and on previously available data on the Mesozoic geological evolution of Calabria and north-east Sicily (the so-called Calabria–Peloritani Arc), provide significant support for the suggestion that the area underwent major strike-slip displacements during its Jurassic history.

INTRODUCTION

The north-eastern margin of the Sila Mountains in North Calabria is made up of Palaeozoic intrusive and metamorphic massifs locally with an unmetamorphosed, transgressive Mesozoic cover sequence. Towards the Ionian coast Miocene sands and conglomerates lie unconformably on this complex. The Mesozoic cover sequence consists of two partly coeval but distinctly different groups named the Longobucco Group and the Caloveto Group by Teale (1985) and Santantonio and Teale (1985) respectively.

The Longobucco Group, described further by Young *et al.* (1986), is a thick (up to 1.4 km) pre-

Marine Clastic Sedimentology © J. K. Leggett & G. G. Zuffa (Graham and Trotman, 1987) pp. 62–74

dominantly terrigenous clastic sequence outcropping over some 170 km^2. The sequence commences with continental clastics then passes, via littoral and shelf, clastic–carbonate sediments, to deep marine turbidites. It ranges in age from Hettangian (or possibly Rhaetic) to Early Toarcian, and appears to document rifting of the crystalline basement here, analogous to and synchronous with the carbonate platform rifting known from throughout the Tethyan region (Bernoulli and Jenkyns, 1974).

The Caloveto Group is, by contrast, a condensed mid-Liassic to early Cretaceous sequence, rarely more than a hundred metres thick, outcropping discontinuously as thrust slices and klippen. It consists of platform carbonates passing into pelagic carbonates and radiolarites, with crystalline detritus of varying grade dispersed more or less throughout. The paucity and extreme fragmentation of the outcrops makes their description and interpretation difficult. However, a detailed study of their lithostratigraphy and ammonite biostratigraphy (Santantonio and Teale, in preparation) has enabled us, in most cases, to date accurately and reassemble the various sections and to interpret their environments of deposition. In this chapter we concentrate on interpretations of the stratigraphy of the Caloveto Group in the regional context, and its tectonic implications. Where appropriate, parallels with and observations from the adjacent Longobucco basin are also made.

Even though siliciclastic input from nearby continental areas occurs throughout both groups, certain parts of the sequences are marked by particularly vigorous input of clastics. These parts are herein referred to as 'detrital episodes', regardless of their thickness or duration. They have a wide variety of causes but seem always to mark important phases of basin evolution, whereas the intervening periods of carbonate and siliceous sedimentation reflect more stable conditions. Hence elucidation of the different causes of these events has proven to be a useful approach to understanding the tectonic evolution of the area.

STRATIGRAPHY AND GEOLOGICAL SETTING

The Mesozoic sedimentary cover of the Hercynian intrusive and metamorphic massifs of Sila (Calabria) (termed the 'Longi Nappe' of the 'Calabride Complex' in Calabria by Ogniben, 1969, 1973; and the calabrian sector of the 'Longobucco–Longi–Taormina Unit' by Amodio-Morelli *et al.*, 1979), outcrops on the north-eastern, Ionian, margin of the chain, which is locally named Sila Greca (Figs 1, 2a).

Previous work

During the last century, geological mapping was carried out in Sila by Cortese (1895), and the attention of students was caught by the rich ammonite faunas yielded by several Jurassic (mainly Liassic) calcareous–terrigenous levels. These were studied by some of the best Italian palaeontologists of those times: Canavari (1891–93), Fucini (1894, 1896), Di Stefano (1900, 1904), and Bonarelli (1896). These authors were often able to correlate their faunas with those from similar sequences in the Peloritani Mountains of eastern Sicily, indicating that they were coeval sequences.

Below the ammonitiferous marls, rich Sinemurian brachiopod faunas were found in the Longobucco and Bocchigliero areas. Other mollusc and brachiopod associations, of Toarcian or even Aalenian age (Greco, 1898), were discovered near Rossano in rather different facies and lithologies, such as 'Ammonitico Rosso' and red crinoidal limestones. Also, the localities yielding these faunas were generally closer to the Ionian coast (Caloveto, Colognati stream valley—Fig. 2b).

Subsequently, several geologists (Quitzow, 1935; Magri *et al.*, 1963–65; Sturani, 1968) made composite sections with the various formations of north-east Calabria from Upper Trias–Hettangian to Upper Jurassic, and Tertiary, assembled as a single succession, only according to their age. In the last decade, however, researchers have advanced the idea that the stratigraphy consists of a number of tectonic slices, comprising different sequences which had been deposited in various palaeogeographic positions, possibly far separated (Amodio-Morelli *et al.*, 1979; Scandone, 1982).

In Teale (1985) and in Santantonio and Teale (1985, in preparation), respectively, the Longobucco Group and the Caloveto Group were formally established, with their distinct formations, as used in this chapter. These units were dated by means of ammonite and calcareous nannofossil biostratigraphy (Young *et al.*, 1986).

The Longobucco Group

The Longobucco Group sequence is essentially terrigenous (Zuffa *et al.*, 1980, Fig. 3). It commences with continental red bed conglomerates and sandstones ('Verrucano' of previous authors) of probable Hettangian age (Torrente Duno Formation). These

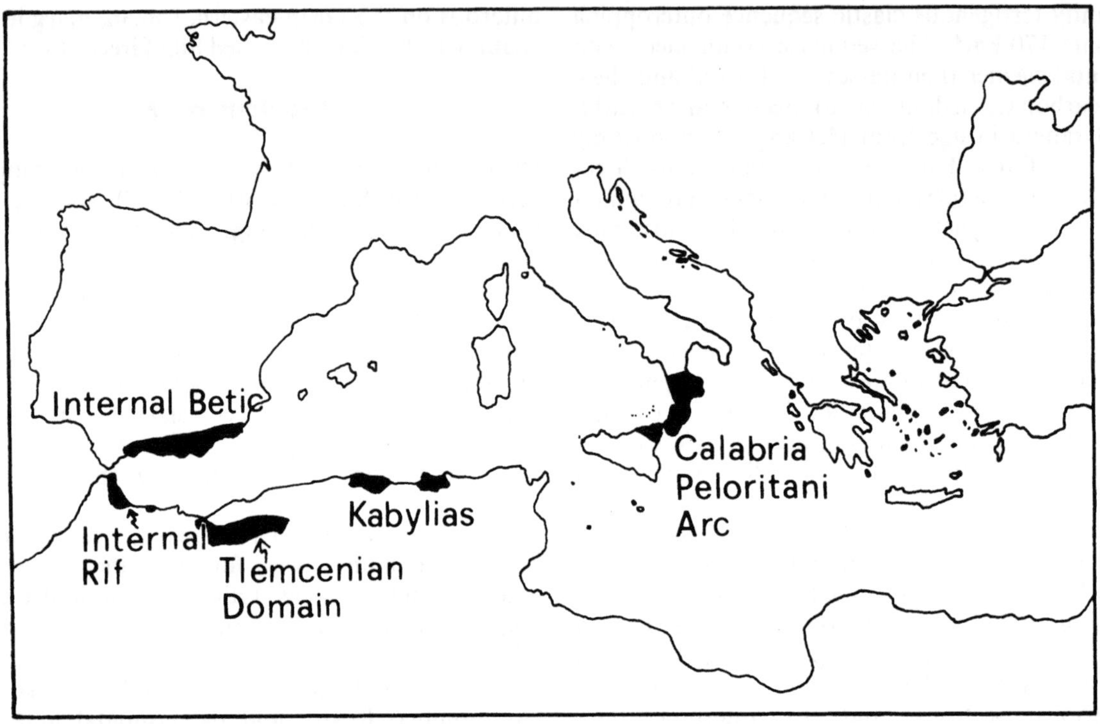

Fig. 1. Map showing the location of the Calabro-Peloritani Arc, and of some Mesozoic successions mentioned in the text. The Tlemcenian Domain belongs to the African foreland. (Partly after Bouillin, 1984; and Elmi *et al.*, 1982.)

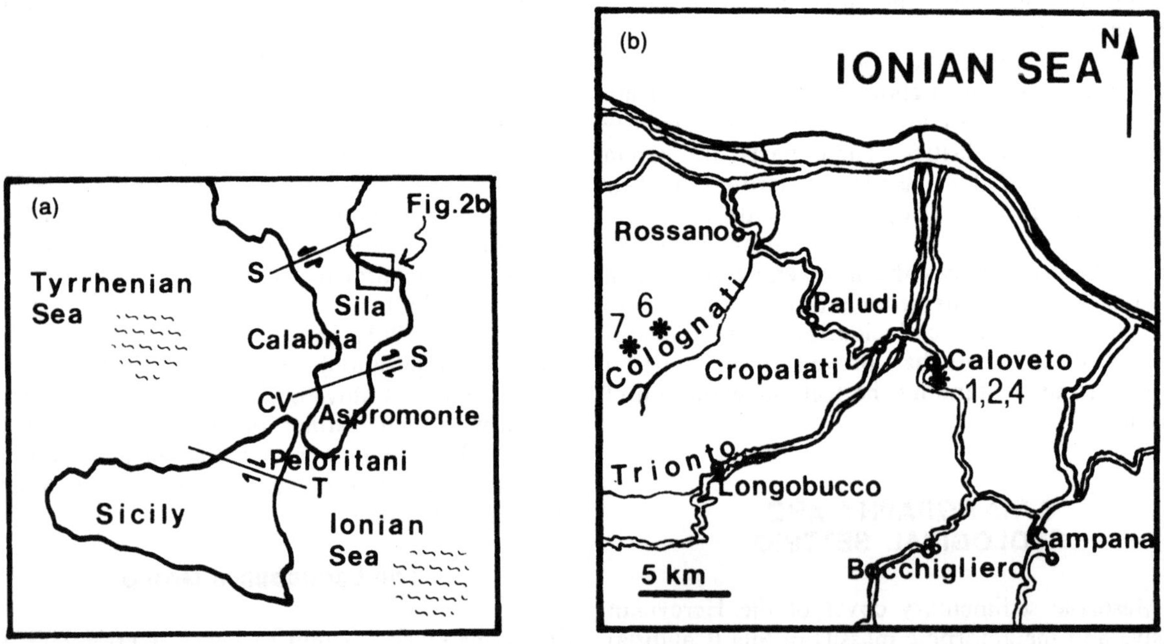

Fig. 2. (a) Location of the study area, and main structural elements of the Calabro-Peloritani Arc. S = Sangineto line; CV–S = Capo Vaticano-Soverato line; T = Taormina line. (b) Study area. The asterisks and numbers indicate the main Caloveto Group outcrops, as mentioned in the text.

rest unconformably on the Hercynian intrusive and metamorphic basement, and pass vertically into a carbonate platform facies, the Sinemurian (*sensu lato*) Bocchigliero Formation ('Calcare nero a Brachiopodi'). This facies, in its turn, passes into shelf-edge or slope marls (Petrone Formation, Carixian–Lower Domerian) and then into a deepening sequence of over 1 km of turbiditic sandstones (Trionto Formation, Upper Domerian–basal Toarcian).

Caloveto Group

The Caloveto Group sequence (Figs 3 and 4) is considerably less thick (maximum about 180 m, Santantonio and Teale in preparation). It commences with white platform carbonates (Middle Lias) containing much coarse crystalline detritus, the Lower Caloveto Formation. These pass into condensed red limestones of 'shallow pelagic platform' facies (Wendt *et al.*, 1984; Wendt and Aigner, 1985) (Lower Toarcian *p.p.*–Middle Toarcian), the Upper Caloveto Formation.

Above the Upper Toarcian the strata become more clay-rich, relating to a deepening environment of deposition. The Upper Toarcian and Aalenian sediments are red marls characteristically crowded with *Zoophycos* burrowing traces, which make them similar to the 'Marnes à Cancellophycos' of French authors. The rest of the Middle Jurassic (Bajocian, Bathonian and Callovian) has not been palaeontologically documented as yet. Only a very few outcrops exist with both Upper Jurassic and Toarcian. In these a very moderate thickness of red marls occurs which could represent this interval. However, even given the lack of any firm evidence for sedimentation during some twenty million years, continuity of deposition can reasonably be inferred since pelagic marls are overlain by pelagic marls and limestones. If there is an hiatus it is likely to reflect changes in pelagic sedimentation rather than a tectonic cause.

The Upper Jurassic is represented by the classic Aptychus limestone–radiolarite–Calpionellid limestone triptych, in various lateral relationships, within which the passage to the Lower Cretaceous occurs. In the Aptychus limestones and radiolarites intercalations of resedimented platform carbonate material (oolites and hermatypic corals) are widespread. The Neocomian sediments are pink to whitish limestones and marly limestones (similar in places to the classic Tethyan Maiolica Formation) displaying thin euxinic organic carbon-rich levels in the upper part; they also contain redeposited clastic beds of various origins. All the pelagic sediments above the Middle Toarcian limestones, up to the Neocomian, have been combined as the Sant' Onofrio Formation, despite the variable lithologies and age, since mapping them as separate units is not consistently practicable.

Both sequences (Longobucco and Caloveto) display unusual features when compared to coeval sequences in the Tethyan Jurassic. The only close parallels are with the diverse successions exposed in the Peloritani Mountains of Sicily (Lentini, 1975; Lentini and Vezzani, 1975), in the Internal Betics (Spain) and, in Africa, in the Kabylias and Internal Rif (Bouillin, 1984; Wildi, 1983) and in the Tlemcenian Trough of the Algerian Atlas (Elmi, 1981; Elmi *et al.*, 1982) (Fig. 1).

TECTONIC SETTING

Since the early 1970s a number of Italian, American and European (mostly French) authors have attempted to explain the structure of the Calabro-Peloritani Arc (CPA, Fig. 1) and to fit it within the framework of Tethyan evolution. This has involved attempts to recognize distinct sectors characterized by the presence or absence, differential development and place of the various tectonic units, and to elucidate the geometry of the contacts between the different allochtonous masses. The resulting models reconstructing the CPA address: (i) the inferred structure of the chain, and the phases of its upbuilding; (ii) the relationships with adjacent mountain belts (Apennines and Sicilian Maghrebides); and (iii) the palaeogeography and the sedimentary environments before orogenic compression.

Points (i) and (ii), which are not the object of this chapter, can be best understood from the recent articles by Scandone *et al.* (1979), Bonardi and Giunta (1982) and Tortorici (1983), and the classic—though now in part out-dated—synthesis of Amodio-Morelli *et al.* (1979). We only remind the reader that it is now widely accepted, though with some exceptions (notably Bouillin, 1984), that the CPA is the result of thrusting of a fragment of Alpine (Cretaceous–Early Palaeogene), Europe-verging terrane over the 'Austroalpine' domains (Longobucco Unit) in the Upper Palaeogene and over the Apennines in Early Miocene (Burdigalian) times. The thrusting took place with an 'African' vergence acquired during the Palaeogene, when mechanisms and directions of subduction changed in the Mediterranean region. This hypothesis was first proposed by Haccard *et al.* (1972). Dubois (1976) has suggested instead that the CPA is the African front thrust over the Apennines

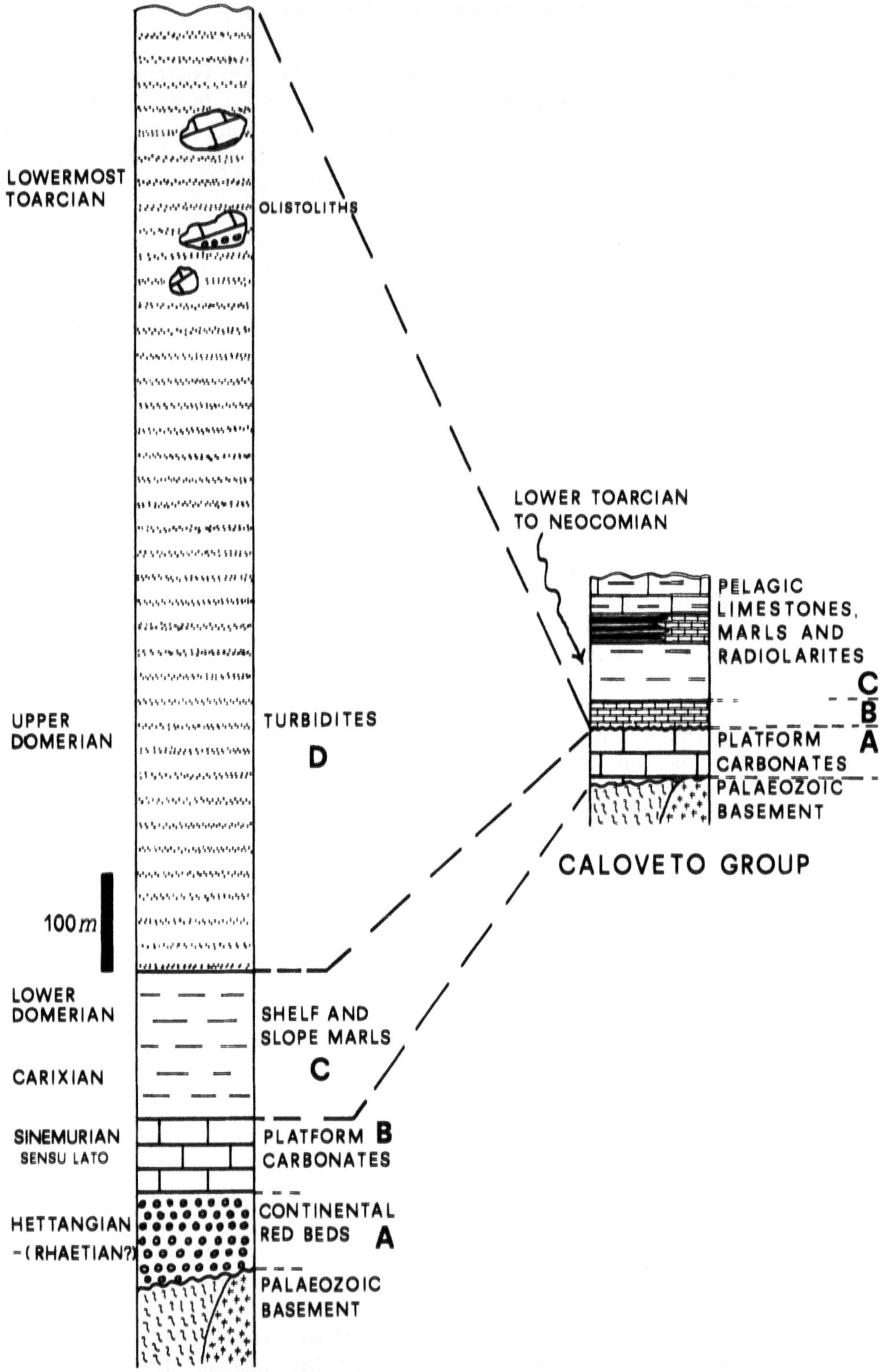

Fig. 3. Simplified stratigraphic columns of the Longobucco and Caloveto Groups: the Caloveto Group appears to be a condensed sequence. Detrital episode number 5 is represented by a deposition of turbidites corresponding to a hiatus in platform areas at the Middle/Upper Lias boundary. Lithostratigraphic subdivisions: Longobucco Group—A, Torrente Duno Fm.; B, Bocchigliero Fm.; C, Petrone Fm.; D, Fiume Trionto Fm. Caloveto Group—A, L. Caloveto Fm.; B, U. Caloveto Fm.; C, Sant' Onofrio Fm.

through deformation of an interposing oceanic belt (Sangineto zone). For other attempts to explain the geology of the CPA in terms of plate tectonics in the Mediterranean region see Dewey *et al.* (1973), Alvarez *et al.* (1974), Carrara and Zuffa (1976), and the discussion in Scandone (1979).

Point (iii), palaeogeography and environments prior to Alpine deformation, has been the object of many stratigraphic studies (Amodio-Morelli *et al.*, 1979; Lanzafame and Tortorici, 1980; Bonardi *et al.*, 1982; Scandone; 1982; Tortorici, 1983). Palaeogeographic reconstructions in these studies were constrained by the tectono–stratigraphic positioning of the crystalline, metamorphic and Mesozoic sedimentary rocks of the CPA. The highest units in the nappe pile were usually thought to be the first deformed, most internal units. The oceanic crust-bearing nappes ('Ligurides'), in this context, are considered to separate 'Europe'-derived terranes (above) from 'Africa'-derived terranes.

Most authors considered the crystalline massifs of the CPA and their sedimentary covers to have held an 'intermediate' position between Africa and Europe (Ogniben, 1969, 1973), belonging, however, to the 'African' margin (Amodio-Morelli, 1979; Channel *et al.*, 1979; Bonardi *et al.*, 1982; Scandone, 1982) on the grounds of both structure and the affinities of Mesozoic facies. The provincial affinities of the ammonite faunas (Santantonio and Teale, in preparation) are fully 'Mediterranean' in character (Cariou *et al.*, 1985).

Virtually all authors regard the Rif/Betic and Kabylian massifs of Spain and N. Africa as the westwards prolongation of the Mesozoic 'Calabrian–Peloritan' palaeogeographic belt (Fig. 1).

Importantly, in Bonardi *et al.* (1980) and Bonardi and Giunta (1982), Scandone (1982) and Tortorici (1983), two distinct sectors are recognized in the CPA, north and south respectively of the Soverato–Capo Vaticano line. The southern region, the Le Serre and Aspromonte massifs (including the Stilo unit), is devoid of any oceanic crust remnant. In contrast the northern, Sila, region includes ophiolitic fragments of the 'Ligurian' ocean. This relationship has been interpreted as the result of a primary ocean boundary related to a major transform fault.

Dissenting opinions on aspects of this analysis have been expressed, however, by various authors. According to Wildi (1983), Calabria was part of the 'Alboran plate' during the Jurassic. Bouillin (1984) regarded the CPA massifs as 'Europe'-derived units. Dercourt *et al.* (1985) considered the CPA as a whole to have been part of the Iberian plate until the Callovian when 'Sila' became incorporated in the southern 'African' continent, being separated from the 'European' 'Stilo' sequences by the Ligurian ocean.

DETRITAL EPISODES

What makes the 'Longobucco' and 'Caloveto' successions particularly interesting in the Mediterranean Tethys is the widespread occurrence of terrigenous debris varying in size from clay, through pebbles, to boulders of Palaeozoic basement and olistoliths (Teale, 1985; Teale and Young, this volume). By contrast, contemporaneous periadriatic sequences are carbonate-dominated with continent-derived material being confined to the clay fraction, sometimes associated with volcanogenic fines.

Six different detrital episodes can be identified in the sedimentary history of the Longobucco and Caloveto Groups. These reflect both the proximity of a tectonically active continental mass and relative sea-level changes. They provide useful tools for interpreting the evolution of the north-east Calabrian sedimentary basins. Four of these (1–4, Fig. 4) are discrete levels of terrigenous input into the Caloveto Group. The other two examples (5, 6, Fig. 4), discussed last, are interpretations in this context of the Longobucco turbidites and of the Paludi Formation.

Carixian–Domerian transgressive conglomerates

In the platform carbonates of the Lower Caloveto Formation crystalline conglomerates and sandstone beds occur extensively. These clastics show many signs of current deposition and remobilization; the conglomerates are commonly imbricated, and individual pebbles commonly show algal coatings. The sandstones usually show sigmoidal or herring-bone cross-stratification indicating tidal deposition. These sedimentary features are clearly not indicative of rapid deposition following influxes of extrabasinal debris. Instead their widespread occurrence in this formation is a result of its transgressive nature. The clastic sediments came directly from the underlying continental substratum, which was being progressively eroded and submerged by the sea. These relationships are best seen in outcrops near the village of Caloveto (locations 2 and 4 as indicated in Fig. 2 and described in detail in Santantonio and Teale, in preparation), where the marine sediments fill fissures and irregularities in the metamorphic

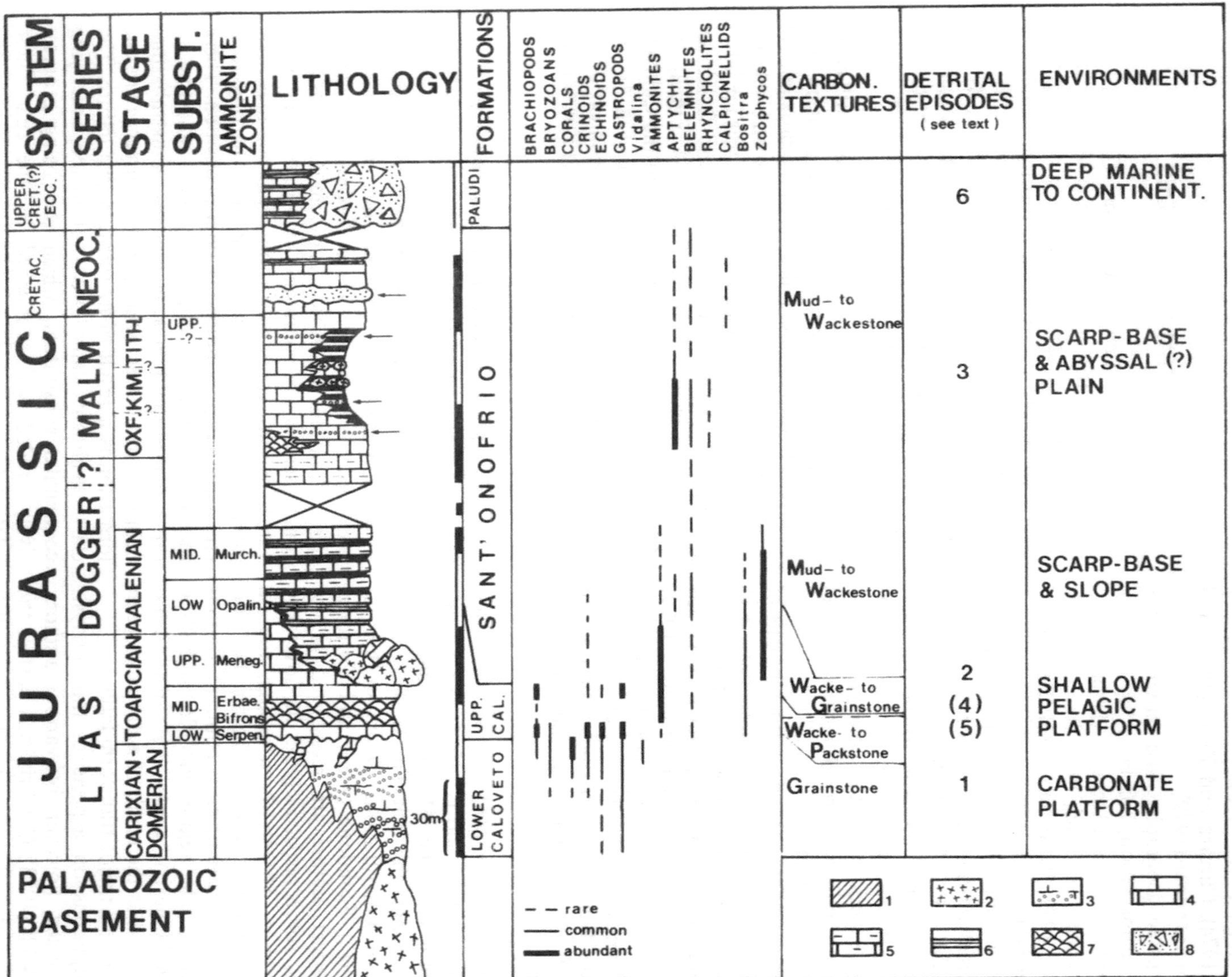

Fig. 4. Idealized stratigraphic column of the Caloveto Group. Arrows in the lithological column indicate resedimented neritic carbonate grains. The Paludi Fm. is not part of the Caloveto Group. Episodes 4 and 5 are not shown.

Key: 1, metamorphites; 2, granite; 3, massive limestone with crystalline detritus; 4, limestone; 5, marly limestone and marls; 6, shale interbeds; 7, nodular limestone; 8, breccia and sandstones. Radiolarites (Malm) are shown by dark bands. The parenthetic question mark after abyssal in the environments column reflects uncertainty as to the depth of deposition of radiolarian cherts and aptychus limestones.

basement, from which angular clasts are clearly derived. Similar features were observed by Truillet (1968) in the Peloritani Mountains.

In the Longobucco Group, the slightly older (Sinemurian *sensu lato*) siliciclast-rich dark limestones of the Bocchigliero Fm., displaying many of the sedimentological features seen in the Lower Caloveto Fm., represent a similar evolutionary stage for that basin.

Upper Toarcian (Meneghini Zone) fault-base conglomerates

At the base of the Sant' Onofrio Formation, blocks of all sizes (up to several metres across) of granite from the basement and limestones from the older Upper Caloveto Formation (less numerous) occur. These coincide with the change to deeper marine facies: the red marls with *Zoophycos* burrows. At Il Torno (location 7) in the Colognati stream valley these conglomerates are particularly well developed, representing the vestiges of a true megabreccia (Colacicchi *et al.*, 1978; Castellarin *et al.* 1978; Castellarin, 1982; Catalano and D'Argenio, 1982; Abate *et al.*, 1982).

The coincidence of the strata of large boulders with deepening to a pelagic domain in which marls could accumulate indicates that the clastic influx is probably due to extension of the basin. Subsidence of the Caloveto Group basin could easily have been coincidental with the development of active normal fault scarps at the margins, and these provide an attractive source for the crystalline debris.

Kimmeridgian–Tithonian debris flows in radiolarites

In the upper part of the Sant' Onofrio Formation angular blocks, and smaller clasts, of granitoids and metamorphites occur as intercalations in the radiolarites, along with gravity flow deposits of shallow marine carbonates (which appear to be partly chertified); these were presumably derived from an adjacent platform. Two distinct (though sometimes mixed) associations can be recognized in these clastic levels, probably reflecting different sources, viz. a granitoid rock, and a phyllitic substratum bearing a sequence of platform carbonates. These deposits can be seen near the Church of Sant' Onofrio (location 6) in the Colognati stream valley below Il Torno.

In this case too, the input of basement-derived crystalline detritus probably reflects an extensional event leading to rejuvenation of fault scarps and deepening of the basin. On a regional scale this is interesting because tensional movements related to spreading in adjacent oceanic areas occur throughout the Tethyan region in the Late Jurassic, though no remnants of ocean floor have ever been reported from the Caloveto area itself.

The occurrence of coarse crystalline detritus (up to tens of centimetres across) and large (some centimetres across) isolated fragments of reef corals (often seen in the Aptychus limestone) in sediments for which a considerable palaeobathymetry is usually inferred (thousands of metres: cf. Garrison and Fisher, 1969; Jenkyns and Winterer, 1982) suggests a basin with very steep margins adjacent to the continent. Inferences on bathymetry of radiolarites should, however, be carefully tested case by case, since these sediments are known to occur in various geological settings (see also Blendinger, 1985; Cecca and Santantonio, 1986). The 'small basin' hypothesis of Jenkyns and Winterer (1982), envisaging raised CCDs for troughs of limited dimensions, could possibly have applied here (see also comments on radiolarite bathymetry in Schlager and Schlager, 1973).

Perhaps surprisingly, this association of clastic debris in radiolarites does have parallels in coeval sequences. Resedimented carbonates of shallow-water origin have been reported from radiolarites overlying oceanic crust in north-west Calabria, belonging to different, but related, tectonic units: the 'Complesso liguride, Unità B' in Lanzafame and Zuffa (1976, p. 239); the 'Malvito Unit' in Dietrich *et al.* (1977); and the 'Upper ophiolitic unit' of Spadea *et al.* (1976). The last work also reports levels rich in sand-size detrital quartz and metamorphic rock debris. Andri and Fanucci (1980) describe similar intercalations of neritic carbonates resedimented in radiolarites from the Ligurian Apennines. Their occurrence, according to the authors, records the limited size of the basin. In their area, however, the authors also report arenaceous beds originated by erosion of ocean-ridge-derived basic rocks. Marini and Terranova (1980a, b) report granite-bearing breccias from the Upper Jurassic cover of the oceanic-crust remnants in the Ligurian Apennines. This led the authors to envisage a 'peri-continental position' for that part of the oceanic basin.

Mid-Toarcian peri-insular conglomerates

Clastic accumulations may occur in certain places while in coeval deposits there is no trace of any disturbance. During the Middle Toarcian, which was a relatively quiet period dominated by deposition of red nodular marly limestones rich in ammonites in most sections, conglomeratic facies are nonetheless locally developed. These occur in places where extremely condensed and hiatus-ridden sequences are exposed (locations 2 and 4, near Caloveto). The latter relationship suggests that these local clastics reflect nearby culminations of the basement, i.e. that they are related to high-energy peri-insular environments. They should not, therefore, be genetically linked to particular geologic events but rather to peculiar environmental conditions; this carbonate-siliciclastic mixing corresponds to the 'facies mixing' of Mount (1984) which can generate 'hybrid arenites' or 'rudites' (Zuffa, 1980).

Longobucco Group: siliciclastic deposition during regression

Under suitable conditions relative sea-level falls can produce marked increases in sedimentation, the extent of the effect depending on the vicinity of a land mass, its areal extent, and ease of denudation (in turn dependent on lithologies exposed, climate, vegetation, etc.). Eustatic control for turbidite deposition by this process has been suggested by Shanmugam and Moiola (1982, 1984) and Mutti *et al.* (1985), among others.

In the Upper Domerian–basal Toarcian (Tenuicostatum Zone), a relatively thick (more than one kilometre) turbiditic succession was deposited in the Longobucco basin. In the coeval carbonate platform succession of Caloveto, a non-depositional surface can be seen, separating the Lower and Upper Caloveto Formations (very well developed at location 1). This coincides with an important global sea-level lowstand documented by Hallam (1978, 1981). As tectonic activity was reaching its acme in Longobucco, regional doming could also easily have played a very effective role in determining this pattern of sedimentation. Unfortunately, the Longobucco sequence is apparently truncated at the Trionto Formation so that the response to the subsequent transgression is unknown. This transgression is responsible, in the Caloveto Group, for the recommencement of sedimentation represented by the Upper Caloveto Fm. limestones (top Lower Toarcian, Serpentinus Zone).

Thus, in this example a hiatus in platform areas is seen to correspond to clastic accumulation in an adjacent turbidite basin. The correlation of these events is well controlled by ammonite biostratigraphy. In the Toarcian and Domerian one ammonite zone is thought to represent an average interval of one million years (Van Hinte, 1976). The Domerian Substage (or Stage, according to some authors) on this basis lasts two million years. The Longobucco basin Trionto Formation turbidites were deposited during the Upper Domerian and Lowermost Toarcian, since Lower Domerian ammonites are documented from the underlying Petrone Formation marls, and *Dactylioceras*- and *Lioceratoides*-bearing beds occur near the top of the turbidites. The mode of preservation and composition of the fauna and sedimentological features, and their occurrence in fine-grained hemipelagic levels, suggest little or no reworking (Young *et al.* 1986). The depositional time-span of this formation is therefore unlikely to exceed two million years.

Paludi Formation: Upper Cretaceous (?)–Eocene 'flysch'

The Paludi Formation (Zuffa and De Rosa, 1980), which is unconformable upon or thrust over the Caloveto Group, displays conglomeratic to turbiditic facies. It is a manifestation of compression of the basin during alpine orogenic movements, since evidence for deformation, possible emergence and erosion of older sedimentary sequences abounds. This includes the composition of the basal conglomerates, with clasts derived from the crystalline basement, platform carbonates, deeper water limestones, and radiolarites—i.e. all the pre-existing formations. Additionally, the conglomerates, many tens of metres thick in places, usually directly overlie the basement and make up deposits for which an 'alluvial fan' origin could be postulated (Teale, unpublished data).

Within the context of the previous discussion this formation can thus be considered as a final detrital episode, reflecting the regional inversion caused by Alpine movements.

TECTONICS AND SEDIMENTATION IN THE JURASSIC OF NORTH-EAST CALABRIA: SPECULATIONS

The importance of strike-slip displacements, as related to major transform boundaries throughout the evolution of the western Tethys (i.e. the Ligurian ocean and non-oceanic sedimentary basins in the western Mediterranean) has recently been emphasized by many authors (Boccaletti, 1982; Bosellini, 1981; Castellarin, 1982; Castellarin *et al.*, 1978; Laubscher and Bernoulli, 1977; Lemoine, 1985; Scandone, 1982). The CPA, which during the Jurassic could have been situated near the south-western edge of the oceanic basin, should on the basis of this work contain clues to a strike-slip mode of deformation.

Rapidly subsiding, and apparently very small basins like the Upper Jurassic radiolarite basin of Caloveto, and the Domerian–Toarcian Longobucco turbidite basin, with their distinctive sedimentary features, seem to match some of the criteria proposed by Reading (1980) for the recognition of ancient strike-slip belts. From this viewpoint a transtensive origin for these basins could be postulated. The main lines of evidence are as follows:

(a) Caloveto Group radiolarite basin (Upper Jurassic). It was a relatively deep basin, but extremely close to the continental land mass. It was presumably bordered by very steep normal faults, responsible for the vigorous multi-directional, and platform-etherometric basement-derived clastic input. Little more can be said about the basin, since it is only known through very limited field exposures.

(b) Longobucco turbiditic basin (Upper Domerian–Lowermost Toarcian). This is a good example of a very rapidly subsiding basin filled by a thick, but not—on a regional scale—laterally extensive sedimentary pile: the classic features of a pull-apart basin, as summarized by Reading (1980).

(c) When the cover sequences of the Peloritani Mountains are compared with these successions, it is clear that during the Domerian and the Toarcian the Longi-Taormina Unit has no features in common with the Longobucco Unit, to which it is usually paralleled. A common starting point (manifested by relatively similar red beds and limestones) can be seen in the lower Liassic sediments, but differences between the two units increase with time. Thus a strike-slip displacement of these two sectors of the CPA from Middle Lias onwards seems plausible, with the Longi-Taormina basin being displaced relative to the main continental clastic sources.

The marine lowstand, whose effects are described above as detrital episode 5, would in this analysis have been superimposed on the regional tectonic trend, the turbidites triggered by the relative sea-level change being piled into a relatively small, and rapidly sinking, pull-apart basin. The synchronous hiatus cannot be ascribed to concurrent transpression, since no trace of compression or unconformity is seen in the Liassic sediments of the Caloveto Group, where only extensional features, notably neptunian dykes, occur (Santantonio and Teale, in preparation). Rifting-related tectonic uplift could, instead, conceivably be an important contributory factor.

DISCUSSION AND CONCLUSIONS

The Caloveto Group, outcropping discontinuously as thrust slices and klippen on the north-east side of Sila Greca (northern Calabria), documents marine sedimentation from Middle Lias to Neocomian on an intrusive and metamorphic Hercynian basement.

The sequences of the Caloveto and Longobucco Groups differ from most coeval successions of the Mediterranean Tethys (both 'African' and 'Apulian' margins) in having been continuously subjected to continental siliciclastic input, and in the absence at their bases of thick shallow marine evaporites and platform carbonates of Late Triassic–Early Liassic age. This was in all probability due to a lack of significant pre-Pliensbachian subsidence, and to rugged topography on the Variscan basement. Those deposits are replaced, in the Longobucco Group, by a moderate thickness (about 150 m) of continental sandstones and conglomerates and shallow marine black limestones rich in crystalline debris.

In the Caloveto Group, which, from its tectonic polarity, must have occupied a more easterly palaeogeographic position (Santantonio and Teale, in preparation), marine sedimentation commenced with pale limestones of Middle Liassic age, directly overlying the Palaeozoic basement. This delayed transgression by the Liassic epicontinental sea was due to the 'horst'-like behaviour of the local basement. It was emergent throughout the Early Lias, becoming a carbonate platform (neritic first, then pelagic) during the Middle and Late Lias (with very local persistences until the Early Aalenian). Subsequently rifting lead to deposition of pelagic sediments in small deep basins during the Late Jurassic and Early Cretaceous.

Although carbonate–siliciclastic mixing is common throughout the sequences, certain detrital episodes can be identified. These, we believe, are genetically linked to the most important stages in basin evolution:

(1) During *transgression* over a crystalline basement and birth of a carbonate platform, pebbles from the underlying basement, commonly displaying oncolitic coatings, were incorporated in the ambient carbonates.

(2) During *reductions* and *drowning* of the carbonate platform, deeper water marls include megabreccias which were deposited at the base of normal fault scarps. These comprise basement-derived and neritic limestone boulders.

(3) During further *basin extension* (and spreading in adjacent oceanic zones), basement-derived clasts of

sand to cobble size were incorporated within siliceous pelagic sediments (radiolarites).

(4) Local *peri-insular environments* generated sporadic clastic accumulations in ambient shallow marine carbonates, and produced locally very condensed sequences.

(5) Enhanced erosion during a *relative sea-level fall* resulted in turbidite sedimentation in the adjacent Longobucco basin.

(6) During *orogenic compression*, a clastic wedge comprising clasts derived from the basement and virtually all pre-existing formations built up. This, the Paludi Formation, caps the Caloveto Group.

In a broader geological context, those events indicating extensional tectonics can be related to the major transform-related strike-slip faulting regime inferred by previous authors to have existed in the Western Mediterranean Jurassic Tethys.

ACKNOWLEDGEMENTS

Professors G. Mariotti (University of Rome), A. Jacobacci (Geological Survey of Italy) and L. Tortorici (University of Calabria) are kindly acknowledged for critically reading the manuscript (Italian draft). Professor M. Sonnino (University of Calabria) and Dr S. Cresta (Geological Survey of Italy—Rome) also provided useful comments. Dr J. K. Leggett and Jeremy Young (Imperial College, London) and Gian Gaspare Zuffa (University of Bologna) reviewed the English draft and made useful remarks. Santantonio publishes through the courtesy of the Director of the Geological Survey of Italy. L. Tortorici and G. Zuffa are also acknowledged for financial support.

References

Abate, B., Catalano, R., D'Argenio, B., Di Stefano, E., Di Stefano, P., Lo Cicero, G., Montanari, L., Pecoraro, C. and Renda, P. (1982). Evoluzione delle zone di cerniera tra piattaforme carbonatiche e bacini nel Mesozoico e nel Paleogene della Sicilia occidentale. *In* R. Catalano and B. D'Argenio (Eds), *Guida alla Geologia della Sicilia Occidentale*, Società Geologica Italiana, Guide Geologiche Regionali, pp. 53–81.

Alvarez, W., Cocozza, T. and Wezel, F. C. (1974). Fragmentation of the Alpine orogenic belt by microplate dispersal, *Nature*, **248,** 309–314.

Amodio-Morelli, L., Bonardi, G., Colonna, G., Dietrich, D., Liguori, V., Lorenzoni, V., Paglionico, A., Perrone, V., Piccareta, G., Russo, M., Scandone, P., Zanettin-Lorenzoni, E. and Zuppetta, A. (1979). L'arco Calabro-Peloritano nell'orogene Appenninico-Maghrebide. *Memorie Società Geologica Italiana*, **17,** 1–60.

Andri, E. and Fanucci, F. (1980). Caratteri sedimentologici e inquadramento paleogeografico di alcune serie pelagiche giurassico-cretacee: (1) I diaspri di Monte Alpe (Liguria orientale), *Memorie Società Toscana di Scienze Naturali*, ser. A, **87,** 39–59.

Bernoulli, D. and Jenkyns, H. C. (1974). Alpine, Meditteranean, and central Atlantic facies in relation to the early evolution of the Tethys. *In* R. H. Dott and H. S. Shaver, Modern and Ancient Geosynclinal Sedimentation, *Soc. Econ. Palaeont. Mineral., Spec. Publ.* **19,** pp. 129–160.

Blendinger, W. (1985). Radiolarian limestones interfingering with loferites (Triassic, Dolomites, Italy), *Neues Jb. Geol. Paläont. Mh.*, 193–202.

Boccaletti, M. (1982). Le catene perimediterranee nel quadro dell'evoluzione della Tetide, *Boll. Soc. Pal. Ital.*, **21,** 2/3, 235–242.

Bonardi, G., Cello, G., Perrone, V., Tortorici, L., Turco, E. and Zuppetta, A. (1982). The evolution of the northern sector of the Calabria-Peloritani arc in a semiquantative palynspastic restoration, *Boll. Soc. Geol. Ital.*, **101,** 259–274.

Bonardi, G., Giunta, G., Perrone, V., Russo, M., Zuppetta, A. and Ciampo, G. (1980). Osservazioni sull'evoluzione dell'arco Calabro-Peloritano nel Miocene inferiore: la Formazione di Stilo-Capo d'Orlando, *Boll Soc. Geol. Ital.*, **99,** 365–393.

Bonardi, G. and Giunta, G. (1982). L'estremità nord orientale della Sicilia nel quadro dell'evoluzione dell'arco calabro-peloritano. *In* R. Catalano and B. D'Argenio (Eds), *Guida alla Geologia della Sicilia Occidentale*, Società Geologica Italiana, Guide Geologiche Regionali, pp. 85–91.

Bonarelli, G. (1896). Sulla età dei calcari marnoso-arenacei varicolori di Pietracutale e Bocchigliero in Calabria, *Rivista Italiana Paleontologia*, **2,** 1–4.

Boselini, A. (1981). The Emilia Fault: a Jurassic fracture zone that evolved into a Cretaceous–Palaeogene sinistral wrench-fault, *Boll. Soc. Geol. Ital.*, **100,** 161–169.

Boulin, J. P. (1984). Nouvelle interprétation de la liaison Apennin–Maghrebides en Calabre; conséquences sur la paléogéographie téthysienne entre Gibraltar et les Alpes. *Revue de Géographie Physique et Géologie Dynamique*, **25/5,** 321–338.

Canavari, M. (1891–93). Conglomerati arenarie e quarziti liassiche di Puntadura in provincia di Cosenza. *Atti Soc. Tosc. Sci. Nat. Proc. Verb.*, **8**.

Cariou, E., Contini, D., Dommergues, D-L., Enay, R., Geyssant, J. R., Mangold, C. and Thierry, J. (1985). Biogéographie des ammonites et évolution structurale de la Téthys au cours du Jurassique, *Bulletin de la Société Géologique de la France*, **8,** 679–697.

Carrara, A. and Zuffa, G. G. (1976). Alpine structures in north-western Calabria, Italy. *Geological Society of America Bulletin*, **87,** 1229–1246.

Castellarin, A. (1982). La scarpata tettonica mesozoica Ballino–Garda (fra Riva e il Gruppo di Brenta). *In* A. Castellarin and G. B. Vai (Eds), *Guida alla Geologica del Sudalpino Centro-orientale*, Società Geologica Italiana, Guide Geologiche Regionali, pp. 79–95.

Castellarin, A., Colacicchi, R. and Praturlon, A. (1978). Fasi distensive, trascorrenze e sovrascorrimenti lungo la 'linea Ancona-Anzio', dal Lias medio al Pliocene. *Geologica Romana*, **17,** 161–189.

Catalano, R. and D'Argenio, B. (1982). Schema geologico della Sicilia, *In* R. Catalano and B. D'Argenio (Eds), *Guida alla Geologia della Sicilia Occidentale*, Società geologica italiana, Guide Geologiche Regionali, pp. 9–41.

Cecca, F. and Santantonio, M. (1986). Le successioni del Giurassico superiore dell'Appenino umbro-marchigiano-sabino: osservazioni sulla geologia e sulla biostratigrafia, *In* G. Pallini (Ed.), *Atti del Convegno 'Fossili, Ambiente, Evoluzione'*, Pergola, 1984, pp. 111–118.

Channell, J. E. T., D'Argenio, B. and Horvarth, F. (1979). Adria, the African Promotory, in Mesozoic Palaeogeography, *Earth Science Review*, **15,** 213–292.

Colacicchi, R., Pialli, G. and Praturlon, A. (1978). Arretramento tettonico del margine di una piattaforma carbonatica e produzione di brecce e megabrecce: l'esempio della Marsica (Appennino centrale). *Quad. Fac. Ing. Univ. di Ancona, 1978*, 295–328.

Cortese, E. (1895). Descrizione geologica della Calabria. *Mem. Descr. Carta Geol. Italia*, **9,** 1–310.

Dewey, J. F., Pitman, W. C., Ryan, W. B. F. and Bonnin, J. (1973). Plate tectonics and the evolution of the Alpine system. *Bulletin of the Geological Society of America*, **84,** 3137–3180.

Dercourt, J., Zonenshain, J. P., Ricou, L. P., Kazmin, V. G., Le Pichon, X., Knipper, A. L., Grandjaquet, C., Sborshchikov, I. M., Boulin, J., Sorokhtin, O., Geyssant, J., Lepvrier, C., Biju-Duval, B., Sibuet, J. C., Savostin, L. A., Westphal, M. and Lauer, J. P. (1985). Présentation de 9 cartes paléogéographiques au 1/20 000 000 s'étendant de l'Atlantique au Pamir pour la période du Lias au l'Actuel, *Bulletin de la Société Géologique de la France*, ser. 8, **1,** 637–652.

Dietrich, D., Lorenzoni, S., Scandone, P., Zanettin-Lorenzoni, E. and Di Pierro, M. (1976). Contribution to the knowledge of the tectonic units of Calabria. Relationships between composition of K-white micas and metamorphic evolution, *Boll. Soc. Geol. Ital.*, **95,** 193–217.

Di Stefano, G. (1900). Il Malm in Calabria. *Rivista Italiana Paleontologica e Stratigrafica*, **6,** 39–47.

Di Stefano, G. (1904). Osservazioni geologiche nella Calabria settentrionale e nel Circondario di Rossano, *Memorie Descr. Carta Geol. Italia*, **9,** 1–310 (Appendix).

Dubois, R. (1976). La suture Calabro-Appéninique et l'ouverture Tyrrhenienne Néogène: étude pétrographique et structurale de la Calabre Centrale. Thèse, Université de M. et P. Curie Paris, pp. 1–567.

Elmi, S. (1981). Comparaison entre l'évolution Jurassique de l'Appénin et la marge nord-africaine, *Memorie Soc. Geol. Ital.*, **21,** 33–40.

Elmi, S., Almeras, Y., Ameur, M., Atrops, F., Brnhamou, M. and Moulan, G. (1982). La dislocation de la plate-forme carbonatée liasique en Méditerranée Occidentale et ses implications sur les échanges fauniques, *Bulletin de la Société Géologique de la France*, sér. 7, **24,** 1007–1016.

Fucini, A. (1894). Due nuovi terreni del Giurassico del circondario di Rossano in Calabria. *Proc. Verb. Soc. Tosc. Nat.* **9,** 164–167.

Fucini, A. (1896). Studio geologico sul circondario di Rossano in Calabria, *Atti Acc. Gioenia Sc. Nat.*, a, **73/9,** 1–87.

Garrison, R. E. and Fisher, A. G. (1969). Deep-water limestones and radiolarites of the Alpine Jurassic, *In* G. M. Friedman (Ed.), *Depositional Environments in Carbonate Rocks, a Symposium*, Society of Economic Palaeontology and Mineralogy, Special Publications 14, 20–56.

Greco, B. (1898). Fauna della zona con *Lioceras opalinum* Rein. sp., *Palaeontographia Italica*, **4,** 93–140.

Haccard, D., Lorenz, C. and Grandjaquet, C. (1972). Essai sur l'évolution tectogénétique de la liaison Alpes-Apénnines de la Liguria à la Calabre), *Memorie della Società Geologica Italiana*, **11,** 309–341.

Hallam, A. (1978). Eustatic cycles in the Jurassic, *Palaeogeography, Palaeoclimatology, Palaeoecology*, **23,** 1–32.

Hallam, A. (1981). A revised sea-level curve for the early Jurassic, *Journal of the Geological Society, London*, **138,** 735–743.

Jenkyns, H. C. and Winterer, E. (1982). Palaeo-oceanography of Mesozoic ribbon radiolarites, *Earth Planetary Science Letters*, **60,** 351–375.

Lanzafame, G. and Tortorici, L. (1980). Le successioni giurassico-eoceniche dell'area compresa tra Bocchigliero Longobucco e Cropalati (Calabria). *Rivista Italiana Paleontologica e Stratigrafica*, **86,** 31–54.

Lanzafame, G. and Zuffa, G. G. (1976). Geologia e petrografia del Foglio Bisignano (Bacino del Crati, Calabria). *Geologica Romana*, **15,** 223–270.

Laubscher, H. P. and Bernoulli, D. (1979). Mediterranean and Tethys, *In* A. E. M. Nairn, W. H. Kanes and F. Stehli (Eds), *The Ocean Basins and Margins*, 4A, New York, Plenum Press, pp. 1–28.

Lemoine, M. (1985). Structuration jurassique des Alpes occidentales et palinspastique de la Téthys ligure, *Bulletin de la Société Géologique de la France*, sér 8, **1**, 126–137.

Lentini, F. (1975). Le successioni mesozoico-terziarie dell'Unità di Longi (Complesso Calabride) nei Peloritani occidentali (Sicilia), *Boll. Soc. Geol. Ital.* **94**, 537–554.

Lentini, F. and Vezzani, L. (1975). Le unità meso-cenozoiche della copertura sedimentaria del basamento cristallino peloritano, *Boll. Soc. Geol. Ital.* **94**, 1477–1503.

Magri, G., Sidoti, G. and Spada, A. (1963–65). Rilevamento geologico sul versante settentrionale della Sila (Calabria), *Mem. Note Ist. Geol. Appl. Napoli*, **9**, 1–60.

Marini, M. and Terranova, R. (1980a). Evoluzione paleogeografica del bacino ligure tra l'Aptiano e il Paleocene, *Memorie Soc. Geol. Ital.* **21**, 143–149.

Marini, M. and Terranova, R. (1980b). I complessi ofiolitiferi dei Monti Aiona e Penna e i loro rapporti con le serie sedimentarie (Appennino ligure-emiliano), *Boll. Soc. Geol. Ital.* **99**, 183–203.

Mount, J. F. (1984). Mixing of siliciclastic and carbonate sediments in shallow shelf environments, *Geology* **12**, 432–435.

Mutti, E., Remacha, E., Sgavetti, M., Rosell, J., Valloni, R. and Zamorano, M. (1985). Stratigraphy and facies characteristics of the Eocene Hecho group turbidites systems, south-central Pyrenees. *In* M. D. Mila and J. Rosell (Eds), 6th Europ. Meeting Int. Assoc. Sedimentol. Lerida, *Excursion Guidebook*, pp. 519–576.

Ogniben, L. (1969). Schema introduttivo alla geologia del confine calabro-lucano, *Memorie Società Geologica Italiana* **12**, 243–585.

Ogniben, L. (1973). Schema geologico della Calabria in base ai dati odierni, *Geologica Romana* **12**, 243–585.

Quitzow, H. W. (1935). Der Deckenbau des Kalabrischen Massivs und Seiner Randgebiete. *Abh. Ges. Wiss. Gottingen Math. Phys. Kl.* **3**, 63–179.

Reading, H. G. (1980). Characteristics and recognition of strike-slip fault systems. *In* P. F. Ballance and H. G. Reading (Eds), *Sedimentation in Oblique-slip Mobile Zones*, Spec. publ. int. Assoc. Sedimentol., **4**, 7–26.

Santantonio, M. and Teale, C. T. (1985). Jurassic condensed sedimentation on Hercynian basement (Calabria, Italy). Abstr. 6th Europ. Meeting Int. Assoc. Sedimentol. Lerida, pp. 459–560.

Santantonio, M. and Teale, C. T. (in preparation). Stratigraphy and sedimentology of the Jurassic–Neocomian Caloveto Group (Southern Italy).

Scandone, P. (1979). Origin of the Tyrrhenian Sea and the Calabrian, Arc, *Boll. Soc. Geol. Ital.* **98**, 27–34.

Scandone, P. (1982). Structure and evolution of the Calabrian Arc, *Earth Evol. Sci.* **3**, 172–180.

Schlager, W. and Schlager, M. (1973). Clastic sediments associated with radiolarites (Tauglboden–Schichten, Upper Jurassic, Eastern Alps), *Sedimentology* **20**, 65–89.

Shanmugan, G. and Moiola, R. J. (1982). Eustatic control of turbidites and winnowed turbidites, *Geology* **10**, 231–235.

Shanmugan, G. and Moiola, R. J. (1984). Eustatic control of calciclastic turbidites. *Marine Geology*, **56**, 273–278.

Spadea, P., Tortorici, L. and Lanzafame, G. (1976). Serie ofiolitifere nell'area tra Tarsia e Spezzano Albanese (Calabria): stratigrafia, petrografia, rapporti strutturali, *Memorie Soc. Geol. Ital.* **17**, 135–174.

Sturani, C. (1968). Il Giurese del Massiccio Calabro-Peloritano. *Geologia dell'Italia*, Vol. Celebr. Cent. Comit. Geol. Ital. Torino, 333–337.

Teale, C. T. (1985). Occurrence and geological significance of olistoliths from the Longobucco Group, Calabria, Southern Italy, Abstr. 6th Europ. Meeting Int. Assoc. Sedimentol. Lerida, pp. 457–458.

Tortorici, L. (1983). Lineamenti geologico-strutturali dell'Arco calabro-peloritano, *Rend. Soc. ital. Min. Petr.* **38**, 927–940.

Truillet, R. (1968). Etude géologique des Péloritains orientaux (Sicile): Thèse, Fac. Sci. Paris, 547 p.

Van Hinte, J. E. (1976). A Jurassic time scale, *Bull. Am. Ass. Petrol. Geol.* **60**, 489–497.

Wendt, J. and Aigner, T. (1985). Facies patterns and depositional environments of Palaeozoic cephalopod limestones, *Sedim. Geol.* **44**, 263–300.

Wendt, J., Aigner, T. and Neugebauer, J. (1984). Cephalopod limestone deposition on a shallow pelagic ridge: the Tafilalt Platform (upper Devonian, eastern Anti-Atlas, Morocco), *Sedimentology* **31**, 601–625.

Wildi, W. (1983). La chaine tello-rifaine (Algérie, Maroc, Tunisie) structure, stratigraphie et évolution du Trias au Miocène. *Revue Geogr. Phys. Géol. Dyn.* **24**, 201–297.

Young, J. R., Teale, C. T. and Bown, P. R. (1986). Revision of the stratigraphy of the Longobucco Group (Liassic, southern Italy); based on new data from nannofossils and ammonites, *Eclog. Geol. Helv.* **79**, 117–135.

Zuffa, G. G. (1980). Hybrid arenites: their composition and classification, *Journal of Sedimentary Petrology* **50**, 21–29.

Zuffa, G. G. and De Rosa, R. (1980). Petrologia delle successioni torbiditiche Eoceniche della Sail Nord-Orientale (Calabria). *Memorie Soc. Geol. Ital.* **18**, 31–55.

Zuffa, G. G., Gaudio, W. and Rovito, S. (1980). Detrital mode evolution of the rifted continental-margin Longobucco Sequence (Jurassic), Calabrian Arc, Italy. *Journal of Sedimentary Petrology* **50**, 51–61.

Chapter 4

Isolated Olistoliths from the Longobucco Basin, Calabria, Southern Italy

C. Tarquin Teale and *Jeremy R. Young*, Geology Department, Imperial College of Science and Technology, London SW7 2BP

ABSTRACT

The Longobucco Group is a cover sequence of Liassic sediments recording rifting of Hercynian crystalline basement. Within turbidites of the upper part of this sequence are numerous olistoliths, up to 250 m long. These are not parts of olistostromes, or other mass movement deposits, but appear to have travelled into the basin independently; hence the term 'isolated olistolith' seems appropriate for them. Soft-sediment deformation structures around the margins of one well-exposed olistolith suggest an analogy with 'outrunner-blocks' recently described from a modern submarine slope-failure complex by Prior *et al.* (1982, 1984). In both cases the blocks appear to have travelled considerable distances across very low slopes without disturbing the underlying sediments to any great extent. Transport probably occurred relatively rapidly with the aid of overpressuring of the underlying sediments.

INTRODUCTION

Isolated olistoliths may be defined as massive blocks of rock occurring out of place, but with sedimentary contact relationships, within unrelated sediments. These relationships, which suggest that the blocks were emplaced as single masses into the basins in which they occur, are conceptually improbable, and only rarely observed. In this paper we describe examples found while mapping the Longobucco Group in Calabria, S. Italy.

The Longobucco Group is a 1500-m thick sequence of sediments which records the formation of a deep extensional basin on Hercynian crystalline basement in the Early Jurassic (Young *et al.*, 1986; Santantonio and Teale, this volume). The olistoliths are colossal blocks, up to 250 m long, of shallow marine limestones and other lithologies from the base of the succession. They occur, however, in the upper part of the sequence, within basin–plain turbidites. Recognition of the blocks as isolated olistoliths helped considerably in the development of our understanding of the history of the basin. Additionally, finely preserved soft-sediment deformation features found around some of the olistoliths provide new evidence for the mechanism of emplacement of such blocks.

After describing the Longobucco olistoliths and their setting, we briefly review analogous cases described from other areas. In particular, comparisons are made between the Longobucco olistoliths and some blocks emplaced during a recent submarine slope failure, described from British Columbia by Prior *et al.* (1982, 1984). This comparison leads us to propose a possible mechanism of emplacement for isolated olistoliths.

Terminology

Some comment on terminology is needed, since virtually every author has used different nomenclature when describing isolated olistoliths (see Table 1). The most common terms are: olistolith; slide block; and glide block. Others include: exotic block; allochthonous block; allochthonous exotic block; sedimentary klippe; olisthothrymma; olisthoplaka; and local names such as Cipit block. The use of the terms 'exotic' and 'allochthonous' seems inappropriate, and potentially confusing, since they have strongly tectonic overtones, whereas the relations are clearly sedimentary. The combination 'allochthonous exotic' is pointless since the two words have essentially the same meaning in this context. 'Sedimentary klippe' evocatively describes a common outcrop mode but is completely misleading in terms of geological relationships. The neologisms 'olisthothrymma' and 'olisthoplaka' were proposed by Richter and Mariolakis (1973) and Richter (1973) for, respectively, slide masses occurring independently, and ones occurring as isolated fragments of gravity nappes. These seem unnecessary and unwieldy names, and they have not been widely used.

'Slide block' and 'glide block' are more attractive terms, and have been increasingly used of late. In current usage (Nardin *et al.*, 1979) slides are mass movements with limited internal deformation. They include both slumps, in which rotation is important, and glides, in which only linear motion occurs. Thus, 'slide block' and 'glide block' have nearly the same meaning and can both legitimately be used. Of them 'slide block' seems preferable as its meaning is clearer, and since some blocks are rotated.

The term is, however, explicitly genetic and so an alternative is needed for cases where the mechanism of emplacement is uncertain, and for field use. Although olistolith (Greek—slide stone) does have genetic roots they are not explicit, so it has been used much more in a descriptive sense. The main problem with its use is that as originally defined, by Flores (1955), it only covered matrix-supported clasts. However, this definition, and that of olistostromes, was clearly emended by Jacobacci (1965) and Abbate *et al.* (1970), so that blocks occurring independently of olistostromes were included. We favour this usage with the addition of the qualifier 'isolated' where appropriate.

THE LONGOBUCCO OLISTOLITHS

Background—The Longobucco Group

The olistoliths discussed in this paper occur within the Liassic sedimentary rocks of the Longobucco Group. These outcrop over some 170 km^2 as a tectonically shortened, but well preserved, sequence on the north-east margin of Calabria, S. Italy (Fig. 1). They probably extend farther beneath the late Tertiary deposits of Calabria, and may occur offshore. Associated with the group are condensed pelagic sediments of Liassic to Early Cretaceous age, the Caloveto Group (Santantonio and Teale, 1985; and in preparation). Together these two groups constitute the Mesozoic cover sequence to Hercynian granitoids and metamorphites of the Longobucco Unit. This unit forms a part of the Calabro-Peloritani Arc, which is generally interpreted as an isolated fragment of the Eo-Alpine chain thrust over the Apenninic carbonate sequences (Scandone 1979, 1982). Palinspastic reconstructions suggest that the Longobucco Unit occupied a position between the Apenninic units to the west and the main Calabrian Alpine units to the east, on the African margin of the Tethys (De Rosa *et al.*, 1980).

The main subsidence and basin formation event recorded by the Longobucco Group occurred during the Late Pleinsbachian–Early Toarcian and so was synchronous with the well-documented late Liassic break-up of Triassic carbonate platforms throughout the Tethyan area (Bernoulli and Jenkyns, 1974). This suggests that it too developed by rifting, although in this case of a crystalline basement. The facies developed in the Longobucco Basin were, however, quite different from the typical Tethyan carbonates, due to the influx of siliciclastic detritus. Santantonio and Teale (this volume) discuss further aspects of the setting of the group. Important studies of its sedimentology and stratigraphy include those of Magri (1963–65), Zuffa *et al.* (1980), and Young *et al.* (1986).

The succession commences with up to 70 m of 'red beds', mixed continental clastics, the Torrente Duno Formation (Fig. 2). These lack age-diagnostic fossils but are probably Hettangian. The base is commonly faulted, but in places the formation can be seen to rest directly on crystalline basement rocks. Upwards, it passes gradationally into variable shallow marine carbonates, commonly sandy up to 60 m thick. These have Sinemurian shelly faunas and constitute the Bocchigliero Formation. The succeeding Petrone Formation is up to 200 m thick and records passage into deeper water with bioturbated marls the dominant lithology; it has yielded quite good Pliensbachian ammonite faunas.

The bulk of the succession is formed of turbidites of the Fiume Trionto Formation, in which the olistoliths occur. Up to 1200 m of the turbidites are preserved, and presumably they were originally thicker, as the top of the formation is everywhere either truncated by erosion, or cut-out by thrust faults. Near the preserved top of the sequence ammonites of the *tenuicostatum* Zone, earliest Toarcian, have been found

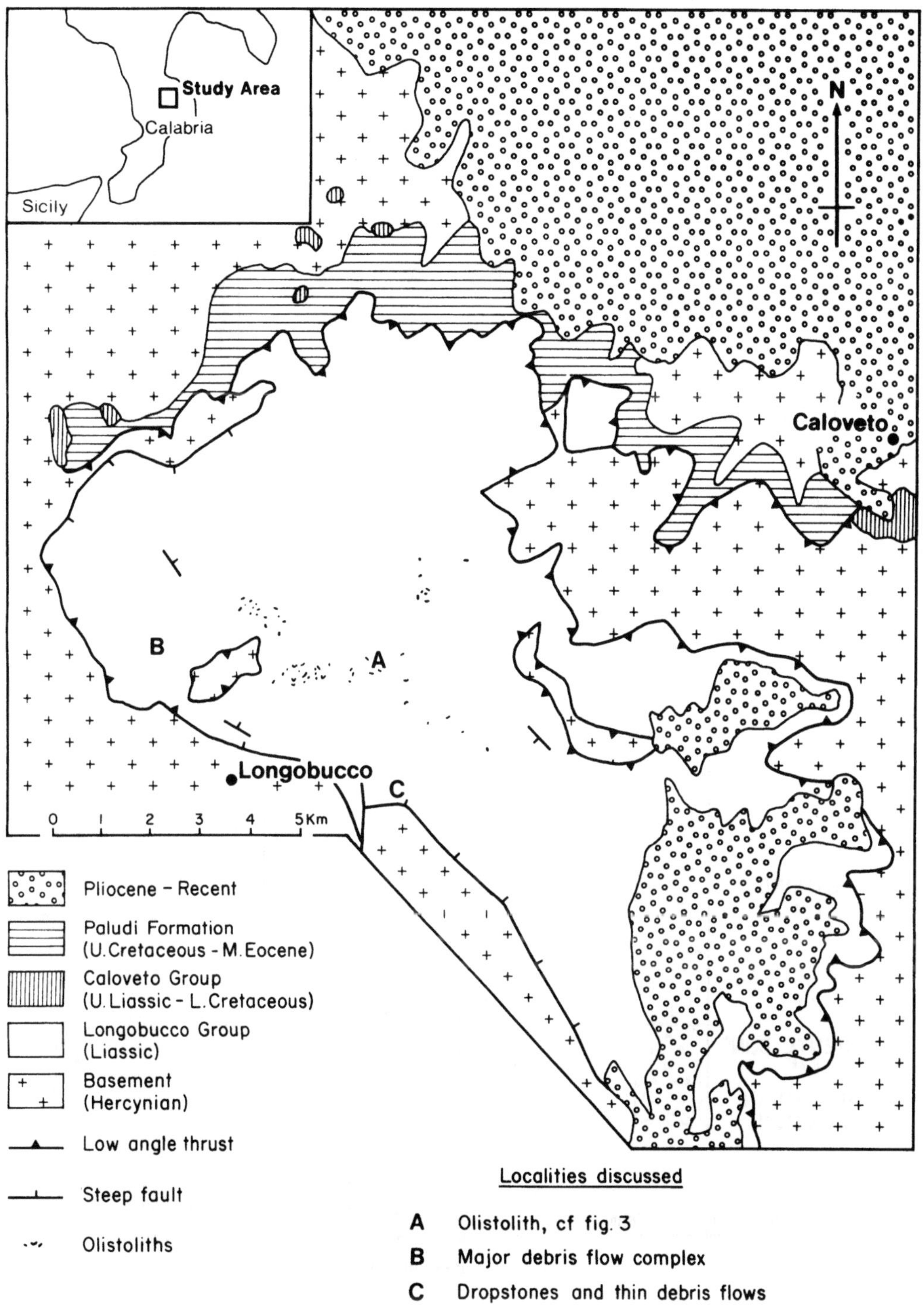

Fig. 1. Sketch map of Longobucco Basin showing extent of the Longobucco and Caloveto Groups and distribution of the olistoliths. Inset, location map.

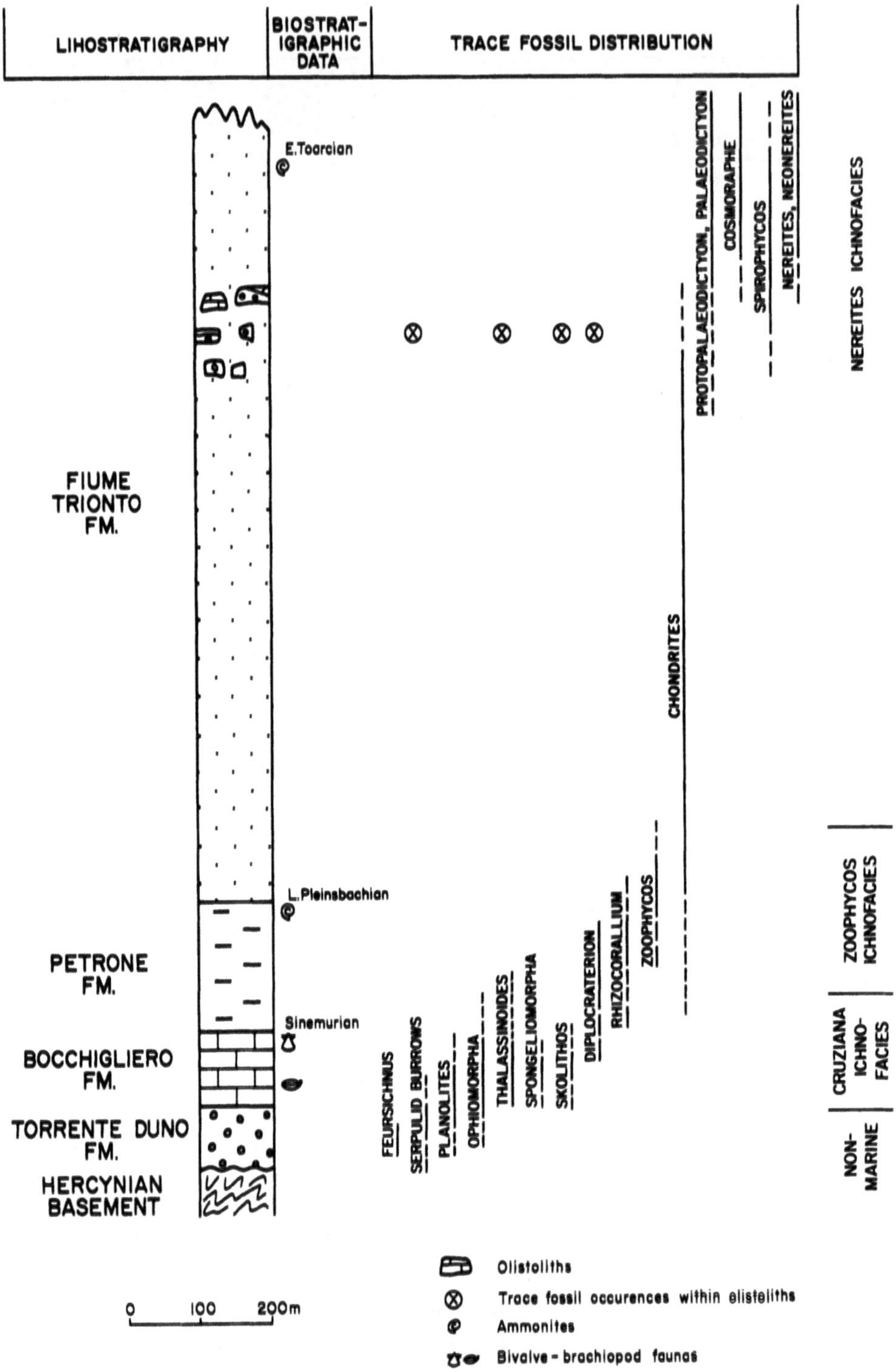

Fig. 2. Composite stratigraphic column of the Longobucco Group indicating the various formations and the main biostratigraphic control points. Nannofossils from throughout the Fiume Trionto Fm. confirm the Pliensbachian–Toarcian age. Also shown is the vertical distribution of trace fossils, illustrating the progressive deepening of the basin.

(Young *et al.*, 1986). Deposition was thus virtually confined to the upper Pliensbachian, and must have been very rapid, *c.* 300 m/Ma. Furthermore, the trace fossil associations in the turbidites indicate progressive deepening (Fig. 2), to the *Nereites* ichnofacies—generally taken to suggest lower bathyal to abyssal sedimentation. So overall subsidence, and presumably rifting, was considerable.

The bulk of the turbidites are quartz-arenites of rather problematic provenance, with palaeocurrents derived from the north-west (CTT unpublished data; Zuffa *et al.*, 1980). They appear to be unchannelized and show the characteristics of the basin plain or outer fan turbidite associations of Mutti and Ricci Lucchi (1972). Subsidiary lithologies include: carbonate turbidites in the lower part; a few very thick beds, up to 17 m, similar to the seismoturbidites of Mutti *et al.* (1984); rare lithic arenite turbidites and debris flows; the olistoliths; and interturbidite hemipelagic marls with common calcareous nannofossils.

The olistoliths

The olistoliths are most common toward the top of the turbidites—about a thousand metres from the base of the succession. They are not, however, grouped at any one horizon, but seem to be randomly distributed through the interval in which they occur. About fifty olistoliths have been identified and mapped. They are typically 50–100 m long and 15–25 m thick, ranging up to 250 m long and 35 m thick. Significantly smaller blocks have only rarely been found, despite intensive mapping and logging of the sequence. The blocks are usually tabular with markedly flat bases and steep sides. With the exception of a few small ones their internal bedding is always sub-parallel to that of the enclosing turbidites—and they are invariably the right way up as indicated by cross-bedding, geopetal fills, etc.

The most common block lithologies are limestones similar to those of the Bocchigliero Formation. This similarity includes: general lithology—siliclastic-rich oolitic and bioclastic limestones; sedimentological features such as trough, sigmoidal, and herring-bone cross-stratification; similar shelly faunas and ichnocoenoses. Perhaps most strikingly, an unusual and distinctive diagenetic fabric seen in the Bocchigliero Fm. also occurs in the olistoliths. This is a mesoscopic banding (up to 5 cm) which occurs in some oolitic limestones due to concentration of quartz into discrete zones, sometimes approximately parallel to bedding, elsewhere forming a reticulate network. The combination of all these features clearly indicates that these olistoliths are composed of the Bocchigliero Formation, and so are indubitably allochthonous.

Other lithologies are also sometimes seen, including the Torrente Duno red beds, rare basement granites, and carbonates, of possibly Pliensbachian age, from the Lower Caloveto Formation. The occurrence of Torrente Duno and Bocchigliero Formation sediments as olistoliths indicates that these formations were originally present over a wider area than the subsequent turbidite basin. Very similar sediments also occur in the Taormina area of Sicily, so they may have been widespread shelf sediments. By contrast, the absence of Petrone Formation marls, and presence of Lower Caloveto Formation limestones, suggests that basin-high differentiation was established by the Pliensbachian.

Olistolith–sediment relations

The Longobucco olistoliths mainly outcrop on valley sides and so their general relations with the enclosing sediments are quite clear, unlike those of olistoliths outcropping in areas of low relief. The details of the block margins are, however, rarely well exposed. The most readily observed feature is turbidite onlap. In almost all cases normal turbidite beds can clearly be seen to be laterally equivalent to the blocks, to butt against them, and lap over them. So, the blocks are not occurring within mass-movement deposits, and must have been introduced to the basin independently.

Details of the block–sediment relationships are particularly well shown around the sides and base of one large, easily accessible block (Fig. 3a). This is located on the north side of the Trionto valley, by the Sila–Rossano road (S.S. 177), 5 km from the village of Longobucco (Fig. 1, locality A). This block is 900 m from the base of the sequence, is composed of Bocchigliero Fm. carbonates, and is about 20 m thick.

As shown schematically in Fig. 4 it lies concordantly on the underlying turbidites, which show no sign of block-related deformation. At various places along the block margins small accumulations of cobble- to boulder-sized limestone clasts occur. These have similar lithologies to the block and are probably talus derived from it. The surrounding turbidites lap onto and around the sides of the block and eventually over its top (Fig. 4).

On the east side of the block is a zone of considerable soft sediment deformation, characterized by recumbent folds, with associated boudinage, accommodation, and injection structures. The folds have an eastward-verging asymmetry—away from the block.

Fig. 3a.

Fig. 3b.

This suggests that they were produced by shortening in front of the block as it slid in, and so that this side was the front of the block.

On the west side of the block, the probable rear, there is another zone of strong soft sediment deformation. Here the structures within the turbidites are chaotic and display no obvious common sense of asymmetry (Fig. 3b). Mixed into them are occasional clasts (from a few millimetres to 3 m long) of comparable lithologies to those of the large block. This

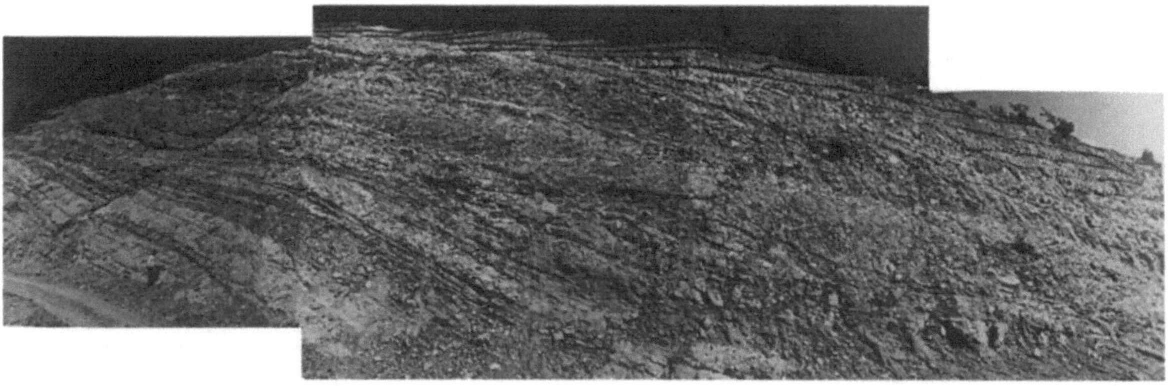

Fig. 3c.

Fig. 3d.

Fig. 3. a. View of the olistolith discussed in the text. b. Block of chaotically deformed sediments from the zone to the west of the olistolith. c. Photomosaic of major debris flow outcrop near the basin margin (at point B, Fig. 1). d. View of a second olistolith.

zone approximately follows bedding, and can be traced some way from the block. Farther west, along the road to Longobucco, a major debris flow is exposed with clasts up to 1.5 m long. The flow is 3.3 m thick at its eastern-most outcrop and can be traced west for 2 km, thickening to 5 m, before exposure is lost. Unfortunately, there is a hundred-metre gap in the outcrop between the block and the debris flow, which means they cannot be directly correlated. However, detailed logging and mapping of the turbi-

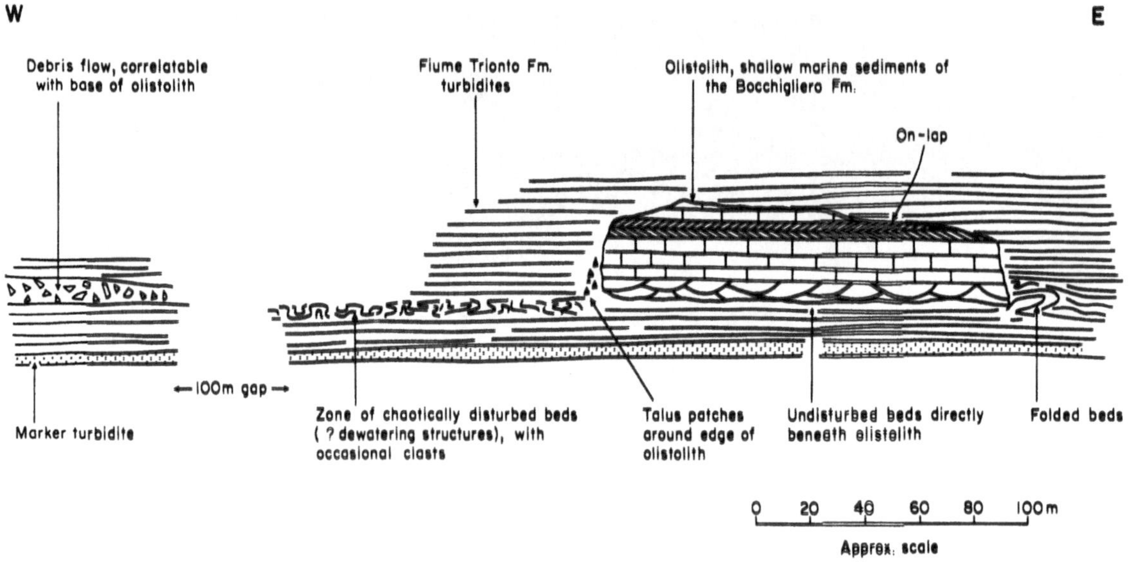

Fig. 4. Schematic diagram of the relationships of the surrounding sediments to the olistolith described in the text.

dite beds above and, most usefully, below the debris flow indicates that it is at the same stratigraphic level as the base of the block. So it seems that the olistolith, chaotic zone and debris flow were all products of the same sedimentation event.

Olistolith setting

The sequence of deposits, the soft sediment structures and the thickening trend of the debris flow all indicate derivation from the west. In this direction the turbidites can be traced continuously for some 4 km until the boundary of the Longobucco Group outcrop is reached. This boundary is a steeply dipping thrust, with basement in the hanging wall, and Longobucco Group sediments in the footwall. The thrusting is probably a result of Alpine compression during emplacement of the Longobucco Unit. However, such thrusting could well have occurred along reactivated normal faults of the original, extensional, margin of the Longobucco turbidite basin. That this is so is suggested by the fact that debris flows and lithic arenite turbidites become more common towards the fault: near it are some spectacular sections almost entirely composed of debris flows (Fig. 3c). Elsewhere along the margin, pebble-sized dropstones of basement schists occur in the turbidites; these must have fallen from a nearby fault scarp.

So it seems reasonable to presume that there was an active fault scarp along this margin, and that the olistoliths were derived from it. This implies transport of 2–5 km for most of the blocks, and probably slightly more if the effects of deformation within the sequence are taken into account.

The bulk of the turbidites, however, are quartz-arenites with NW–SW-directed palaeocurrents, as indicated by widespread sole marks. Neither the composition nor the palaeocurrents suggest derivation from the basin margin to the south-west, which was the source of the lithic arenite turbidites, debris flows, and olistoliths. Instead, the quartz-arenites were probably derived by recycling of sandstones from a source to the north-west, now lost. Hence the main axis of the turbidite basin, and so the regional palaeoslope, must have been oriented NW–SE, parallel to the south-west margin. Furthermore, the unchannelized basin–plain type facies of the turbidites, and the rarity of slumping in the sequence, suggest that this slope was gentle. So the olistoliths must have been emplaced obliquely across a very shallow slope, as shown in Fig. 5. This poses intriguing problems for their mechanism of emplacement, problems which we address in the final section.

OTHER CASES

Isolated olistoliths do not seem to be common and there are only a few well-described examples. The main features of a number of these olistoliths—from a variety of areas, geological periods, and settings—are summarized in Table 1. They are listed in order

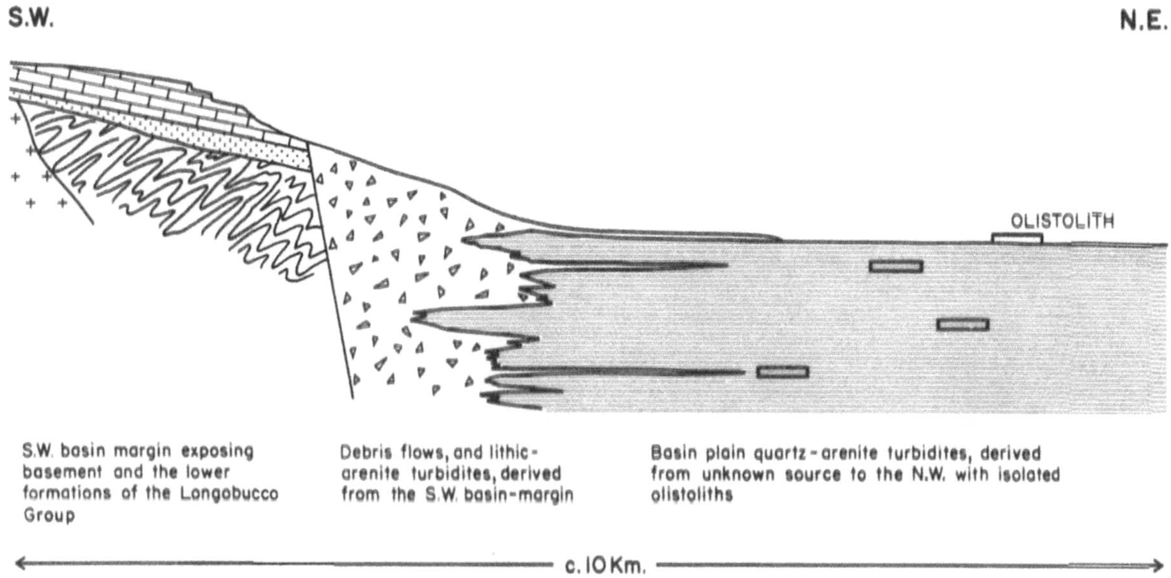

Fig. 5. Cartoon illustrating the inferred facies relationships across a section of the south-west margin of the Longobucco Basin, during the early Toarcian.

of block size and include, for comparison, examples probably best considered as fallen boulders (the Cipit Blocks), and as small gravity nappes (the Oman Exotics). Some other possibly analogous cases were excluded from the table since the blocks have debris flow deposits around their bases. These include the 'fallen stack' and other massive blocks in the Helmsdale boulder beds of north-east Scotland (Bailey and Weir, 1932; Pickering, 1984), and the celebrated Cretaceous blocks occurring in Oligocene lake sediments near Alès, S. France (Heim, 1923; Koop, 1952; Arène *et al.*, 1978).

It is clear from the table that the nature of the enclosing sediments is highly variable—and so it is presumably not a critical factor. Block lithology is also variable, but with carbonates dominating.

The predominance of extensional tectonic settings is interesting. This is doubtless in part an indication that extensional tectonics are likely to produce suitable conditions for block formation and transport, but may also in part reflect the greater potential of extensional settings for eventual preservation of decipherable relationships. In thrust settings subsequent deformation is liable to destroy block–sediment relations producing melanges of similar appearance whatever the genetic relations of the blocks to the matrix.

The olistoliths range in size from tens of metres to a couple of kilometres. They thus grade down into boulders, and up into sedimentary nappes—such as the Oman exotics described by Glennie *et al.* (1973) and Wilson (1969). Robertson (1977) and Richter (1973) have described blocks of similar size apparently formed by the detachment of fragments from the leading edge of larger sedimentary nappes, or slides. Distances travelled are rarely unambiguously constrainable, but in all the cases in the table it is clear that a suitable fault-scarp or other source occurred within a few kilometres of the blocks. Ten kilometres is about the maximum indicated distance of transport.

The cases most closely analogous to the Longobucco olistoliths seem to be those described from Antarctica by Ineson (1985), and from the Northern Apennines by Naylor (1981, 1982) and by Abbate *et al.* (1970). Both Naylor (1982) and Ineson (1985) discuss block–sediment relations. As with the Longobucco olistoliths they found that the blocks typically had planar bases which rested concordantly on the sediments below. Above the base the sediments onlapped or draped over the block. Naylor also recorded block-derived debris and breccias around the sides of the blocks, analogous to the talus accumulations seen around the Longobucco olistoliths. In addition, he listed a range of deformation features occurring in the sediments at the bases of the blocks which he ascribed to settling of the blocks after emplacement—including de-watering and liquefaction structures, small slump structures, and crumpling and contortion of bedding. Ineson did not find

TABLE 1. Listing of the salient features of various olistoliths comparable with the Longobucco olistoliths, organized in approximate order of size. The Oman Exotics are probably better considered as gravity nappes, but are included for comparison.

Author	*Location*	*Setting*	*Enclosing sediments*	*Olistoliths*	*Size*	*Distance*	*Terms used*
Fursich and Wendt (1977), Cros (1967)	Cassinian Fm. Dolomites, Northern Italy	Depositional carbonate margin	Triassic slope carbonates	Triassic reef carbonates	1–20 m	0.5–5 km*	Cipit blocks
Prior *et al*. (1982, 1984)	Kitimat Arm Fjord, B.C., Canada	Head of fjord	Modern fjord bottom clays and silts	Slabs, probably of delta front sands and silts	to 75 m	to 6 km*	Slide blocks Outrunner blocks
Cronan *et al*. (1974), Fisher and Bunce (1974)	DSDP site 232. Gulf of Aden	Newly formed oceanic rift	Upper Miocene nannofossil ooze	Upper Miocene lithified shallow marine sandstone	25 m T	—	Slide blocks
Teale (1985), Young *et al*. (1986)	Longobucco Basin, Calabria, S. Italy	Small extensional basin	Upper Liassic basin–plain turbidites	Lower Liassic shelf carbonates and clastics	10–250 m	2–6 km*	Olistoliths
Ineson (1985), Farquharson *et al*. (1984)	James Ross Island, Antarctic Peninsula	Back-arc basin	Lower Cretaceous lower slope clastics	Upper Jurassic laminated radiolarian mudstones	200–800 m	5–10 km	Glide blocks Slide blocks
Conaghan *et al*. (1976)	Nubrigyn Field, New South Wales, Australia	Depositional carbonate margin	Lower Devonian slope carbonates	Lower Devonian reef and platform carbonates	4–1400 m	to 10 km	Allochthonous blocks
Carrasco-V. (1973, 1977)	Valles-San Luis Potosi platform, Mexico	Bypass carbonate margin	Albian slope micrites and debris flows	Lower Cretaceous platform carbonates	to 95 m T	to 3.5 km	Allochthonous exotic blocks
Naylor (1981/2, 1984), Abbate *et al*. (1970)	Casanova Complex, N. Apennines, Italy	Passive continental margin	Upper Cretaceous–P'gene turbidites and debris flows	Ophiolitic igneous rocks and pelagic sediments	to 2 km	? a few km	Slide blocks Olistoliths
Stoneley (1981)	Kuh-e-Dalneshin area, S. Iran	Extensional basin	Upper Cretaceous siliceous pelagics	Cenomanian neritic limestones	to 1 km	? a few km	Exotic blocks
Richter and Mariolakis (1973)	Gavrovo Massif, Epiros, Greece	Intra-basinal ridge	Oligocene turbidites	Upper Cretaceous reef limestones	1+ km	?12 km*	Olisthothrymma
Wood (1981)	Lagonegro Basin, S. Apennines, Italy	Extensional basin	Middle Triassic (Ladinian) shales	Middle Triassic (Anisian) neritic lsts	to 20 km	—	Olistoliths
Glennie *et al*. (1973), Wilson, (1969)	Oman Mountains	Newly formed oceanic rift	Upper Cretaceous radiolarites	Permo-Triassic platform carbonates	to 20 km	?10 s of km	Exotic blocks Oman exotics

Author: where two references are given the main source of information is listed first.

Size: in most cases this is range of lengths of olistoliths. Where size of smallest olistoliths is not stated in references only size of largest is quoted. T denotes thickness, as opposed to length.

Distance: this is the distance from present position of block to probable source. *—our estimate, from maps etc. in the reference.

Abbreviations: P'gene—Palaeogene; lsts—limestones.

talus deposits or soft sediment deformation in the enclosing rocks but did describe 'glide-induced shear brecciation and folding' in the base of one block. Overall, then, there does seem to be a characteristic suite of emplacement-related structures. Which of these are seen in any one case will depend on such factors as the lithology of the block and of the enclosing sediment, and their subsequent deformation and diagenesis.

The best description of a modern analogue for these deposits seems to be that given by Prior *et al.* (1982, 1984) from the Kitimat Arm Fjord in British Columbia, Canada. They used a variety of oceanographic techniques to study a major slope-failure complex affecting a delta at the head of the fjord.

Side-scan sonar mapping of the fjord/bottom revealed isolated blocks occurring in front of the main landslide/debris flow. These so-called 'outrunner' blocks were found to occur up to one kilometre down-fjord from the front of the debris flow. The blocks appear to have continued to move downslope after the underlying debris flow, on which the blocks had been moving, had lost momentum and stopped. The blocks, lying on the fjord-bottom muds, are up to 50 m × 75 m in surface area, and 4 m high and are lying on a slope of less than 0.5°. Prior and his co-workers were also able to identify glide tracks left by the blocks on the fjord bottom mud as they moved over it. Surrounding these glide tracks, detritus was found which had been carried and shed by the blocks during their passage.

MODE OF OCCURRENCE

There are two basic problems in explaining isolated olistoliths: the mechanisms of formation of the blocks, and mechanisms of transport. These are discussed separately below.

Mechanisms of formation

This is the lesser of the two problems, and the clear association of olistoliths with active extensional regimes suggests that one common process is likely to be detachment of blocks from submarine fault scarps during earthquakes. This should occur most readily where strata in the fault scarp dip towards the basin. Detachment can then occur by sliding along bedding planes. Assuming rotation does not occur during emplacement, this would produce olistoliths with internal bedding subparallel to that of the surrounding sediments—as is the case with the Longobucco olistoliths, and those from the Antarctic Peninsula (Ineson, 1985). However, normal faulting is usually accompanied by back-rotation of the footwall block. This means that dips in the fault scarp are much more likely to be away from the basin than towards it. Under these circumstances slope-failure would be most likely to occur by rotational collapse. Blocks produced in this way would have internal bedding at high angles to the enclosing sediments—as described by Wood (1981) from the Lagonegro Basin, and by Naylor (1982).

Lithology is also obviously important: only massive rocks are liable to produce large blocks. This is reflected in the predominance of carbonates among the block lithologies in Table 1. It is also clear in the particular case of Longobucco, where the olistoliths are formed predominantly of the shallow marine limestones of the Bocchigliero Formation. The associated debris flows and lithic turbidites, by contrast, are dominated by material derived from the basement granites and schists, even though like the olistoliths they appear to have been derived from the south-west margin of the basin. Evidently these rocks were more common in the source area of the olistoliths but were less liable to form massive blocks.

Mechanisms of emplacement

The basic method of emplacement which has been suggested for most of the examples in the table is slow downslope sliding, similar to that of subaerial block slides, with motion occurring when the downslope component of gravity is sufficient to overcome the frictional forces. The principal problem with this model is that a large block resting on soft sediments should sink into them, or at least compact them. So, downslope motion would involve ploughing through a layer of sediment. This should in turn considerably impede movement, making forward motion unlikely on gentle slopes, and should create a wake of severely deformed sediments.

The blocks in Kitimat Fjord described by Prior *et al.* (1982, 1984) occur on a slope of less than 0.5 degrees and appear to have sunk somewhat into the surrounding sediment. In consequence, Prior *et al.* suggested a rather different mechanism of emplacement: relatively rapid 'skidding' of slabs over the sediment, with support in part from high pore-fluid pressures generated by the weight of the slabs. This overpressuring effect would also produce a low-strength layer able to reduce the resistance to motion—an effect analogous to the lubrication generated by the water layer produced beneath an ice-skate. This mechanism might enable a block to travel considerable distances over negligible, or even reversed, slopes, until it slowed to a speed where it began to sink into the underlying sediments. The additional resistance to motion caused by the deformation of these sediments would then abruptly brake the block.

This mechanism seems to be applicable to the Longobucco olistoliths. It provides an explanation for their occurrence several kilometres from the basin margin on what would otherwise be expected to be a negligible slope. The field relations, discussed above, fit well with this model: the folding in front of the block could have been caused by it nosing into the sediments as it came to rest; similarly, the disruption of the sediments behind the block may have been produced by the rapid escape of overpressured pore waters after the passage of the block. Prior *et al.* also suggested this as a cause for the glide tracks they observed, the sediments of which appeared to have lower pore-water contents than the surrounding fjord-bottom muds.

A major difference between the two cases is that in Kitimat the blocks are associated with a massive slope-failure complex which seems to have transported the blocks a considerable distance before they outran it under their own momentum, as the main flow came to a halt. In the Longobucco case, by contrast, there is only a relatively thin debris flow deposit in direct association with the block. Thus the independent motion of the block is liable to have been much more important in this case.

CONCLUSIONS

Isolated olistoliths are phenomena intermediate between fallen boulders and gravity nappes in scale and in emplacement process, in that important effects arise from the overpressuring of sediments beneath them. The Longobucco olistoliths provide unusually clear evidence for this and for the ability of such blocks to move long distances down gentle slopes independently of other mass-movement processes. This evidence may be worth considering when the origin of blocks with more ambiguous relations are considered—for instance olistoliths associated with debris flows, or blocks in melanges.

Whatever their mode of origin, olistoliths can be useful in elucidating basin histories, and deserve special attention wherever their occur. In the particular case of the Longobucco Basin they have given information on the basin geometry, on the original extent of the various lithofacies, on the diagenetic history of the rocks, and on synsedimentary tectonics.

ACKNOWLEDGEMENTS

The research behind this paper was, of course, entirely Tarquin's. I was privileged both to work with him on the biostratigraphy of the Longobucco Group, and to be asked to prepare his work on the olistoliths for publication, by the amalgamation of two of his draft manuscripts. Any mistakes in the final result are probably mine (J. R. Y.).

Olistoliths form a good subject for discussion and we both benefited greatly from the help of many colleagues and friends, including particularly: Emiliano Mutti, Gian Gaspare Zuffa, Jerry Leggett, Massimo Santantonio and John MacKenzie. R. H. Campbell and D. B. Prior gave useful comments on an earlier manuscript.

References

Abbate, E., Bortolotti, V. and Passerini, P. (1970). Olistostromes and olistoliths. *Sedim. Geol.* **4,** 521–557.

Arène, J., Berger, G-M., Gras, H., Poidevin, J-L. and Sauvel, C. (1978). *Alès. Notice explic. carte géol. France* 1/50 000, Bur. Rech. géol. min., 58 p.

Bailey, E. B. and Weir, J. (1932). Submarine faulting in Kimmeridgian times: East Sutherland, Transactions of the Royal Society, Edinburgh, **47,** 429–468.

Bernoulli, D. and Jenkyns, H. C. (1974). Alpine, Mediterranean, and central Atlantic Mesozoic facies in relation to the early evolution of the Tethys. *In* R. H. Dott and H. S. Shaver *Modern and Ancient Geosynclinal Sedimentation*, Soc. Econ. Palaeont. Mineral., Spec. Publ. 19, 129–160.

Carrasco-V, B. (1973). Exotic blocks of forereef slope, Cretaceous Valles—San Luis Potosi Platform (Mexico). (Abstract). Bull. Am. Ass. Petrol. Geol., **57,** 772.

Carrasco-V, B. (1977). Albian sedimentation of submarine autochthonous and allochthonous carbonates, East edge of the Valles–San Luis Potosi Platform, Mexico, *In* H. E. Cook and P. Enos (Eds), *Deep-water Carbonate Environments*, Soc. Econ. Palaeont. Mineral., Spec. Publ. **25,** 263–272.

Conaghan, P. J., Mountjoy, E. W., Edgecombe, D. R., Talent, J. A. and Owen, D. E. (1976). Nubrigyn algal reefs (Devonian), eastern Australia: allocthonous blocks and megabreccias, *Bull. Geol. Soc. Am.* **87,** 515–530.

Cronan, D. S., Damiani, V. V., Kinsman, D. J. J. and Thiede, J. (1974). Sediments from the Gulf of Aden and Western Indian Ocean. *In* R. L. Fisher and E. T. Bunce, Init. Reps. deep Sea drill. Proj. **24,** 1047–1110.

Cros, P. (1967). Origine de certain blocs calcaires de 'Cipit' du Trias dolomitique, *C.r. hebd. Séanc. Acad. Sci. Paris*, Sér. D, **264,** 556–559.

De Rosa, P., Zuppetta, A., Cavaliere, S., La Fratta, R., Marmolino, R. and Turco, E. (1980). I rapporti tra 1 unità Longobucco e le unità della catena alpina Europa-vergente nella finestra tettonica del Crati (Calabria nord-orientale), *Boll. Soc. Geol. It.* **99,** 129–138.

Farquharson, G. W., Hamer, R. D. and Ineson, J. R. (1984). Proximal volcaniclastic sedimentation in a Cretaceous back-arc basin, northern Antarctic Peninsula, *In* B. P. Kokelaar, M. F. Howells and R. D. Roach (Eds.), *Marginal Basin Geology*, Spec. publ. geol. Soc. Lond., **16,** 219–229.

Fisher, R. L. and Bunce, E. T. (1974). Site 232, *In* R. L. Fisher and E. T. Bunce, Init. Reps. deep Sea drill. Proj. 24, 127–196.

Flores, G. (1955). Discussion. World Petroleum Congress, Proceedings, 4th, Rome, **A2,** 120–121.

Fursich, F. T. and Wendt, J. T. (1977). Biostratinomy and palaeo-ecology of the Casinian Formation (Triassic) of the Southern Alps. *Palaeogeography, Palaeoclimatology, Palaeoecology* **22,** 257–323.

Glennie, K. W., Boeuf, M. G. A., Hughes-Clarke, M. W., Moody-Stuart, M., Pilaar, W. H. F. and Reinardt, B. M. (1973). Late Cretaceous nappes in Oman Mountains and their geological evolution, *Bull. Am. Ass. Petrol. Geol.* **57,** 5–27.

Heim, A. (1923). La prétendue nappe de recouvrement du bassin d'Alès (Gard) et l'origine des brèches urgoniennes dites mylonitiques, *Eclog. Géol. Helv.* **17,** 531–539.

Ineson, J. R. (1985). Submarine glide blocks from the Lower Cretaceous of the Antarctic Peninsula, *Sedimentology* **32,** 659–670.

Jaccobacci, A. (1965). Frane sottomarine nelle formazioni geologiche. Interpretazione dei fenomeni olistostromici e degli olistoliti nell'Appennino e in Sicilia, *Boll. Serv. Geol. Ital.* **86,** 65–85.

Koop, O. J. (1952). A megabreccia formed by sliding in southern France, *American Journal of Science* **250,** 822–828.

Magri, G., Sidoti, G. and Spada, A. (1963–65). Rilevamento geologico sul versante settentrionale della Sila (Calabria), *Memorie Note Ist. Geol. Appl. Napoli*, **9,** 1–60.

Mutti, E. and Ricci Lucchi, F. (1972). Le torbiditi dell'Appennino settentrionale: introduzione all'analisi di facies, *Memorie Soc. Geol. Ital.* **11,** 161–199.

Mutti, E., Ricci-Lucci, F., Seguret, M. and Zanzucchi, G. (1984). Seismoturbidites: a new group of resedimented deposits, *Mar. Geol.* **55,** 103–116.

Nardin, T. R., Hein, F. J., Gorsline, D. S. and Edwards, B. D. (1979). A review of mass-movement processes, sediment and acoustic characteristics, and contrasts in slope and base-of-slope systems versus canyon–fan–basin–floor systems, *In* L. J. Doyle and O. H. Pilkey (Eds), *Geology of Continental Slopes*, Soc. Econ. Palaeont. Mineral., Spec. Publ., 27, 61–73.

Naylor, M. A. (1981). Debris flows (olistostromes) and slumping on a distal passive continental margin: the Palombini limestone–shale sequence of the northern Apennines, *Sedimentology* **28,** 837–852.

Naylor, M. A. (1982). The Casanova Complex of the Northern Apennines: a melange formed on a distal passive continental margin, *Journal of Structural Geology* **4,** 1–18.

Pickering, K. T. (1984). The Upper Jurassic 'Boulder Beds' and related deposits: a fault-controlled submarine slope, north-east Scotland, *Journal of the Geological Society, London* **141,** 7–74.

Prior, D. B., Bornhold, B. D., Colemans, J. M. and Bryant, W. R. (1982). Morphology of a submarine slide, Kitimat Arm, British Columbia, *Geology* **10,** 588–592.

Prior, D. B., Bornhold, B. D. and Johns, M. W. (1984). Depositional characteristics of a submarine debris flow, *Journal of Geology* **92,** 707–727.

Richter, D. (1973). Olisthostrom, Olistholith, Olisthothrymma und Olisthoplaka als Merkmale von Gleitungs—und Resedimentationsvorganen infolge synsedimentärer tektogenetischer Bewegungen in Geosynklinalbereichen, *Neues Jb. Geol. Palaont. Abh.* **143,** 304–344.

Richter, D. and Mariolakis, I. (1973). Olistothrymma, ein bisher nicht bekanntes tekto-sedimentologisches Phanomen in Flysch-Ablagerungen, *Neues. Jb. Geol. Palaont. Abh.* **142,** 165–190.

Robertson, A. H. F. (1977). The Moni Mélange—Cyprus: an olistostrome formed at a destructive plate margin, *Journal of the Geological Society, London* **133,** 447–466.

Santantonio, M. and Teale, C. T. (1985). Jurassic condensed sedimentation on Hercynian basement (Calabria, Italy), Abstr. 6th Europ. Meeting int. Assoc. Sedimentol. Lerida, 459–560.

Santantonio, M. and Teale, C. T. (1987). An example of the use of detrital episodes in elucidating complex basin histories: the Caloveto and Longobucco Groups of north-east Calabria, S. Italy, this volume.

Santantonio, M. and Teale, C. T. (in preparation). Stratigraphy and sedimentology of the Jurassic–Neocomian Caloveto Group (Southern Italy).

Scandone, P. (1979). Origin of the Tyrrhenian Sea and the Calabrian Arc, *Boll. Soc. Geol. Ital.* **98,** 27–34.

Scandone, P. (1982). Structure and evolution of the Calabrian Arc. *Earth Evoln. Sci.* **3,** 172–180.

Stoneley, R. (1981). The geology of the Kuh-e Dalneshin area of Southern Iran, and its bearing on the evolution of southern Tethys, *Journal of the Geological Society, London* **138,** 509–526.

Teale, C. T. (1985). Occurrence and geological significance of olistoliths from the Longobucco Group, Calabria, Southern Italy. Abstr. 6th Europ. Meeting int. Assoc. Sedimentol. Lerida, 457–458.

Wilson, H. H. (1969). Late Cretaceous eugeosynclinal sedimentation, gravity tectonics, and ophiolite emplacement in Oman Mountains, south-east Arabia. *Bull. Am. Ass. Petrol. Geol.* **53,** 629–671.

Wood, A. W. (1981). Extensional tectonics and the birth of the Lagonegro Basin (Southern Italian Apennines), *Neues Jb. Geol. Palaont. Abh.* **161,** 93–131.

Young, J. R., Teale, C. T. and Bown, P. R. (1986). Revision of the stratigraphy of the Longobucco Group (Liassic, southern Italy); based on new data from nannofossils and ammonites, *Eclog. Geol. Helv.* **79,** 117–135.

Zuffa, G. G., Gaudio, W. and Rovito, S. (1980). Detrital mode evolution of the rifted continental-margin Longobucco Sequence (Jurassic), Calabrian Arc, Italy, *J. Sedim. Petrol.* **50,** 51–61.

Chapter 5

The *Griestoniensis* Zone Turbidite System, Welsh Basin

R. D. A. Smith, Department of Earth Sciences, Downing Street, Cambridge, CB2 3EQ

ABSTRACT

The early Silurian *griestoniensis* Zone turbidite system developed in the ensialic, tectonically active Welsh Basin. The preserved remnants of the system are interpreted as deposits of the largely unchannelled outer region of a moderately large, high-efficiency, confined turbidite system with a minimum length of 100 km. Large-scale conglomerate and sandstone-filled erosional structures are present in the most proximal areas of exposure. Palaeocurrent patterns and inferred onlap relationships reflect strong lateral confinement by structurally controlled slopes. Growth of the system appears to have been essentially aggradational and vertical stacking of lobe sandstone packets is ascribed to pulses of rapid tectonic uplift at source. The sediment source was probably a large delta located on the uplifting southern basin margin.

INTRODUCTION

Remnants of the early Silurian *griestoniensis* Zone turbidite system are discontinuously exposed over a distance of about 100 km in a NNE–SSW direction in Wales (Fig. 1). They consist of parts of a sequence up to about 400 m thick of turbidites, other sediment gravity flow deposits, and hemipelagites of late Llandovery *griestoniensis* Zone age (Fig. 2). The sequence is part of an Upper Ordovician and Lower Silurian turbidite-dominated succession, approximately 3 km thick, deposited in the Welsh Basin.

Early work on the *griestoniensis* Zone basinal facies focused on lithostratigraphy, biostratigraphy and structure. Rocks of this age were described under various names: 'Pysgotwr Grits' in the south (Davies, 1933), 'Moelfre Group' in the central area (Jones, 1945), and 'Talerddig Group' in the north (Bassett, 1955; Wood, 1906). Long (1966), prior to the advent of deep-sea fan models, interpreted the sequence as the deposits of a turbidite system lacking channel elements, sourced from a delta located on an uplifting 'geanticlinal ridge' to the south-west.

Several features of the *griestoniensis* Zone system make it an excellent study example of a moderately large, confined turbidite system developed in a tectonically active ensialic basin:

1. Rocks of this age are widely exposed in Wales and the Welsh Borderland (Fig. 1).
2. Good biostratigraphic control is available using graptolites in basinal facies, bachiopods in shelf facies, and both groups in distal shelf facies (Benton and Gray, 1981; Cocks *et al.*, 1984; Wood, 1906; Ziegler, 1965).
3. Detailed studies of benthic fossil assemblages in the adjacent shelf facies provide constraints on relative water depths and sea-level fluctuations (Cocks and McKerrow, 1984; McKerrow, 1979; Ziegler *et al.*, 1968b).
4. Recent advances in the understanding of the tectonic framework of the Welsh Basin (reviews in Kokelaar *et al.*, 1984; Woodcock 1984a) enable improved evaluation of tectonic controls on sedimentation.

Marine Clastic Sedimentology © J. K. Leggett & G. G. Zuffa (Graham and Trotman, 1987) pp. 89–107

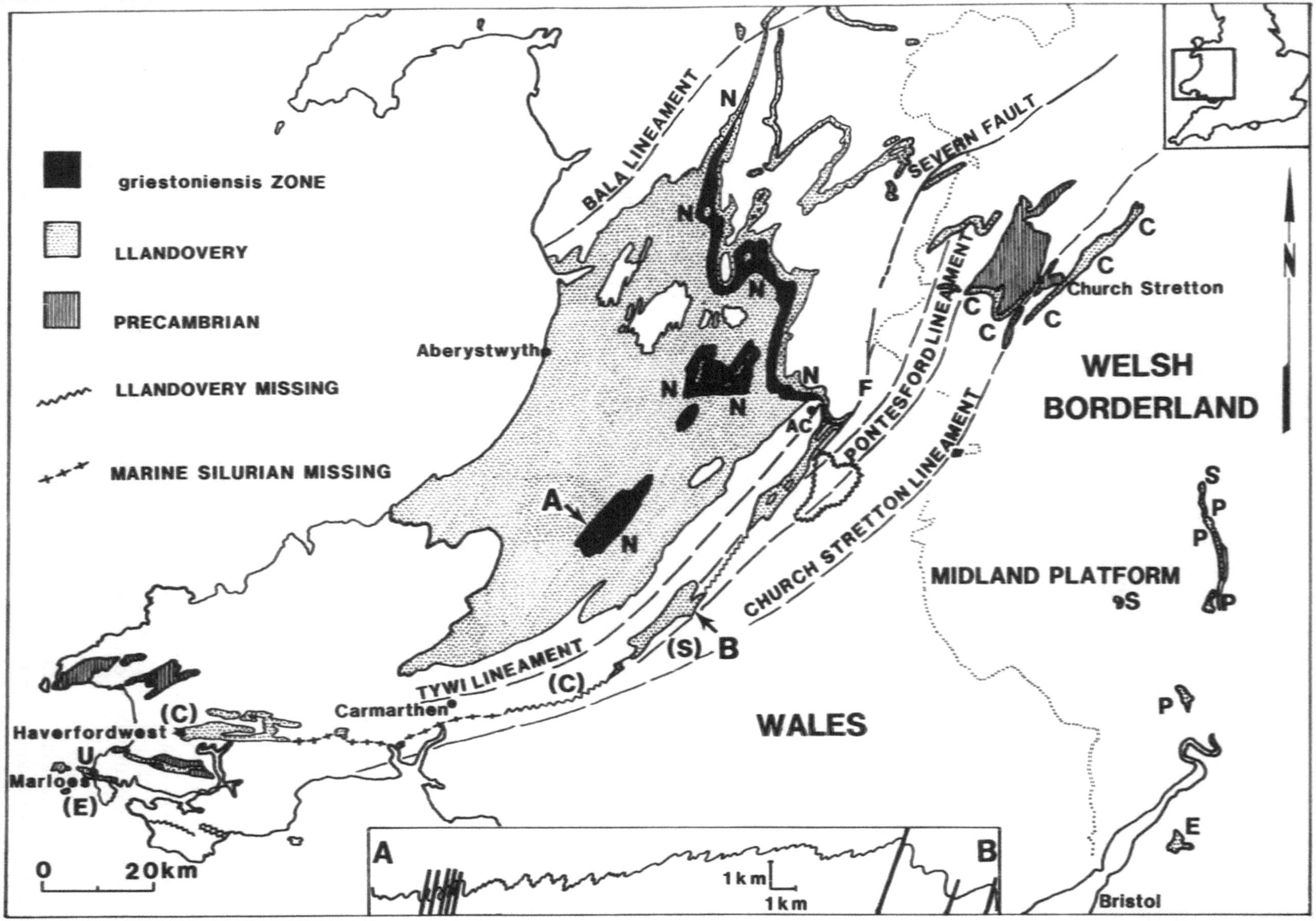

Fig. 1. Location map showing approximate areas of *griestoniensis* Zone, Llandovery Series, and Precambrian outcrop in Wales and the Welsh Borderland, and place names mentioned in the text (AC, Abbey Cwmhir). The *griestoniensis* Zone outcrop pattern is compiled from Bassett (1955), Jones (1945), Long (1966), Roberts (1929), Smith (1987) and Wood (1906). Simplified structural cross-section is compiled from Smith (1987), A. Mackie (unpublished data), Smallwood (1986), and Woodcock (1987). Distribution of brachiopod-dominated communities of approximately *griestoniensis* Zone age is after Ziegler *et al.* (1968b): [illegible] *Eocoelia*; P, *Pentameroides*; S, *Costistricklandia*; C, *Clorinda*. Locations of communities occurring in rocks of approximately *crenulata* Zone age,

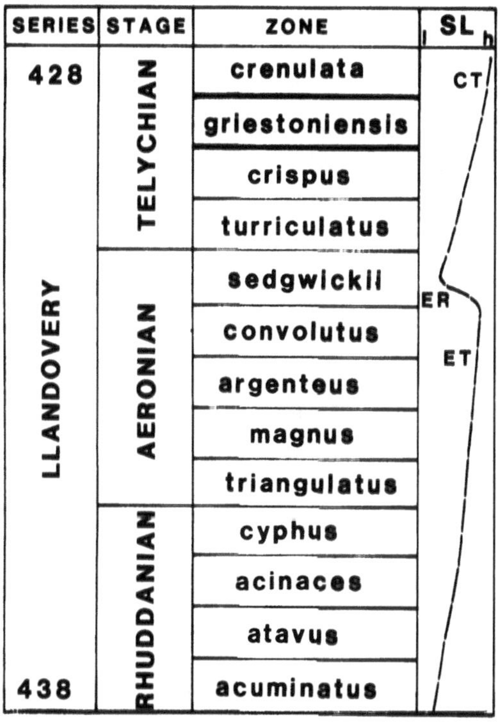

Fig. 2. Llandovery Series stratigraphy (after Cocks *et al.*, 1984) and sea-level curve generalized from McKerrow (1979). ET, eustatic transgression; ER, eustatic regression; CT, continental transgression. Dates of beginning and end of the Series, in millions of years before present, are from Harland *et al.* (1980).

In this paper new sedimentological data are presented and, together with information on the regional setting, are used to infer important aspects of the geometry of the turbidite system and the dominant factors controlling its pattern of development.

GENERAL GEOLOGICAL SETTING

Tectonic setting

The early Palaeozoic Welsh Basin was a relatively rapidly subsiding area of continental crust separated from the more stable Midland (or London) Platform by an arcuate array of steep (at the surface) fault belts (Fig. 1, Woodcock, 1984a). The basin is thought to have been located close to the leading edge of a northward-drifting microcontinental fragment, known as Armorica (André *et al.*, 1986; Perroud *et al.*, 1984) or Cadomia (Soper and Hutton, 1984) on the southern margin of the Iapetus Ocean. Soper (1986; see also Burrett, 1985) suggests that a still smaller terrane, Eastern Avalonia (including the Welsh Basin and Midland Platform), may have been separate from the Armorican Craton to the south, although André *et al.* (1986) argue against there having been a large ocean in this position.

Thorpe (1979) argued that the continental crust underlying the basin is largely the product of late Precambrian–earliest Cambrian arc accretion events (see also Thorpe *et al.*, 1984). He also suggested that several of the major lineaments in Wales may represent sutures between accreted arc and fore-arc terranes. The large-scale anastomosing pattern of the important Welsh Lineaments may, at least in part, be a reflection of strike-slip displacements during oblique accretion episodes (cf. Woodcock, 1986; see also Gibbons (1983), (1985) and Pauley (1986) for discussion of evidence for major late Precambrian–earliest Cambrian strike-slip in North Wales and the Welsh Borderland). The nature of the lineaments at depth is unknown, although one major reflector, the South Irish Sea Lineament, dipping west-north-west at approximately 25° to about 15–18 km depth, has been recorded by the BIRPS deep seismic reflection WINCH traverse (Brewer *et al.*, 1983). It is not known which onshore structure this lineament corresponds to. Repeated reactivation of such early structures characterized the ensuing complex early Palaeozoic history of the basin (e.g. James, 1972; James and James, 1969; Kokelaar *et al.*, 1984 and references therein; Smith, 1987; Woodcock, 1984b).

The petrochemistry of Welsh volcanic rocks is thought to record a late Tremadoc episode of arc volcanism, followed by a phase of Arenig to Caradoc back-arc extension (Kokelaar *et al.*, 1984). This is consistent with the location of a relict Ordovician arc in the Leinster–Lake District Zone (Fig. 4). André *et al.* (1986) interpret the coeval igneous rocks of the Brabant and Ardennes massifs of Belgium as the products of south-directed subduction to the east of the Midland Platform.

The detailed Ashgill and Silurian setting of the basin remains obscure due to uncertainties concerning the exact timing of final closure of Iapetus, the extent of strike-slip displacements associated with the Armorica/Cadomia-Laurasia collision (Murphy and Hutton, 1986; Soper, 1986; Soper and Hutton, 1985) and the importance of other convergence zones to the south and north-east (Soper, 1986). Apparent eastward migration of depocentres during the Silurian (Cummins, 1969; Smith and Long, 1969) led Okada and Smith (1981) to suggest a fore-arc setting. Their model involves compressional uplift of the Irish Sea High (Fig. 4) as a result of south-eastward

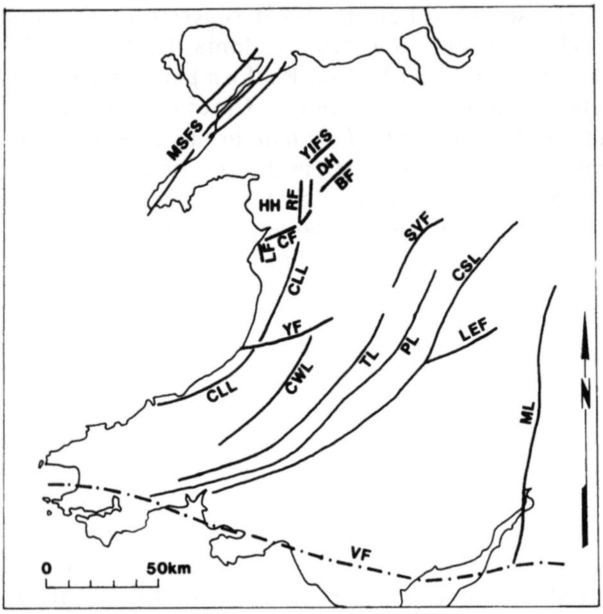

Fig. 3. The 'Variscan Front' and structures inferred to have been active during the early Palaeozoic history of the Welsh Basin: BF, Bala Fault (James and James, 1969; Fitches and Campbell, 1987); CLL, Corris–Llangranog Lineament (James and James, 1969; James, 1972; Craig, 1985; Craig, 1987); CF, Ceunant Fault (Fitches and Campbell, 1987); CSL, Church Stretton Lineament (Holland and Lawson, 1963); CWL, Central Wales Lineament (new name for core of Central Wales Synclinorium: Smith, 1987); DH, Derwen Horst (Fitches and Campbell, 1987); HH, Harlech Horst (Fitches and Campbell, 1987); LEF, Leinthall Earls Fault (Lawson, 1973); LF, Llanegryn Fault (Fitches and Campbell, 1987); ML, Malvern Line (Hurst *et al.*, 1978); MSFS, Menai Straits Fault System (Kokelaar *et al.*, 1984 and references therein); PL, Pontesford Lineament (Woodcock, 1984b); RF, Rhobell Fracture (Kokelaar *et al.*, 1984 and references therein; Fitches and Campbell, 1987); SVF, Severn Valley Fault (James and James, 1969); TL, Tywi Lineament (Smallwood, 1986); YIFS, Yspytty Ifan Fault System (Fitches and Campbell, 1987); YF, Ystwyth Fault (James and James, 1969).

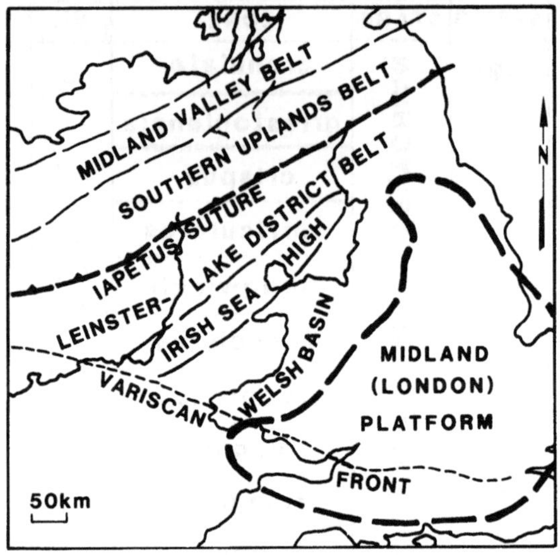

Fig. 4. Regional tectonic framework of the Welsh Basin (modified after Ziegler, 1984).

migration of a southeast-dipping subduction zone. Possible Lower Silurian volcanic arc rocks are exposed near Marloes (Ziegler *et al.*, 1969) and in the Tortworth and Mendips inliers near Bristol (Hurst *et al.*, 1978; Stillman and Francis, 1979). Abundant bentonites of late Llandovery to early Ludlow age record more widespread volcanism during this period of time, but Teale and Spears (1986) suggest that the source for the bentonites may have been a considerable distance away, possibly in north-east Europe. A potential problem with the fore-arc model, noted by Woodcock (1984a) and Kelling *et al.* (1985), is the lack of evidence for a subduction-related accretion zone in the Leinster–Lake District Belt. A recently proposed alternative possibility for the origin of the Silurian igneous rocks of south-west Wales and the Welsh Borderland is that they are the products of northward subduction to the south, between the Midland/London Platform and the Ardennes (Soper, 1986; but see also André *et al.* (1986), Perroud *et al.* (1984); Ziegler (1984) for discussion of the nature and history of the 'mid-European Caledonides' located between the Midland/London Platform and the Armorican Craton). Murphy and Hutton (1986) contend that Iapetus oceanic crust had been entirely consumed by early Silurian time. Their model sites the Silurian sequence of the Southern Uplands Belt (Fig. 4) in a successor (or remnant) basin setting, between a relict (now no longer preserved) arc, 'Cockburnland', to the north and the relict Leinster–Lake District arc to the south. In this context the Silurian Welsh Basin may have been a remnant marginal basin which underwent compressional modification as the Southern Uplands Belt imbricate stack, or along-strike equivalents, overthrust southwards onto Armorican/Cadomian crust (see Leggett *et al.*, 1983). Evidence for early Silurian tilting of basement fault blocks bounded by normal faults (Smith, 1987) suggests active extension in the basin during early Silurian time, but the driving mechanism behind this extension is unknown. Silurian crustal shortening in the Welsh Basin is suggested by post-Wenlock uplift west of the Tywi

Lineament (Smith, 1987; Woodcock, 1976), and tectonic uplift in south-west Wales (Cummins, 1969; Hurst *et al.*, 1978; Smith and Long, 1969). The relative mildness of collision-related deformation, such as the absence of large-scale thrust sheet translation to the south of the Iapetus Suture, has been ascribed to the slow and oblique nature of the Armorica/Cadomia-Laurasia collision (Murphy and Hutton, 1986).

The essential feature of the basin history is the reutilization of early structural discontinuities in a variety of regional stress regimes. The Silurian Welsh Basin is closest to a Type C basin in the classification scheme of Mutti and Normark (this volume).

Late Ordovician and early Silurian turbidite systems in the Welsh Basin

During Ashgill to mid-Llandovery (Aeronian) times turbidite systems in the Welsh Basin were mainly laterally fed from the eastern basin margin with a minor component of axial supply from the south (Cave, 1979; James, 1972, 1983a; Kelling and Woollands, 1969). These systems were strongly influenced by glacioeustatic sea-level fluctuations and fault-controlled basin topography (Cave, 1979; James, 1972, 1983a; James and James, 1969; Smith, 1987).

The importance of the eastern margin as a source area decreased dramatically during the late Llandovery transgression of the Midland Platform (Bridges, 1975) and the late Llandovery to Wenlock turbidite systems of the basin were predominantly axially fed (Cave, 1979; Cummins, 1969; Smith and Long, 1969).

The Upper Llandovery (Telychian) to Wenlock basinal facies record a progressive eastward migration of the eastern margins of successive turbidite systems (Cummins, 1969). This may represent a tectonic migration of basin axis eastward with time since:

1. Palaeocurrents are consistently unidirectional with low variance reflecting confinement of flows this study; Cummins, 1957; Wood and Smith, 1959). This confinement was probably the product of a corrugated basin floor topography resulting from draping of the sedimentary cover sequence over tilted basement fault blocks (cf. Smith, 1987).

2. A recent white mica crystallinity study Robinson and Bevins, 1986), indicating a primary burial depth control on metamorphic grade although strain-related grade variations have also been documented in mid-Wales: Angus Mackie, personal communication, 1986), shows equally low grades in the *griestoniensis* Zone rocks and the older (*turriculatus* Zone) Aberystwyth Grits exposed along the coast to the west. This suggests that, if depth of burial is the dominant control on grade, the Aberystwyth Grits were not overlain by any significantly greater thickness of later Silurian sediments than was the *griestoniensis* Zone sequence.

3. Eastward-directed slumps of early Ludlow (Gorstian) age in the region of the Tywi Lineament (Woodcock, 1976) suggest reactivation of this structure with a reverse sense of movement. Since the Pontesford and Church Stretton lineaments still appear to have had basin-down throws at this time (Bailey, 1969; Holland and Lawson, 1963; Woodcock, 1976; Woodcock, 1984b), this would be consistent with west-to-east reactivation of basement lineaments as reverse faults forcing foreland-verging monoclines in the cover (cf. Smith, 1987). Details of the timing of inversion of previously subsiding blocks to the west of the Tywi Lineament are unknown.

Other possibly significant controls on the pattern of apparent eastward migration of turbidite systems with time are:

1. The influence of the depositional topography of each preceding lobe (Cave, 1979).

2. Preferential breaching of the lower eastern levees of feeder channels. Since the Welsh Basin was situated south of the equator in early Palaeozoic time (Briden *et al.*, 1974; Cocks and Fortey, 1982; Perroud *et al.*, 1984), the Coriolis Force would have had the effect of causing overbanking flows in channels to be deflected to the left, leading to the development of higher western levees (Komar, 1969; Nelson and Kulm, 1973).

3. Sinistral strike-slip along the basin-margin fault systems removing basinal systems to the west of an important point source.

However:

1. Depositional lobe topography in systems of this scale with distal sand depocentres would probably be extremely gentle (e.g. Ghibaudo, 1981; Nelson, 1985) and it seems unlikely that successive flows would be consistently deflected to the right of each preceding lobe deposit.

2. The proximal parts of all these systems have been removed by erosion and are not available for study.

3. There is no unequivocal evidence for intra-Silurian sinistral strike-slip in the basin.

A schematic cross-section of the pre-shortening Telychian to Wenlock basin-fill, in which eastward

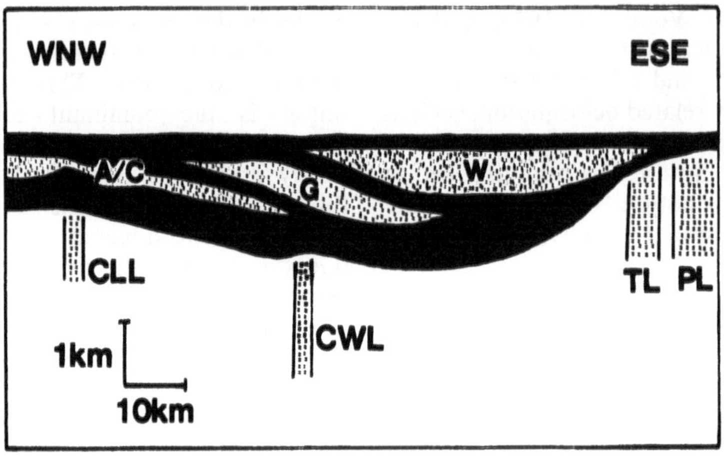

Fig. 5. Schematic WNW–ESE cross-section of the Upper Llandovery (Telychian) to Wenlock basin-fill. A/C, Aberystwyth–Cwmystwyth turbidite system lobe deposits (*turriculatus-crispus* Zones); G, *griestoniensis* Zone turbidite system lobe deposits; W, Wenlock turbidite system lobe deposits. Black shading represents fine-grained, thin-bedded turbidites. Eastern migration of western margins of systems has been assumed. Lineament abbreviations as in Fig. 3.

migration of the western margins of successive turbidite systems has been assumed, is shown in Fig. 5.

THE *GRIESTONIENSIS* ZONE TURBIDITE SYSTEM

Structure of the *griestoniensis* Zone sequence

The present structural style is dominated by strongly asymmetrical, generally tight, southeast-verging folds (Fig. 1), the product of an important component of northwest–southeast compression, probably culminating in early to early mid-Devonian time (Woodcock, 1984a, 1987). Common clockwise transection of folds by cleavages suggests that basin inversion involved sinistral transpression (Smith, 1987). The trend of the folds swings from northeast–southwest in the south to approximately north–south in the north. Coward and Siddans (1979) and Woodcock (1984a) argue that the arcuate pattern is probably not a secondary bending effect, preferring the influence of rigid basement structures, such as volcanic centres, during compression. The extent of rotations in the sedimentary cover sequence associated with the deformation is unknown. Cleavages are well developed in the mudstones and more poorly sorted sandstones of the sequence (Smith, 1987) and metamorphic grade ranges from diagenetic Zone to upper anchizone (Robinson and Bevins, 1986; Smith 1987). Faults are mainly concentrated into narrow belts localized by long-lived basement structures (Smith, 1987).

Biofacies framework

Studies of the Llandovery Series shallow water shelf faunas (Cocks and McKerrow, 1984; Ziegler, 1965; Ziegler *et al.*, 1968a, 1968b) have documented near-shore to offshore variations in the composition of fossil benthic associations (mainly brachiopods) related to palaeo-water depth (justified by Cocks and McKerrow, 1984). Four 'communities' have been recognized in shelf facies of roughly equivalent age to the *griestoniensis* Zone basinal sequence. From shallowest to deepest they are the *Eocoelia*, *Pentameroides*, *Costistricklandia*, and *Clorinda* communities. Their distribution in rocks of this age is shown in Fig. 1. Further offshore from the *Clorinda* community band, skeletonized benthos disappears entirely and records of benthic animal activity are provided by trace fossils. *Cruziana* ichnofacies (Seilacher, 1967) trace fossil assemblages, such as the one associated with *Clorinda* community faunas near Church Stretton, described by Benton and Gray (1981), are replaced offshore by an apparently low-diversity trace fossil assemblage in fine-grained slope facies near Abbey Cwmhir (e.g. at SO08756878). The low apparent diversity may be a result of reduced preservation potential of many trace fossils (especially graphoglyptids) in fine-grained sediments (e.g. Seilacher, 1977). Dominant forms are *Chondrites*, *Planolites* and *Paleodictyon*. In the basinal facies interturbidite hemipelagites are either micro-bioturbated and associated with a diverse *Nereites* ichnofacies (Seilacher, 1967) trace fossil assemblage

(Smith, in preparation) or finely laminated and unbioturbated. Clearly, bottom waters in the deepest part of the basin were low in dissolved oxygen (fluctuating from dysaerobic to anaerobic) due to stratification of the water column (cf. Ettensohn and Elam, 1985). Leggett *et al.* (1981) suggest that the early Silurian corresponded to a period of a 'greenhouse' state earth. During such 'G' state periods sea levels were high and ocean waters below the zone of surface overturning poorly aerated. This is a result of the concentration of primary productivity in shelf areas, such that there is a greater drain on sea-water oxygen levels when shelf areas are transgressed.

Absolute water depths are unknown, but Cocks and McKerrow (1984) estimated that the total depth range for the shelf communities was probably no more than 200 m. Assuming this value and a lateral basin-slope gradient of between 1 and 3 (cf. Thornton, 1984) gives a depth range of the order of 500–1000 m for the turbidite system.

Lithofacies associations

Lithofacies of the *griestoniensis* Zone sequence can be interpreted in terms of environments within a relatively deep-water turbidite system for the following reasons:

1. The sequence consists of turbidites, related sediment gravity-flow deposits and hemipelagites with no evidence for wave reworking, and characteristic turbidite system facies (facies A to G of Mutti and Ricci Lucchi, 1972) are present.
2. In the regional palaeogeographical context the sequence was deposited offshore, across a slope, from a shelf over 100 km wide with laterally extensive sub-wave base storm deposits located near its distal edge (Benton and Gray, 1981; Bridges, 1972, 1975; Ziegler *et al.*, 1968b).
3. Hemipelagic facies and trace fossil assemblages indicate that the turbidites were deposited in the oxygen-deficient deeper parts of a stratified water column.
4. Distinct down-palaeocurrent and across-palaeocurrent facies changes can be demonstrated, and large-scale erosional features are preserved in the most proximal parts of the sequence.

Four major lithofacies associations have been recognized which represent the deposits of the outer area of a moderately large turbidite system and its lateral bounding slope.

Channel–Lobe transition zone facies association

This facies association is present in the most south-westerly area of exposure (Fig. 6) and is characterized by the presence of coarse-grained, often conglomeratic facies, with very frequent amalgamation of beds due to extensive scouring at their bases. In this area conglomerate and sandstone-filled erosional features of the order of one kilometre wide and ten metres or more deep occur incised into packets of lobe sandstones. Available exposures allow reconstruction of oblique profiles, but do not enable reconstruction of the down-palaeocurrent geometry of these features. The best exposed fill of one of these huge scours (almost continuously exposed for 1200 m on Craig Twrch, SN6548) has an up to 4.2 m thick basal unit consisting of a clast- to matrix-supported pebble conglomerate with occasional intrabasinal boulders which grades abruptly up into coarse sandstone. An inverse-graded layer roughly five centimetres thick is discontinuously developed along the base of the conglomerate. The conglomerate base cuts down through thick (to 2 m) coarse tail-graded granule conglomerate–sandstone units with strongly scoured bases into typical lobe sandstones. The basal pebble conglomerate–sandstone unit is succeeded by thinner (150–15 cm) coarse-tail graded conglomerate–sandstone units which are overlain by thick-bedded sandstones. The crude vertical thinning- and fining-upwards pattern is interpreted as the product of deposition from a surging high-density turbidity current (cf. Hendry, 1973; Lowe, 1982; Surlyk, 1984) with traction carpet development at the flow base, followed by deposition from smaller volume flows. The giant erosional structures are seen as features eroded by the heads of large-volume high-density turbidity currents and then rapidly infilled. They may represent either the terminal reaches of feeder channels or, alternatively, exceptionally large scours unattached to channels, excavated by large-volume, high-velocity flows undergoing a hydraulic jump in the channel–lobe transition zone (cf. Mutti and Normark, this volume; Normark *et al.*, 1979). The facies immediately underlying and overlying the lenticular conglomerates are strikingly similar to channel–lobe transition facies described by Cazzola *et al.* (1981). They consist of amalgamated, coarse-tail graded units, several with near-horizontal stratification (see Hiscott and Middleton, 1979) and some with sets of cross-stratification up to 40 cm thick developed at their tops.

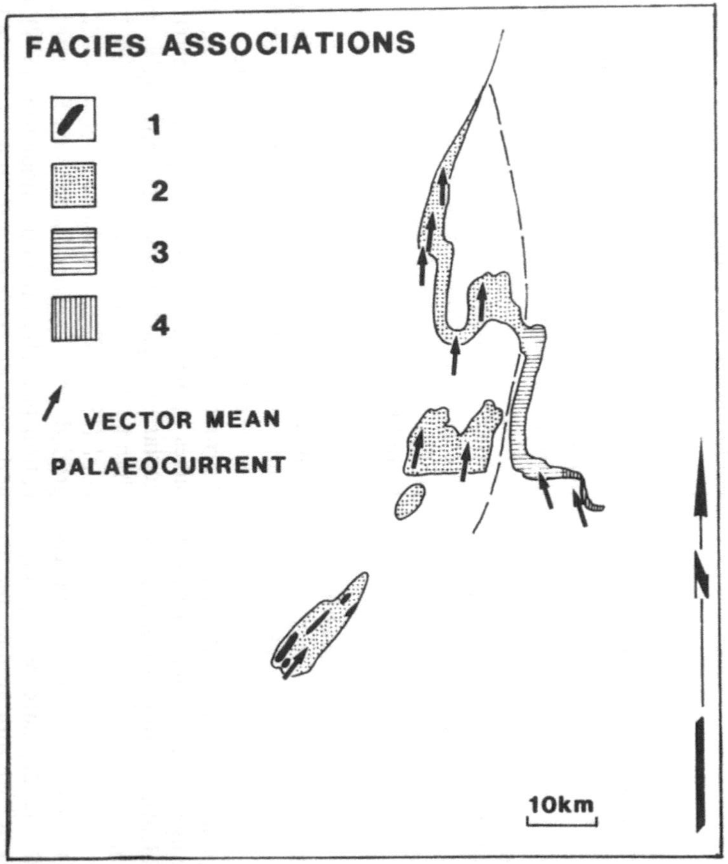

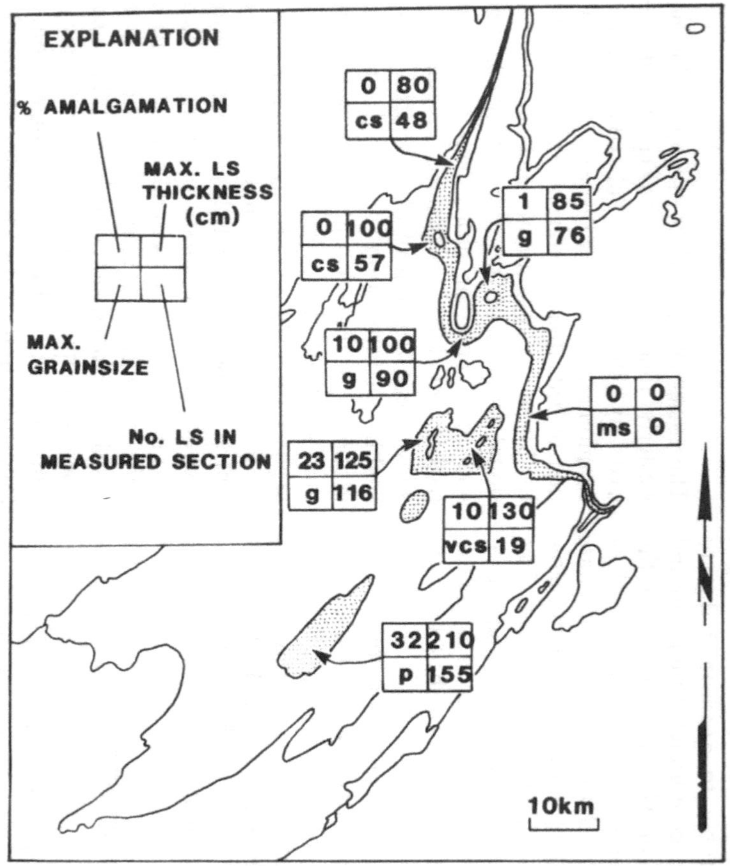

Fig. 6 (left). Distribution of *griestoniensis* Zone lithofacies associations and vector mean palaeocurrents. 1, channel–lobe transition zone facies association (approximate extent); 2, sandstone lobe facies association (maximum lateral extent); 3, lobe fringe facies association; 4, lateral basin-slope facies association.

Fig. 7 (right). Lateral variation in the *griestoniensis* Zone sandstone lobe facies association illustrated by percentage amalgamation, maximum bed thickness, and maximum grain size for selected measured sections. LS, 'lobe sandstone', including ungraded mud-rich 'slurried' beds. p, pebble; g, granule; vcs, very coarse sand; cs, coarse sand; ms, medium sand.

Extrabasinal clast types in the conglomerates include rhyolites, albite granites, granophyres, tuffs and tuffites, vein quartz and various sedimentary rock fragments. The igneous clasts closely match late Precambrian rock types presently exposed in south-west Wales (Bridges, 1972; Cummins, 1957).

Sandstone lobe facies association

Packets of non-channelized thin- to very thick-bedded sandstones punctuate the thin-bedded turbidite background of the *griestoniensis* Zone sequence (Fig. 11c, Fig. 12a, b). The packets range from about 5 to 30 m in thickness in the most proximal area, but are thinner and less well defined in more distal (to the north) and lateral (to the east) areas. These packets consist of predominantly normally graded, very poorly sorted sandstones with basal grain sizes from medium sand occasionally up to small pebble grade, and bed thicknesses ranging from (rarely) as thin as 5 cm up to 2 m. Bouma (1962) $T_{a(d)e}$, T_{ac-e} and T_{abc-e} sequences occur. T_c to T_e transitions commonly show features of the Stow and Shanmugam (1980) fine-grained turbidite model including fading ripples, graded alternating silt and mud laminae, and graded and ungraded mud. Observed water-escape structures in the sandstones include diffuse dish structures and large (to 15 cm × 6 cm) pillar structures. The sandstones are interpreted as the deposits of turbidity currents of various degrees of hydraulic maturity, attributable mostly to Facies C of Mutti and Ricci Lucchi (1972) and partly to Facies B. Also present are very mud-rich ungraded 'slurried' units (up to 20% of all 'lobe sandstones' near the eastern limit of the sandstone lobe facies association, e.g. near Talerddig SH9300) which frequently contain abundant large (to more than 1.5 m maximum dimension), often highly deformed, siltstone and mudstone rip-up clasts. These are interpreted as the deposits of cohesive laminar flows transformed from fully turbulent flows during deceleration (cf. Enos, 1977; Hiscott and Middleton, 1979; Lowe, 1982; Postma, 1986).

Runs tests (using the procedure of Heller and Dickinson, 1986) show that the sandstones occur in random bed thickness patterns. Minor thickening- and thinning-upwards sequences do occur, however, and possibly represent local progradational and re-gradational phases and/or processes of compensation for gentle depositional topography (Hiscott, 1981; Mutti and Normark, this volume; Mutti and Sonnino, 1981). There is no large-scale coarsening- or thickening-upwards of the lobe sandstone packets throughout the *griestoniensis* Zone sequence.

Individual sandstones within packets are either amalgamated with adjacent beds or separated by thin mudstones or thin-bedded turbidite sequences. Confident correlation of individual beds over distances larger than about five kilometres has not proved possible, but general down-palaeocurrent and across-palaeocurrent variations in this facies association are illustrated in Fig. 7 by values of percentage amalgamation, maximum bed thickness, and maximum grain size for selected measured sections. These consistently decrease distally and towards the eastern lobe fringe area. Lobe sandstone packets are only present near the top of the *griestoniensis* Zone sequence in the north (Wood, 1906) and east (Jones, 1945).

Lobe–Fringe facies association

'Thin-bedded turbidites' (actually very thin- to medium-bedded: Ingram, 1954) form a monotonous background to the *griestoniensis* Zone sequence. They enclose lobe sandstone packets (Fig. 12b) and fringe them to the east and north (Fig. 6). Thicknesses of thin-bedded turbidite sequences between lobe sandstone packets range from about one to forty metres. They are predominantly coarse silt-mud turbidites, but basal grain sizes sometimes range up to medium sand grade. Beds are either parallel-sided or markedly lenticular, with thicknesses ranging from the order of one to twenty centimetres. Bouma (1962) T_{b-e}, T_{c-e}, T_{d-e} or just T_e sequences are represented. Climbing ripple cross-lamination (both subcritical and supercritical varieties) and convolute lamination are frequently present. The T_c to T_e transition may be sharp with no development of T_d, or gradual, via fading ripples, convolute laminae, irregular laminae, graded alternating silt and mud laminae through graded mud to ungraded mud. These turbidites correspond to Facies D of Mutti and Ricci Lucchi (1972), but the finer-grained units are better described by the fine-grained turbidite model of Stow and Shanmugam (1980). They are interpreted as the deposits of low-density turbidity currents (cf. Stow and Bowen, 1980).

Two distinct types of hemipelagite (Facies G of Mutti and Ricci Lucchi, 1972) occur between the thin-bedded turbidites. The first is a dark grey, finely laminated facies consisting of alternating carbonaceous laminae and silt-mud laminae. The carbonaceous laminae probably represent seasonal phytoplankton blooms in the surface waters, possibly annual (cf. Soutar and Crill, 1977). Zooplankton are represented by abundant graptolites. Many of the silty laminae may be the deposits of very low-density turbidity currents since small-scale, very low-angle scours are occasionally present at their bases, and associated graptolites are commonly aligned. An alternative is that these features may have been produced by extremely oxygen-depleted bottom currents. The second type of hemipelagite is microbioturbated and olive to pale olive–grey in colour. Fine-grained turbidites associated with this facies are frequently reworked by a diverse assemblage of trace fossils attributable to the deep-water *Nereites* ichnofacies of Seilacher (1967).

Lateral basin–slope facies association

This facies association is preserved in a thin strip of outcrop around the axis of the north-east-plunging Tywi anticlinorium (Fig. 6). It grades into lobe fringe thin-bedded turbidites to the west. The dominant facies present are thin mud and silt-mud turbidites with laminated or unbioturbated hemipelagites, unbioturbated hemipelagites becoming increasingly rarer towards the east. The location of the base of slope cannot be clearly defined, but a gradual slope decrease to the west is reflected by increasing frequency of siltstone units from east to west. Near the axis of the anticlinorium (at SO05677290) wet-sediment normal faults with basin-down throws of up to 2 m occur (Fig. 12c). Very thin (2–3 cm) pebbly mudstones consisting of rounded pebbles (to 1.5 cm diameter) floating in mudstone are also present at this locality and are interpreted as thin debris flow deposits (cf. Hill *et al.*, 1982). Further east (*c.* 6 km south-east of this locality at SO10436838) slump sheets consisting of tightly folded mud turbidites and sparsely bioturbated hemipelagites with abundant micro-reverse faults (Facies F of Mutti and Ricci Lucchi, 1972) are exposed. Clearly the Telychian shelf–slope break did not correspond to the axis of the Tywi anticlinorium as previously thought (e.g. James, 1983b). An important fault (labelled F in Fig. 1), which appears to have been active during the Wenlock (Dimberline and Woodcock, 1987), is a better candidate for a tectonic control on the Telychian slope in this area. Another possibility is that the slope was locally cut back in the Abbey Cwmhir district as a result of pre-Telychian large-scale slope failure events (probably associated with late Ordovician or late Aeronian regressive phases, Fig. 2).

The *griestoniensis* Zone lateral basin–slope facies association is closely analogous to recent slope facies of the Santa Barbara Basin, California (Thornton, 1984).

Distinctive slope facies associations have not been documented from areas further to the north. The absence of recorded mass movement or mass flow deposits in fine-grained sediments separating distal shelf facies with shelly faunas from basinal turbidite-dominated facies (King, 1928) indicates that slopes were very gentle in this area.

Source area

Probably the best candidate for the source area for the turbidite system is the Carmarthen region (Fig. 1). In this area Old Red Sandstone rests unconformably on Ordovician. Much of the erosion represented by the unconformity is probably of intra-Silurian age, given the huge volumes of material resedimented in the basin (a conservative estimate for the *griestoniensis* Zone alone is 1000 km^3). Cope (1979) interprets the structure of the area as a system of north-dipping thrusts which is consistent with inferred tectonic uplift on the southern basin margin (Cummins, 1969; Smith and Long, 1969), but these structures may have had a later, end Caledonian, origin. Also, the presence of late Precambrian igneous clasts in the sandstones and conglomerates of the *griestoniensis* Zone sequence is consistent with Cope's evidence that late Precambrian source rocks were available at the surface in the area in early Palaeozoic time. An alternative possibility might be that the turbidite system was sourced from a land area located to the north of Haverfordwest (see Bridges, 1972, for evidence for the existence of such a land area).

No rocks of proven *griestoniensis* Zone age have been documented in south-west Wales, but shelf facies of *crenulata* Zone age have been recorded, containing *Clorinda* community faunas near Haverfordwest and *Eocoelia* and *Costistricklandia* community faunas near Marloes (Fig. 1; Bassett, 1982; Hurst *et al.*, 1978). A sub-*crenulata* unconformity is present near Marloes (Fig. 1), probably a product of subaerial erosion (Ziegler *et al.*, 1969).

PALAEOGEOGRAPHY

Turbidite system geometry

The large lateral extent of the sandstone lobe facies association (Fig. 6) suggests analogy with elongate high efficiency (Mutti, 1979), Type I (Mutti, 1985) turbidite systems. The minimum dimensions of the depositional sandstone lobe area are about 90 km (down-palaeocurrent) by 40 km (across palaeocurrent, unshortened).

Palaeocurrents in the central and northern areas are very consistent with low circular variance (Figs 6 and 8), reflecting confinement of flows by the lateral basin-slope (Figs 9 and 10). The wider scatter of palaeocurrents in the south may reflect proximity to feeder channel mouths with the associated lateral spreading of flows or may simply be a function of the larger number of measurements obtained from this area. In the southern and central areas palaeocurrents are roughly parallel to the probable main slope-controlling lineaments (the Tywi and Pontesford Lineaments, Fig. 1). Farther north, lineaments with north–south trends, parallel to palaeocurrents, have been identified from a Landsat imagery study (Craig, 1985). There may be important north–south trending basement structures in this area which generated gentle monoclinal flexures in the sedimentary cover above buried fault tiplines. This would be consistent with the inferred gentleness of the bounding slopes to the turbidite system in the northern area. A marked thinning of Llandovery Series strata towards the Bala Fault (Fig. 1, Fig. 3) suggests a control of differential subsidence by this Lineament, which, as one margin of the Derwen Horst (Fitches and Campbell, 1987), also significantly influenced later Silurian sedimentation (Cummins, 1957, 1959, 1969). The presence of lobe sandstone packets only in the highest parts of the *griestoniensis* Zone sequence in the east and north suggests a degree of passive onlap of bounding slopes. The lack of any large-scale vertical coarsening-up sequence, together with generally random bed thickness patterns, indicates that the system grew in an essentially aggradational manner without significant lateral migration of channel and depositional lobe areas.

Deposits of the more proximal parts of the *griestoniensis* Zone system have been removed by erosion, and their characteristics can only be inferred. If the system was sourced from the Carmarthen area the length of the missing portion would be of the order of 50 km if the shelf–slope break was localized by the continuations of the Tywi and Pontesford Lineaments in this region. The scale of the depositional sandstone lobe area implies the presence of a major stable feeder channel system with well developed levees (Mutti 1979; Stow *et al.*, 1985; Figs 9 and 10).

Marginal areas

Development of large-volume turbidite systems (high-efficiency, Type I systems of Mutti 1979, 1985) appears to require prograding mud-rich river-delta sources (Mutti, 1985; Nelson, 1985; Stow *et al.*, 1985). Hence, such a delta was probably active on the southern margin of the Welsh Basin in late Llandovery time (Fig. 9). In view of the large volumes of material resedimented in the basin, the major river system feeding the delta must have drained an extensive area of land. Resedimented clast types in the basinal facies suggest that late Precambrian basement was exposed at the surface.

The Telychian transgression of the Midland Platform to the east of the turbidite system submerged a broad shelf area (Fig. 9). Nearshore sandstones in the east grade westwards into mudstones, with thin siltstones and sandstones representing storm events (Benton and Gray, 1981; Bridges, 1975).

To the north of the turbidite system, Telychian rocks are poorly exposed in north Wales, but are uniformly fine-grained (Warren *et al.*, 1984 and references therein). The *griestoniensis* Zone is represented by silt–mud turbidites with laminated and bioturbated hemipelagites, and the sequence of this age appears to be only a few metres thick. The late Llandovery north Wales area was clearly a wide, relatively deep (sub- to intrapycnocline) area receiving very low sediment input at this time.

FACTORS CONTROLLING TURBIDITE SYSTEM DEVELOPMENT

Turbidite system development is influenced by many factors which can be broadly grouped into tectonic setting, sediment source, and sea level (Stow *et al.*, 1985) which are to varying extents interrelated. The good lateral outcrop control and biofacies and tectonic frameworks for the *griestoniensis* Zone system enable identification of the dominant factors shaping its geometry and growth pattern.

Tectonic setting

Tectonic settings of turbidite systems affect the type of system formed through determining basin shape, size, and gradients and rates of uplift of basin margins, which in turn affect local relative sea level and rates of sediment supply.

In the *griestoniensis* Zone system the influence of basin shape was manifested in confinement of depositional lobes by tectonically controlled slopes, documented by palaeocurrent patterns and inferred onlap relationships. The lateral slope-controlling structures are steep at the surface, their nature at depth being presently unknown.

Tectonic uplift of the southern basin margin generated unconformities in south-west Wales (near Marloes, Fig. 1) and caused relative sea-level falls during which large volumes of detritus were supplied to the basin. The nature of the structures in this area

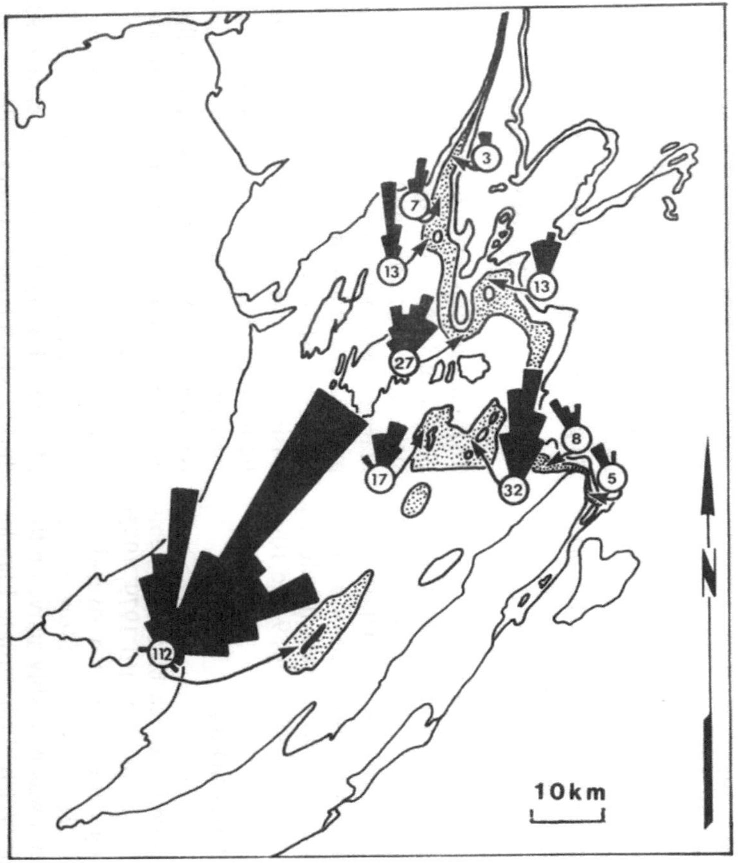

Fig. 8

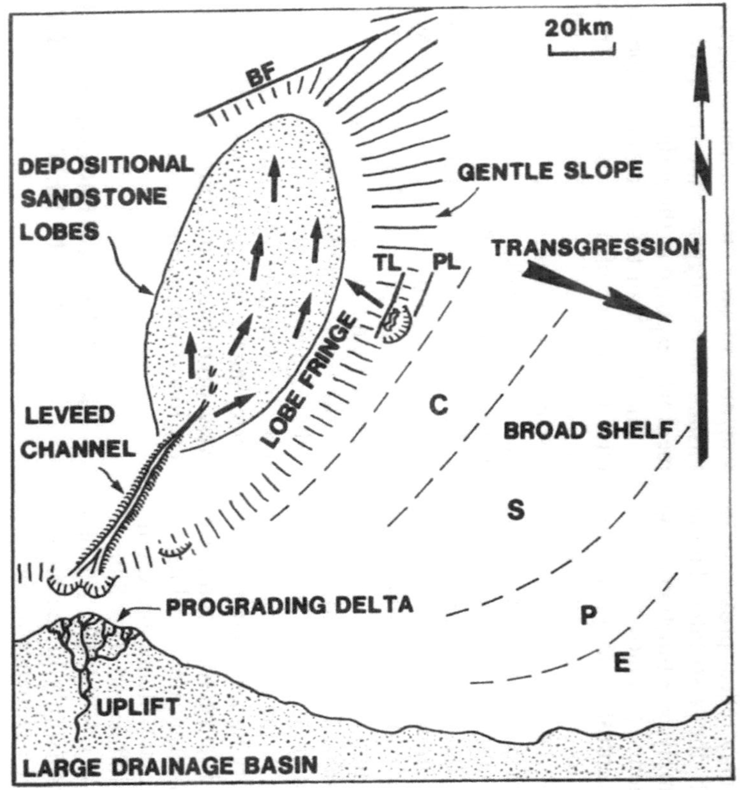

Fig. 9

Fig. 8. Rose diagrams of palaeocurrents recorded from selected measured sections.

Fig. 9. Cartoon of *griestoniensis* Zone palaeogeography. Community abbreviations as in Fig. 1, lineament abbreviations as in Fig. 3.

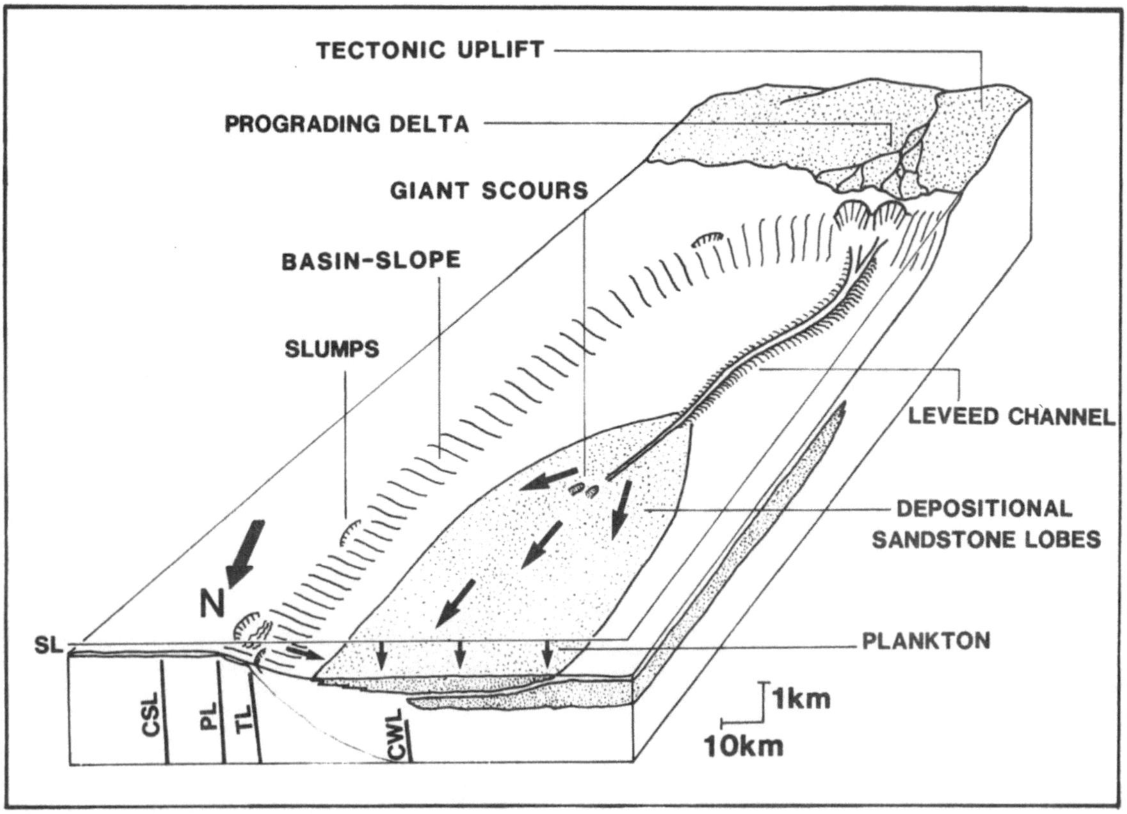

Fig. 10. Depositional model for the *griestoniensis* Zone turbidite system. Arrows indicate directions of sediment transport. Lineament abbreviations as in Fig. 3.

during the Silurian is very poorly constrained, partly as a result of Variscan overprint. However, mapped pre-Variscan northward-dipping thrusts may possibly have been active in early Silurian times. Seismicity may have been an important trigger for slope failure during periods of rapid uplift. Oversteepening of slopes due to structural uplift is another, possibly major, factor in the initiation of large slope failure events (Nagel *et al.*, 1986).

Sediment source

Sediment source type and volume play perhaps the most important part in controlling turbidite system shape and dimensions. The two dominant parameters are (a) composition of source sediment (emphasized by Mutti, 1979; Nelson 1985; Stow *et al.*, 1985), and (b) the volumes of failure-prone sediment available for resedimentation (emphasized by Mutti, 1985). Large-volume mud-rich sediment sources result in elongate turbidite systems with distal sand depocentres, since stable leveed feeder channels can be constructed and sand can be efficiently transported to distal depositional sites. Small-volume, mud-poor sediment sources result in small radial systems (in the absence of strong basin shape influence), or, if there are multiple sources, slope aprons (Nelson, 1985; Stow *et al.*, 1985).

The *griestoniensis* Zone system with its extensive depositional sandstone lobe area is clearly closely comparable with modern medium-sized systems fed by river deltas (Stow *et al.*, 1985). Hence, a mud-rich point source which supplied large volumes of material to the southern shelf can be inferred. The volumes and composition of material supplied to the shelf will have fluctuated with relative sea-level changes, maximum delta progradation occurring during phases of sea-level fall (Coleman *et al.*, 1983) and coarsest grain sizes being supplied during sea-level low stands.

Sea level

The importance of sea level on turbidite system activity is dramatically demonstrated by the effects of the Plio-Pleistocene glacioeustatic fluctuations on

Fig. 11. (a) Base of conglomeratic scour-fill, channel–lobe transition zone facies association, Craig Twrch (SN653481), coloured section of pole is 50 cm long; (b) stacked conglomerate–sandstone units overlying thick basal conglomerate–sandstone unit, interpreted as the product of deposition from a surging high-density turbidity current (same locality as (a)); (c) laterally continuous non-channelized packets of lobe sandstones (same general location as (a) and (b)).

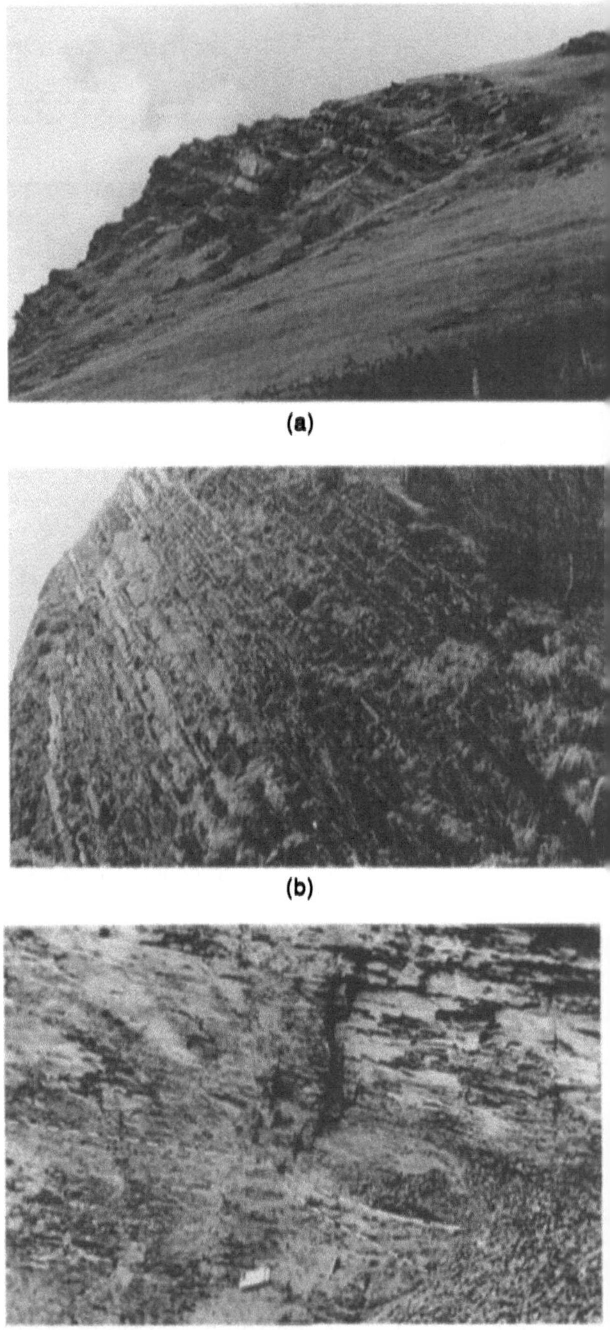

Fig. 12. (a) Lobe sandstone packet, Cwm Ystwyth (SN850759); (b) lobe fringe thin-bedded turbidites enclosing a lobe sandstone packet, Talerddig (SH9300); (c) fine-grained turbidites and hemipelagites of the Lateral basin-slope facies association, near Abbey Cwmhir (SO05677290). Wet-sediment normal fault and its draped scar are emphasized by dashed lines, mapping case is 30 cm long.

recent submarine fans (Shanmugam *et al.*, 1985). Loading of shelves with huge volumes of detritus during progradation of shoreline systems associated with sea-level falls led to large-scale slope failure events and rapid fan growth (Coleman *et al.*, 1983). In contrast, the present high sea level has resulted in removal of major river mouths from shelf edges and negligible growth of fans.

The *griestoniensis* Zone system developed at a time of widespread transgression (Bridges, 1975; McKerrow, 1979, Fig. 2; Ziegler *et al.*, 1968b) which resulted in abandonment of the eastern basin margin as an important sediment source area for the basin. The packets of lobe sandstones in the basinal facies of the *griestoniensis* Zone are interpreted as representing phases of depositional sandstone lobe growth during periods of local relative sea-level fall caused by pulses of rapid tectonic uplift on the southern basin margin.

By *crenulata* Zone times the rate of sea-level rise exceeded uplift rates to the south with the result that the *griestoniensis* turbidite system was abandoned and draped by 'Pale Shales' (Smith and Long, 1969), a sequence of very fine-grained turbidites and hemipelagites.

CONCLUSIONS

The *griestoniensis* Zone turbidite system was a high-efficiency Type I turbidite system probably fed from a mud-rich point source located on the southern Welsh Basin margin. A combination of factors resulted in the development of a moderately large turbidite system with an extensive relatively distal sand depocentre. These factors include:

(a) Storage of large volumes of unlithified or poorly lithified sediment on the southern shelf;

(b) a high proportion of mud in the stored sediment; and

(c) large volumes involved in slope failure events during local relative sea-level falls caused by tectonic uplift.

In common with many other modern and ancient turbidite systems developed in active margin settings the depositional sandstone lobes of the system were strongly confined by tectonically controlled slopes. In the case of the *griestoniensis* Zone turbidite system the main basin shape-controlling structures were an array of long-lived, repeatedly reactivated basement lineaments.

ACKNOWLEDGEMENTS

Discussions with Dick Cave, Andrew Dimberline, Dave Johns, Angus Mackie, Nick McCave, Stewart Smallwood, John-Joe Traynor and Nigel Woodcock are gratefully acknowledged. Improvements to the manuscript were suggested by Andrew Dimberline, John-Joe Traynor, Jerry Leggett and Stewart Smallwood. This work was carried out whilst in receipt of a NERC studentship, Cambridge Earth Science Series No. 789.

References

André, L., Hertogen, J. and Deutsch, S. (1986). Ordovician–Silurian magmatic provinces in Belgium and the Caledonian Orogeny in middle Europe, *Geology* **14**, 879–882.

Bailey, R. J. (1969). Ludlovian sedimentation in south Central Wales. *In* A. Wood (Ed.), *The Precambrian and Lower Palaeozoic Rocks of Wales*, Cardiff, University of Wales Press, pp. 283–304.

Bassett, D. A. (1955). The Silurian rocks of the Talerddig district, Montgomeryshire, *Quarterly Journal of the Geological Society, London*, **111**, 239–265.

Bassett, M. G. (1982). Silurian rocks of the Marloes and Pembroke peninsulas. *In* M. G. Bassett (Ed.), *Geological Excursions in Dyfed, South-west Wales*, National Museum of Wales, Cardiff, pp. 103–122.

Benton, M. J. and Gray, D. I. (1981). Lower Silurian distal shelf storm-induced turbidites in the Welsh Borders: sediments, tool marks and trace fossils, *Journal of the Geological Society, London* **138**, 675–694.

Bouma, A. H. (1962). Sedimentology of some flysch deposits, Amsterdam, Elsevier, 168 pp.

Brewer, J. A., Matthews, D. H., Warner, M. R., Hall, J., Smythe, D. K. and Whittington, R. J. (1983). BIRPS deep seismic reflection studies of the British Caledonides, *Nature* **305**, 206–210.

Briden, J. C., Drewry, G. E. and Smith, A. G. (1974). Phanerozoic equal area world maps, *Journal of Geology* **82**, 555–574.

Bridges, P. H. (1972). The sedimentology of a marine transgression: as exemplified by the Llandovery transgression in Wales and the Welsh Borderland, Unpublished Ph.D. Thesis, University of Reading.

Bridges, P. H. (1975). The transgression of a hard substrate shelf: the Llandovery (Lower Silurian) of the Welsh Borderland. *Journal of Sedimentary Petrology* **45**, 79–94.

Burrett, C. F. (1985). Comment on 'Paleozoic evolution of the Armorica plate on the basis of paleomagnetic data', *Geology* **13,** 380.

Cave, R. (1979). Sedimentary environments of the basinal Llandovery of mid-Wales. *In* A. L. Harris, C. H. Holland and B. E. Leake (Eds), *Caledonides of the British Isles, Reviewed,* Special Publication of the Geological Society, London **8,** pp. 517–526.

Cazzola, C., Fonnesu, F., Mutti, E., Rampone, G., Sonnino, M. and Vigna, B. (1981). Geometry and facies of small, fault-controlled deep-sea fan systems in a transgressive depositional setting (Tertiary Piedmont Basin, northwestern Italy), *In* F. Ricci Lucchi (Ed.), *Excursion Guidebook*: 2nd European Regional Meeting of International Association of Sedimentologists, Bologna, pp. 5–56.

Cocks, L. R. M. and Fortey, R. A. (1982). Faunal evidence for oceanic separation in the Palaeozoic of Britain. *Journal of the Geological Society, London* **139,** 465–478.

Cocks, L. R. M. and McKerrow, W. S. (1984). Review of the distribution of the commoner animals in the Lower Silurian marine benthic communites. *Palaeontology* **27,** 663–670.

Cocks, L. R. M., Woodcock, N. H., Lane, P. D., Rickards, R. B. and Temple, J. T. (1984). The Llandovery Series of the type area, *Bulletin of the British Museum* (*Natural History*).

Coleman, J. M., Prior, D. B. and Lindsay, J. F. (1983). Deltaic influences on shelf edge instability processes. *In* D. J. Stanley and G. T. Moore (Eds), *The shelf break, critical interface on continental margins,* S.E.P.M. Spec. Publ. **33,** 121–137.

Cope, J. C. W. (1979). Early history of the southern margin of the Tywi Anticline in the Carmarthen area, South Wales. *In* A. L. Harris, C. H. Holland and B. E. Leake (Eds), *Caledonides of the British Isles, Reviewed,* Special Publication of the Geological Society, London **8,** pp. 187–198.

Coward, M. P. and Siddans, A. W. B. (1979). The tectonic evolution of the Welsh Caledonides. *In* A. L. Harris, C. H. Holland and B. E. Leake (Eds), *Caledonides of the British Isles, Reviewed,* Special Publication of the Geological Society, London **8,** pp. 187–198.

Craig, J. (1985). Tectonic evolution of the area between Borth and Cardigan, Dyfed, West Wales. Unpublished Ph.D. Thesis, University College of Wales, Aberystwyth.

Craig, J. (1987). The structure of the Llangranog Lineament, West Wales: a Caledonian transpression zone, *Geological Journal,* **22,** 167–181.

Cummins, W. A. (1957). The Denbigh Grits; Wenlock greywackes in Wales, *Geological Magazine* **94,** 433–451.

Cummins, W. A. (1959). The Nantglyn Flags: Mid-Salopian basin facies in Wales, *Liverpool and Manchester Geological Journal* **2,** 159–167.

Cummins, W. A. (1969). Patterns of sedimentation in the Silurian rocks of Wales, *In* A. Wood (Ed.), *The Precambrian and Lower Palaeozoic rocks of Wales,* Cardiff, University of Wales Press, pp. 219–237.

Davies, K. A. (1933). The geology of the country between Abergwesyn (Breconshire) and Pumpsaint (Carmarthenshire), *Quarterly Journal of the Geological Society, London* **89,** 172–201.

Dimberline, A. J. and Woodcock, N. H. (1987). The south-east margin of the Wenlock turbidite system, Mid-Wales, *Geological Journal,* in press.

Enos, P. (1977). Flow regimes in debris flows, *Sedimentology* **24,** 133–142.

Ettensohn, F. R. and Elam, T. D. (1985). Defining the nature and location of a late Devonian–early Mississippian pycnocline in eastern Kentucky. *Bulletin of the Geological Society of America* **96,** 1313–1321.

Fitches, W. R. and Campbell, S. D. G. (1987). Tectonic evolution of the Bala Lineament in the Welsh Basin, *Geological Journal,* **22,** 131–153.

Ghibaudo, G. (1981). Deep-sea fan deposits in the Macigno Formation (Middle–Upper Oligocene) of the Gordana Valley, northern Appenines, Italy—reply, *Journal of Sedimentary Petrology* **51,** 1021–1026.

Gibbons, W. (1983). Stratigraphy, subduction and strike-slip faulting in the Mona Complex of North Wales: a review, *Proceedings of the Geologists' Association* **94,** 147–163.

Gibbons, W. (1985). Geology and paleobiology of islands in the Ordovician Iapetus Ocean: review and implications: discussion and reply, *Bulletin of the Geological Society of America* **96,** 1225–1226.

Harland, W. B., Cox, A. V., Llewellyn, P. G., Pickton, C. A. G., Smith, A. G. and Walters, R. (1980). A geologic time scale, Cambridge, Cambridge University Press.

Heller, P. L. and Dickinson, W. R. (1985). Submarine ramp facies model for delta-fed sand-rich turbidite systems, *Bulletin of the American Association of Petroleum Geologists* **69,** 960–976.

Heller, P. L. and Dickinson, W. R. (1986). Submarine ramp facies model for delta-fed, sand-rich turbidite systems: reply to Waldron Discussion. *Bulletin of the American Association of Petroleum Geologists* **70,** 1747–1748.

Hendry, H. E. (1973). Sedimentation of deep-water conglomerates in Lower Ordovician rocks of Quebec—composite bedding produced by progressive liquefaction of sediment? *Journal of Sedimentary Petrology* **43,** 125–136.

Hill, P. R., Aksu, A. E. and Piper, D. J. W. (1982). The deposition of thin-bedded subaqueous debris flow deposits. *In* S. Saxov and J. K. Nieuwenhuis (Eds), *Marine Slides and Other Mass Movements,* New York and London, Plenum Press, pp. 273–287.

Hiscott, R. N. (1981). Deep-sea fan deposits in the Macigno Formation (Middle–Upper Oligocene) of the Gordana Valley, Northern Appenines, Italy—discussion, *Journal of Sedimentary Petrology* **51,** 1015–1033.

Hiscott, R. N. and Middleton, G. V. (1979). Depositional mechanics of thick-bedded sandstones at the base of a submarine slope, Tourelle Formation (Lower Ordovician), Quebec, Canada, *In* L. J. Doyle and O. H. Pilkey (Eds), *The Geology of Continental Slopes*, S.E.P.M. Special Publication No. 27, 307–326.

Holland, C. H. and Lawson, J. D. (1963). Facies patterns in the Ludovian of Wales and the Welsh Borderland, *Liverpool and Manchester Geological Journal* **3,** 269.

Hurst, J. M., Hancock, N. J. and McKerrow, W. S. (1978). Wenlock stratigraphy and palaeogeography of Wales and the Welsh Borderland, *Proceedings of the Geologists' Association* **89,** 197–226.

Ingram, R. L. (1954). Terminology for the thickness of stratification and parting units in sedimentary rocks, *Bulletin of the Geological Society of America* **65,** 937–938.

James, D. M. D. (1972). Sedimentation across an intrabasinal slope: the Garnedd-wen Formation (Ashgillian), west central Wales, *Sedimentary Geology* **7,** 291–307.

James, D. M. D. (1983a). Sedimentation of deep-water slope base and inner-fan deposits—the Drosgol Formation (Ashgill), west central Wales, *Sedimentary Geology* **34,** 21–40.

James, D. M. D. (1983b). Observations and speculations on the north-east Towy 'axis', mid-Wales, *Geological Journal* **18,** 283–296.

James, D. M. D. and James, J. (1969). The influence of deep fractures on some areas of Ashgillian–Llandoverian sedimentation in Wales, *Geological Magazine* **106,** 562–582.

Jones, W. D. V. (1945). The Valentian succession around Llanidloes, Montgomeryshire, *Quarterly Journal of the Geological Society, London* **100,** 309–332.

Kelling, G., Phillips, W. E. A., Harris, A. L. and Howells, M. F. (1985). The Caledonides of the British Isles: a review and appraisal, *In* D. G. Gee and B. A. Sturt (Eds), *The Caledonian Orogen—Scandinavia and Related areas*, Chichester, John Wiley and Sons, 1133 p.

Kelling, G. and Woollands, M. A. 1969. The stratigraphy and sedimentation of the Llandoverian rocks of the Rhayader district, *In* A. Wood (Ed.), *The Precambrian and Lower Palaeozoic Rocks of Wales*, University of Wales Press, Cardiff, 255–282.

King, W. B. R. (1928). The geology of the district around Meifod (Montgomeryshire), *Quarterly Journal of the Geological Society, London* **84,** 671–702.

Kokelaar, B. P., Howells, M. F., Bevins, R. E., Roach, R. A. and Dunkley, P. N. (1984). The Ordovician marginal basin of Wales, *In* B. P. Kokelaar and M. F. Howells (Eds), *Marginal Basin Geology*, Special Publication of the Geological Society, London, **16,** 245–271.

Komar, P. D. (1969). The channelised flow of turbidity currents with application to Monterey deep-sea fan channel, *Journal of Geophysical Research* **74,** 4544–4558.

Lawson, J. D. (1973). Facies and faunal changes in the Ludlovian rocks of Aymestry, Herefordshire, *Geological Journal* **8,** 247–278.

Leggett, J. K., McKerrow, W. S., Cocks, L. R. M. and Rickards, R. B. R. (1981). Periodicity in the early Palaeozoic marine realm, *Journal of the Geological Society, London* **138,** 167–176.

Leggett, J. K., McKerrow, W. S. and Soper, N. J. (1983). A model for the crustal evolution of southern Scotland, *Tectonics* **2,** 187–210.

Long, G. H. (1966). Investigations into the sedimentation and sedimentary history of the Talerddig Grits (upper Llandoverian) and their lateral equivalents in Central Wales. Unpublished Ph.D. Thesis, University of London.

Lowe, D. R. (1982). Sediment gravity flows: II. Depositional models with special reference to the deposits of high-density turbidity currents, *Journal of Sedimentary Petrology* **52,** 279–297.

McKerrow, W. S. (1979). Ordovician and Silurian changes in sea level, *Journal of the Geological Society, London* **136,** 137–145.

Murphy, F. C. and Hutton, D. H. W. (1986). Is the Southern Uplands of Scotland really an accretionary prism? *Geology* **14,** 354–357.

Mutti, E. (1979). Turbidites et cones sous-marines profonds, *In* P. Homewood (Ed.), *Sédimentation Détritique* (*Fluviatile, Littorale et Marine*), Institut de géologie, Université de Fribourg, Fribourg, Switzerland, pp. 353–419.

Mutti, E. (1985). Turbidite systems and their relations to depositional sequences, *In* G. G. Zuffa (Ed.), *Provenance of Arenites*, NATO-ASI Series, Dordrecht, D. Reidel, pp. 65–93.

Mutti, E. and Sonnino, M. (1981). Compensation cycles: a diagnostic feature of turbidite sandstone lobes. Abstracts of the 2nd European Regional Meeting IAS, Bologna, pp. 120–123.

Mutti, E. and Ricci Lucchi, F. (1972). Le torbiditi dell' Appenino settentrionale: introduzione all'analisi di facies, *Memorie della Società Geologica Italiana* **11,** 161–199.

Nagel, D. K., Mullins, H. T. and Green, H. G. (1986). Ascension Submarine Canyon, California—evolution of a multi-head canyon system along a strike-slip continental margin, *Marine Geology* **73,** 285–310.

Nelson, C. H. (1985). Astoria Fan, Pacific Ocean, *In* A. H. Bouma, W. R. Normark and N. E. Barnes (Eds), *Submarine Fans and Related Turbidite Systems*, New York, Springer-Verlag, pp. 45–50.

Nelson, C. H. and Kulm, L. D. (1973). Submarine fans and channels, *In* G. V. Middleton and A. H. Bouma (Eds), *Turbidites and Deep-water Sedimentation*, S.E.P.M. Pacific Section Short Course, Anaheim, pp. 39–78.

Normark, W. R., Piper, D. J. W. and Hess, G. R. (1979). Distributary channels, sand lobes, and mesotopography of Navy submarine fan, California Borderland, with applications to ancient fan sediments, *Sedimentology* **26,** 769–774.

Okada, H. and Smith, A. J. (1980). The Welsh 'geosyncline' of the Silurian was a fore-arc basin, *Nature* **288,** 352–354.

Pauley, J. C. (1986). The Longmyndian Supergroup: facies, stratigraphy and structure. Unpublished Ph.D. Thesis, University of Liverpool.

Perroud, H., Van der Voo, R. and Bonhommet, N. (1984). Paleozoic evolution of the Armorica plate on the basis of paleomagnetic data, *Geology* **12,** 579–582.

Postma, G. (1986). Classification for sediment gravity flow deposits based on flow conditions during sedimentation, *Geology* **14,** 291–294.

Roberts, R. O. (1929). The geology of the district around Abbey-Cwmhir (Radnorshire), *Quarterly Journal of the Geological Society, London,* **85,** 651–676.

Robinson, D. and Bevins, R. E. (1986). Incipient metamorphism in the Lower Palaeozoic marginal basin of Wales, *Journal of Metamorphic Geology* **4,** 101–113.

Seilacher, A. (1967). Bathymetry of trace fossils, *Marine Geology* **5,** 413–428.

Seilacher, A. (1977). Pattern analysis of *Paleodictyon* and related trace fossils, *In* T. P. Crimes and J. C. Harper (Eds), *Trace Fossils 2*, Geological Journal Special Issue No. 9, pp. 289–334.

Shanmugam, G., Moiola, R. J. and Damuth, J. E. (1985). Eustatic control of submarine fan development, *In* A. H. Bouma, W. R. Normark and N. E. Barnes (Eds), *Submarine Fans and Related Turbidite Systems*, New York, Springer-Verlag.

Smallwood, S. D. (1986). Sedimentation across the Tywi Lineament, mid-Wales, *Philosophical Transactions of the Royal Society, London* **A 317,** 279–288.

Smith, A. J. and Long, G. H. (1969). The Upper Llandovery sediments of Wales and the Welsh Borderlands, *In* A. Wood (Ed.), *The Precambrian and Lower Palaeozoic Rocks of Wales*, Cardiff, University of Wales Press. pp. 239–254.

Smith, R. D. A. (1987). Structure and deformation history of the Central Wales Synclinorium, north-east Dyfed: evidence for a long-lived basement structure, *Geological Journal*, **22,** 183–198.

Soper, N. J. (1986). The Newer Granite Problem: a geotectonic view, *Geological Magazine* **123,** 227–236.

Soper, N. J. and Hutton, D. H. W. (1985). Late Caledonian sinistral displacements in Britain: implications for a three-plate collision model, *Tectonics* **3,** 781–794.

Soutar, A. and Crill, P. A. (1977). Sedimentation and climatic patterns in the Santa Barbara Basin during the 19th and 20th centuries, *Bulletin of the Geological Society of America* **88,** 1161–1172.

Stillman, C. J. and Francis, E. H. (1979). Caledonide volcanism in Britain and Ireland. *In* A. L. Harris, C. H. Holland and B. E. Leake (Eds), *Caledonides of the British Isles—Reviewed*, Special Publication of the Geological Society, London, **8,** 557–577.

Stow, D. A. V. and Bowen, A. J. (1980). A physical model for the transport and sorting of fine-grained sediment by turbidity currents, *Sedimentology* **27,** 31–46.

Stow, D. A. V. and Shanmugam, G. (1980). Sequence of structures in fine-grained turbidites; comparison of recent deep-sea and ancient flysch sediments, *Sedimentary Geology* **25,** 23–42.

Stow, D. A. V., Howell, D. G. and Nelson, C. H. (1985). Sedimentary, tectonic, and sea level controls. *In* A. H. Bouma, W. R. Normark and N. E. Barnes (Eds), *Submarine Fans and Related Turbidite Systems*, New York, Springer-Verlag, pp. 15–23.

Surlyk, F. (1984). Fan-delta to submarine fan conglomerates of the Volgian–Valanginian Wollaston Forland Group, East Greenland. *In* E. H. Koster and R. J. Steel (Eds), *Sedimentology of Gravels and Conglomerates*, Canadian Society of Petroleum Geologists, Memoir 10, 359–382.

Teale, C. T. and Spears, D. A. (1986). The mineralogy and origin of some Silurian bentonites, Welsh Borderland, *UK Sedimentology* **33,** 757–765.

Thornton, S. E. (1984). Basin model for hemipelagic sedimentation in a tectonically active continental margin: Santa Barbara Basin, California Continental Borderland, *In* D. A. V. Stow and D. J. W. Piper (Eds), *Fine-grained Sediments, Deep-water Processes and Facies*, Special Publication of the Geological Society, London, **15,** 377–395.

Thorpe, R. S. (1979). Late Precambrian igneous activity in southern Britain. *In* A. L. Harris, C. H. Holland and B. E Leake (Eds), *Caledonides of the British Isles—Reviewed*, Special Publication of the Geological Society, London, **8,** 579–584.

Thorpe, R. S., Beckinsale, R. D., Patchett, P. J., Piper, J. D. A., Davies, G. R. and Evans, J. A. (1984). Crustal growth and late Precambrian–early Palaeozoic plate tectonic evolution of England and Wales, *Journal of the Geological Society London* **141,** 521–536.

Warren, P. T., Price, D., Nutt, M. J. C. and Smith, E. G. (1984). Geology of the country around Rhyl and Denbigh *Memoir of the British Geological Survey*, sheets 95 and 97.

Wood, E. M. R. (1906). The Tarannon Series of Tarannon. *Quarterly Journal of the Geological Society, London* **62,** 644–701.

Wood, A. and Smith, A. J. (1959). The sedimentation and sedimentary history of the Aberystwyth Grits (Upper Llandoverian), *Quarterly Journal of the Geological Society, London* **114,** 163–195.

Woodcock, N. H. (1976). Ludlow Series slumps and turbidites and the form of the Montgomery Trough, Powys, Wales, *Proceedings of the Geologists' Association* **87,** 169–182.
Woodcock, N. H. (1984a). Early Palaeozoic sedimentation and tectonics in Wales. *Proceedings of the Geologists' Association* **95,** 323–335.
Woodcock, N. H. (1984b). The Pontesford Lineament, Welsh Borderland, *Journal of the Geological Society, London* **141,** 1001–1014.
Woodcock, N. H. (1986). The role of strike-slip fault systems at plate boundaries, *Philosophical Transactions of the Royal Society*, **A 317,** 13–29.
Woodcock, N. H. (1987). Structural geology of the Llandovery Series in the type area, Dyfed, Wales, *Geological Journal* **22,** 199–209.
Ziegler, A. M. (1965). Silurian marine communities and their environmental significance, *Nature* **207,** 270–272.
Ziegler, A. M., Cocks, L. R. M. and Bambach, R. K. (1968a). The composition and structure of Lower Silurian Marine communities, *Lethaia* **1,** 1–27.
Ziegler, A. M., Cocks, L. R. M. and McKerrow, W. S. (1968b). The Llandovery transgression of the Welsh Borderland. *Palaeontology* **11,** 736–782.
Ziegler, A. M., McKerrow, W. S., Burne, R. V. and Baker, P. E. (1969). Correlation and environmental setting of the Skomer Volcanic Group, Pembrokeshire, *Proceedings of the Geologists' Association* **80,** 409–439.
Ziegler, P. A. (1984). Caledonian and Hercynian crustal consolidation of western and central Europe—a working hypothesis, *Geologie en Mijnbouw* **63,** 93–108.

Chapter 6

Mineralogical Consequences of Organic Matter Degradation in Sediments: Inorganic/Organic Diagenesis

Charles Curtis, Department of Geology, Beaumont Building, The University, Sheffield S3 7HF, UK

ABSTRACT

A significant fraction of the products of organic matter diagenesis is water soluble. Some products react directly to precipitate sedimentary minerals: carbonates, sulphides and phosphates. Low-temperature reactions are mostly microbial whereas later (deeper) reactions are thermally induced. Most of the soluble products are acidic and do not, by themselves, stimulate carbonate mineral precipitation. Reduction of Fe(III) by organic matter, however, causes massive increase in alkalinity. The balance between this and the microbial reactions is a major controlling influence on patterns of mineral diagenesis. Organic acids are produced during catagenesis and can cause mineral dissolution, both of carbonates and silicates. The permeability and porosity of sedimentary rocks may be modified with obvious consequences for petroleum geology. The amounts and types of acids generated are not well known. Equally poorly documented are the Fe(III)–organic matter reactions, yet these must affect both mineral authigenesis and organic maturation.

INTRODUCTION

The commercial success of the petroleum industry has ensured that research into the formation of petroleum hydrocarbons has been both detailed and extensive. Consequently our knowledge of these compounds is excellent. Hydrocarbons are hydrophobic and interact little with the mineral constituents of sedimentary rocks.

Water-soluble organic molecules are also produced, however, during organic-matter diagenesis and catagenesis. These do react with minerals, and in several different and important ways. The consequences of these interactions are important for the petroleum industry in particular and also for sedimentology in general.

In the discussion that follows, the terms 'diagenesis' and 'catagenesis' will be used to describe processes occurring respectively below and above (approximately) 50 °C. Inorganic geochemists and sedimentologists tend to use diagenesis to cover everything between sedimentation and metamorphism: some confusion is possible!

SHALLOW BURIAL PROCESSES: ORGANIC MATTER DIAGENESIS

Mineralization and isotope fractionation

The simplest water-soluble degradation product of organic matter is formed by its complete oxidation to

Marine Clastic Sedimentology © J. K. Leggett & G. G. Zuffa (Graham and Trotman, 1987) pp. 108–123

carbon dioxide. Carbon dioxide dissolves readily in water to give HCO_3^- and subsequently may be incorporated into carbonate minerals such as calcite ($CaCO_3$) and siderite ($FeCO_3$). One of the most useful techniques for following carbon mineralization is stable isotope geochemistry. Strong and sometimes specific isotopic fractionations have been observed in a variety of metabolic processes and the evidence of fractionation may then be preserved in minerals.

Mineralization takes place on land as well as in buried sediments and is there more amenable to experimental investigation. Plants fix carbon from atmospheric CO_2 via two photosynthetic pathways: C_3 and C_4. Both selectively take up the light isotope ^{12}C. The stable isotope composition of the resulting biomass ($\delta^{13}C$ approximately $-27‰$ and $-13‰$* for the two pathways respectively) is enriched relative to atmospheric CO_2 ($\delta^{13}C$ approximately $-7‰$). Within the soil, root respiration and microbial litter degradation reoxidize organic matter to CO_2. The isotopic composition of soil CO_2 thus reflects the proportion of C_3 and C_4 plants in the biomass and the net respiration rate (low rates favour mixing with atmospheric CO_2). Soils in dry climates are characterized by accumulation of authigenic $CaCO_3$ within the profile. Cerling (1984) has demonstrated that this preserves variable isotope ratios which reflect the counterplay of biological and environmental controls.

Soil studies are important in another way. Detrital sediments are redistributed soils. The materials of soil profiles are thus the starting point for diagenetic alteration. This applies equally to mineral constituents and variably degraded organic matter.

Microbial processes in buried sediments and links with depositional environment

A stratified 'ecological succession' can be recognized in recently buried sediments. Evidence from many different investigations is evaluated and summarized by Claypool and Kaplan (1974). Beneath oxygenated waters, oxygen is rapidly eliminated from organic-rich sediments by the activities of aerobic respirers. The depth of penetration of oxygen, which defines the base of the respirative zone, is limited by its diffusion rate in porewater and by the ventilative activities of burrowing organisms. Eventually consumption of oxygen matches supply and no oxygen penetrates beneath this depth.

Beneath the respiration zone, anaerobic conditions prevail and, in waters containing dissolved SO_4^{2-}, organic matter is degraded by sulphate-reducing bacteria. Two important products are H_2S and CO_2. Little carbon isotope fractionation appears to accompany either respiration or sulphate reduction and isotopic composition of the dissolved bicarbonate is similar to the precursor organic matter ($\delta^{13}C$, $-25‰_{PDB}$). If the sulphate concentration of porewater is at very low levels (either because of consumption by sulphate reducers or because there was little in the first place—as in freshwater sediments) organic matter is degraded by microbial methanogenesis: CO_2 and CH_4 being principal products. This time there is a very pronounced fractionation with $\delta^{13}C$ approximately $-75‰$ (CH_4) and $\sigma^{13}C$ up to $+15‰$ (CO_2). Extreme values like these were observed directly in porewaters from DSDP sites 102, 147, 174A and 180 (Claypool and Kaplan, 1974). The lower limit of the sulphate reduction zone is again determined by the balance between SO_4^{2-} diffusing down from the depositional waters above and its consumption by bacteria.

Of course, the organisms referred to above are not the only ones to inhabit shallow buried sediments. There are many different species and it is probable that few have been fully characterized. Their varied and complex metabolic pathways break down biopolymers to water-soluble monomers which, if they survive utilization by heterotrophic bacteria, eventually condense to form the geopolymers that constitute kerogen (Tissot and Welte 1979; Hunt 1979). Recognition that the same microbial processes, operating within distinctly stratified depth intervals, generate important water-soluble mineralizing agents (HCO_3^-, H_2S, H_3PO_4) has been very helpful to workers investigating diagenetic mineral assemblages in ancient sediments (Irwin *et al.*, 1977).

The microbial degradation pattern in any particular sediment, however, will depend strongly on the depositional environment.

In Fig. 1 four possibilities are considered: SO_4^{2-}-rich depositional waters (marine) with and without anoxic bottom waters and the same two alternatives for fresh-water deposition. In the first case, sulphate reducers inhabit both bottom waters and uppermost sediment layers. Below the depth of maximum diffusive penetration by SO_4^{2-}, microbial methanogenesis becomes the dominant process. Sediment passing downwards through the column would encounter first sulphate reduction then methanogenesis and, apart from suspended transport, would not encounter molecular oxygen.

In fresh waters with anoxic bottom conditions, the only important organic matter degradation pathway would be microbial methanogenesis. Under normal

* All isotope ratios refer to the PDB standard

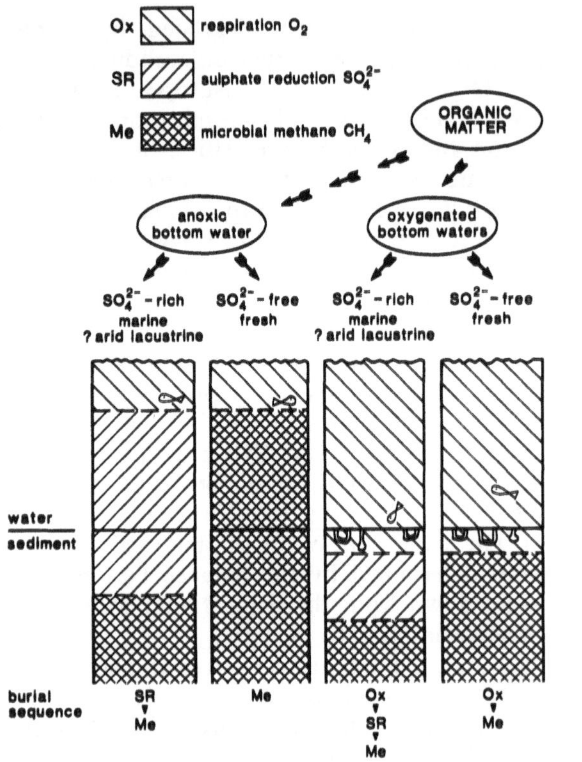

Fig. 1. The dependence of geomicrobial zonation pattern on depositional environment.

oxidizing conditions (cases 3 and 4 in Fig. 1), organic matter would be degraded by molecular oxygen and by the activities of respirative organisms at, and just below, the sediment–water interface. The difference between marine and fresh-water depositional environments is seen in the absence of sulphate reduction in the latter: sulphate reduction rates are dependent on SO_4^{2-}-concentrations at low levels (Boudreau and Westrich, 1984). The most complicated diagenetic sequences are thus seen in 'normal' marine environments where, during burial, sediments successively encounter and are modified by oxygen (inorganic oxidation and respiration), sulphate reduction and microbial methanogenesis. All these processes degrade organic matter and, indirectly, modify the mineral constituents of sediments. Figure 1 details the successions encountered in sediments accumulating in a variety of depositional environments.

It should be noted that the term 'fresh water' here is used in a fairly literal (not littoral) sense. Fresh waters are very low in solutes, including SO_4^{2-}. 'Non-marine' is a much more general descriptive term for environments and can include high-salinity lacustrine environments (with or without SO_4^{2-}) as well as fresh-water deposition.

Energy release during organic matter degradation and 'Preservation Potential'

A very simple and plausible model for organic diagenesis is that organic matter is oxidized by the available oxidant liberating the most molar free energy (Froelich *et al.*, 1979, and references therein). The respiration zone, sulphate reduction zone and microbial methanogenesis zone are in vertical sequence which parallels free energy release. In future discussion, these zones (and the principal chemical processes occurring in them) will be referred to as Ox, SR and Me, following the scheme suggested by Coleman (1985). This is less likely to cause confusion than schemes using numbers—as adopted by several authors previously, including myself.

Froelich *et al.* (1979) compare degradation of organic matter of average composition for marine productivity—the 'Redfield' molecule:

Ox zone—respiration

$$(CH_2O)_{106}(NH_3)_{16}(H_3PO_4) + 138O_2$$
$$= 106CO_2 + 16HNO_3 + H_3PO_4 + 122H_2O$$
$$\Delta G_r^0 = -3190 \text{ kJ mole}^{-1} \text{ (glucose)}$$

(HNO_3 is reduced to N_2 with a similar energy release: -3030 kJ mole^{-1}.)

SR zone—sulphate reduction

$$(CH_2O)_{106}(NH_3)_{16}(H_3PO_4) + 53SO_4^{2-}$$
$$= 106CO_2 + 16NH_3 + 53S^{2-}$$
$$+ H_3PO_4 + 106H_2O$$
$$\Delta G_r^0 = -380 \text{ kJ mole}^{-1} \text{ (glucose)}$$

Me zone—microbial methanogenesis

$$(CH_2O)_{16}(NH_3)_{16}(H_3PO_4)$$
$$= 53CO_2 + 53CH_4 + 16NH_3 + H_3PO_4$$
$$\Delta G_r^0 = -350 \text{ kJ mole}^{-1} \text{ (glucose)}$$

Within many sediments, these three zones are fairly clearly separated, although local perturbations occur in the vicinity of burrows or where there is a

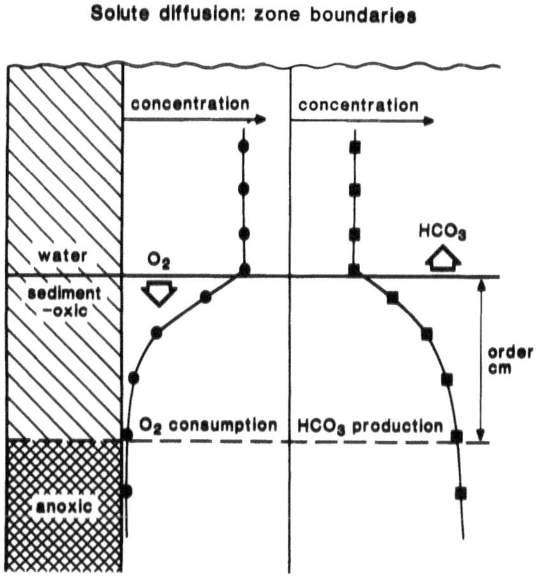

Fig. 2. Diffusional control on microbial zone boundaries: lower boundary of the Ox and SR zones being determined by the limit of downward movement of O_2 and SO_4^{2-} respectively (diffusional transfer being augmented by bioturbation).

concentration of organic matter. In pelagic sediments, a thin zone where nitrate reduction (NR) is important has been recognized, just beneath the Ox zone. Mineral oxidants (Mn(IV) and Fe(III)) are reduced and also occupy discrete depth zones (MnR and FeR) in sequence of decreasing molar free energy release. Within organic-rich sediments, however, such fine resolution is not observed; Ox, SR and Me are the only obvious zones.

Microbial processes thus degrade organic matter and produce potentially mineralizing solutes. It seems unlikely, however, that much in the way of mineral precipitation will occur in the Ox zone. This is because increased concentrations of HCO_3^- in the vicinity of the sediment–water interface will be dissipated by upward diffusion and dispersal in depositional waters. Much longer diffusion pathways (i.e. reactions deeper in the sediment) would be required to establish high solute concentrations relative to depositional waters. This is illustrated in Fig. 2.

The energy release data support the intuitive view that degradation within the Ox zone is likely to be much more dramatic than in the SR zone, with Me (not strictly involving an 'external' oxidant, therefore not depth-limited in the way that Ox and SR are by supply from above) somewhat less reactive still. If this is so, then the depositional environments considered in Fig. 1 have different potentials for organic matter preservation. This is illustrated diagrammatically in Table 1. The importance of anoxic environments in favouring organic matter preservation has been clearly identified (Demaison and Moore, 1980) but the considerable extent to which microbial sulphate reduction can degrade and modify organic matter during diagenesis has been less often evaluated.

TABLE 1 The link between prospect for organic matter preservation and depositional environment.

Degradation	*Environment*	*Burial sequence*
Minimum	fresh-water anoxic	Me
Intermediate	marine anoxic	SR → Me
Intermediate	fresh-water oxygenated	Ox → Me
Maximum	marine oxygenated	Ox → SR → Me

The influence of sedimentation (burial) rate

The total production of potentially mineralizing solutes within each microbial zone depends on the rate of organic matter degradation and the time for which the particular zonal reaction operates. The latter is a simple function of zone thickness and rate of burial. Degradation within one zone influences subsequent zones by reducing the amount and attractiveness to bacteria of organic substrate. Factors influencing the kinetics of individual mechanisms (particularly SR) have been quantified by Berner (1974) and the "carry over" effect of alteration in one zone influencing subsequent reactions has been explored by Curtis (1977). Rate of sedimentation is critical where respiration and, to a lesser extent, sulphate reduction are concerned. This is because the Ox zone is not often more than about 10 cm deep and the SR zone seldom more than a metre or two.

Extremely slow rates of sediment accumulation favour massive degradation of, or even complete elimination of, organic matter from sediments beneath oxygenated waters. Slow accumulation beneath anoxic marine waters, however, favours extensive degradation of organic matter by sulphate reduction. This generates very large amounts of CO_2 and H_2S which, in turn, might favour mineralization. By contrast, rapid burial takes sediment quickly to the Me zone where degradation is least intensive. These points are illustrated in Fig. 3. (Note that the thickness of each zone in Fig. 3 is not entirely independent of sedimentation rate.)

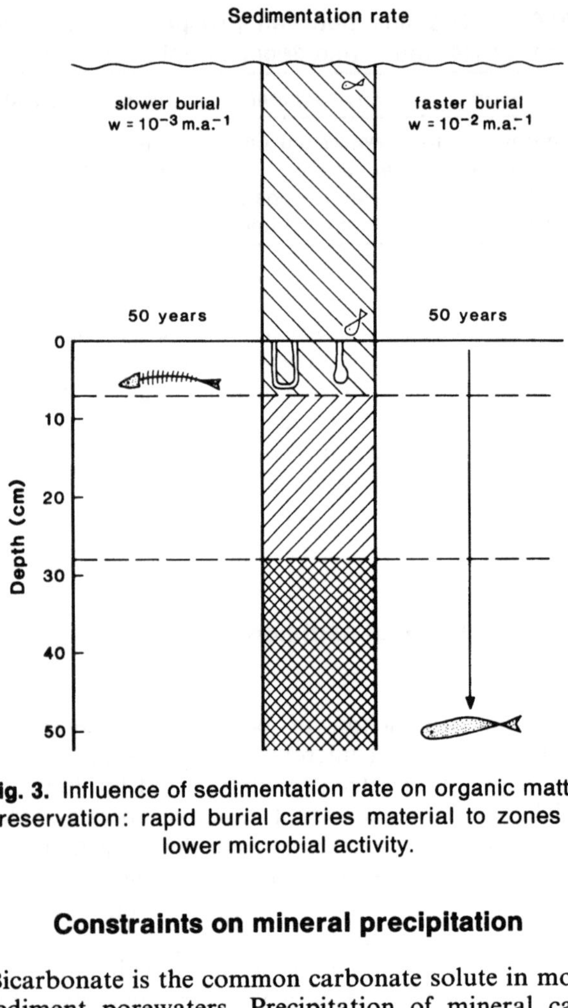

Fig. 3. Influence of sedimentation rate on organic matter preservation: rapid burial carries material to zones of lower microbial activity.

Constraints on mineral precipitation

Bicarbonate is the common carbonate solute in most sediment porewaters. Precipitation of mineral carbonates can thus be represented by:

$$Ca^{2+} + HCO_3^- = CaCO_3 + H^+ \quad (1)$$

Any chemical reaction that introduces HCO_3^- to the porewater system will increase HCO_3^- activity and encourage precipitation of authigenic carbonate minerals. Introduction of H^+ will lower pH and destabilize detrital carbonates. Acidification can also destabilize detrital aluminosilicate minerals:

$$NaAlSi_3O_8 + 2H^+ + 6H_2O = Al(OH)_2^+ + Na^+ + 3Si(OH)_4^0 \quad (2)$$

The pH-dependence of aluminosilicate solubility, however, is more complicated than this single equation suggests. This is because other solutes (Al^{3+}, $AlOH^{2+}$, $Al(OH)_4^-$, etc.) occur within the pH range likely to be encountered within sediment porewaters (Curtis, 1983a).

Returning to the three important microbial zones, each can be approximated by a simple reaction (assuming full oxidation to CO_2 followed by dissolution in porewater):

$$\text{Ox:} \quad CH_2O + O_2 = H^+ + HCO_3^- \quad (3)$$

$$\text{SR:} \quad 2CH_2O + SO_4^{2-} = H^+ + 2HCO_3^- + HS^- \quad (4)$$

$$\text{Me:} \quad 2CH_2O + H_2O = H^+ + HCO_3^- + CH_4 \quad (5)$$

Respiration and microbial methanogenesis thus introduce HCO_3^- and H^+ in roughly equal proportions. The *net* effect would certainly be to cause detrital carbonate mineral dissolution rather than authigenic mineral precipitation (Garrels and Christ, 1965, Chapter 3).

The situation with respect to sulphate reduction is not so clear because more bicarbonate than hydrogen is introduced. Coleman (personal communication, 1983; 1985) has pointed out that saturation with respect to H_2S gas and its loss from the system would almost certainly lead to carbonate precipitation since alkalinity is increased without H^+ addition. From equation (4):

$$H^+ + HS^- + 2HCO_3^- \rightarrow 2HCO_3^- + H_2S \quad (6)$$

Diffusive loss of HS^- upwards (or migration of H_2S as a gas) into the Ox zone followed by oxidation by O_2 (microbial or direct) would acidify porewaters by introducing sulphuric acid (Coleman, 1985).

These considerations collectively suggest that microbial organic degradation reactions, far from promoting precipitation of authigenic mineral cements, are likely to cause detrital carbonate (and aluminosilicate) minerals to dissolve. The possible exception is vigorous sulphate reduction accompanied by loss of H_2S as a gas. Since the solubility of calcium phosphates also increases in acid solution, authigenic phosphate cements are similarly unlikely.

Reduction of iron and manganese minerals

In the pelagic sediments described by Froelich *et al.* (1979), Mn(IV) and Fe(III) are reduced to Mn^{2+} and Fe^{2+}. In near-shore muddy sediments of the Rio Ameca Basin, Mexico, Drever (1971) showed that Fe(III) oxides had been progressively lost with depth. The particular significance of these observations is that organic matter is the only available reducing agent and, no matter how reactions are formulated, the alkalinity of the sediment porewater must be drastically modified (Curtis, 1977):

$$2Fe_2O_3 + CH_2O + 3H_2O = 4Fe^{2+} + HCO_3^- + 70H^- \quad (7)$$

This reaction would dramatically favour carbonate precipitation by increasing bicarbonate activity *and raising pH*. Within organic-rich sediments, iron reduction probably occurs in both SR and Me zones (Fe(III) oxides are stable in the Ox zone). Only in slowly deposited, pelagic sediments can separate manganese reduction (MnR) and iron reduction (FeR) zones (the 'suboxic' zones) be recognized.

Early diagenesis in clastic sediments where iron is available must thus be described by combinations of equations (4) and (5) with equation (7). Such an analysis was undertaken by Curtis (1983b) and, in a much more detailed way, by Coleman (1985).

One obvious criticism of equation (7) is that the mechanism is completely unknown: Fe(III) reduction is simply seen to take place. Organic matter must be oxidized, but not necessarily to CO_2. Numerous other possibilities exist and the evolution of kerogen, particularly during early maturation, could well be influenced by the amount and reactivity of iron (III) minerals in the sediment. I think that organic geochemists could usefully conduct experiments in this direction.

Mineral authigenesis in the SR zone

A number of iron sulphides form in marine muddy sediments, but pyrite, FeS_2, is normally most stable. It will form at the expense of almost any other iron mineral in the SR zone (Berner, 1964, 1970; Curtis and Spears 1968). Simultaneous reduction of iron (III) compounds and SO_4^{2-} yields Fe^{2+} and HS^- in solution which quantitatively precipitate as pyrite (the reaction is slightly more complicated than this since a polysulphide, S_2^{2-}, is involved). For balanced reduction to produce pyrite:

$$15CH_2O + 2Fe_2O_3 + 8SO_4^{2-} = 4FeS_2 + 7H_2O + 15HCO_3^- + OH^- \quad (8)$$

Provided sufficient iron is available, therefore, the reactions which are responsible for pyrite formation inject very large amounts of HCO_3^- into sediment porewaters and also raise pH. From equation (1) it follows that carbonate precipitation is almost inevitable. If the sulphate reduction rate balances or exceeds Fe(III) reduction, the carbonate phase will be iron-poor (calcite, dolomite). In the opposite case, iron-rich carbonates (ferroan calcite, ferroan dolomite, ankerite, siderite) will coprecipitate with pyrite.

In summary, reduction of iron III compounds within the SR zone will be conducive to simultaneous carbonate and sulphide mineral authigenesis. The composition of the carbonate phase will reflect the relative rates of Fe(III) and SO_4^{2-} reduction. Figure 4 outlines the likely mineralization trends within the SR zone for Fe(III)-poor and Fe(III)-rich sediments respectively. Only in the former case is it probable that sulphur will combine with more reactive organic molecules, subsequently to be incorporated in kerogen. In these shallow burial environments, the sediment as a whole is not in equilibrium. Mention has been made of different situations arising depending on essentially kinetic controls (relative rates of sulphate and iron reduction). Coleman (1985) has taken this analysis much further and shown that the rate of organic matter degradation by any process relative to iron reduction is probably the most significant factor determining the trend of mineral authigenesis.

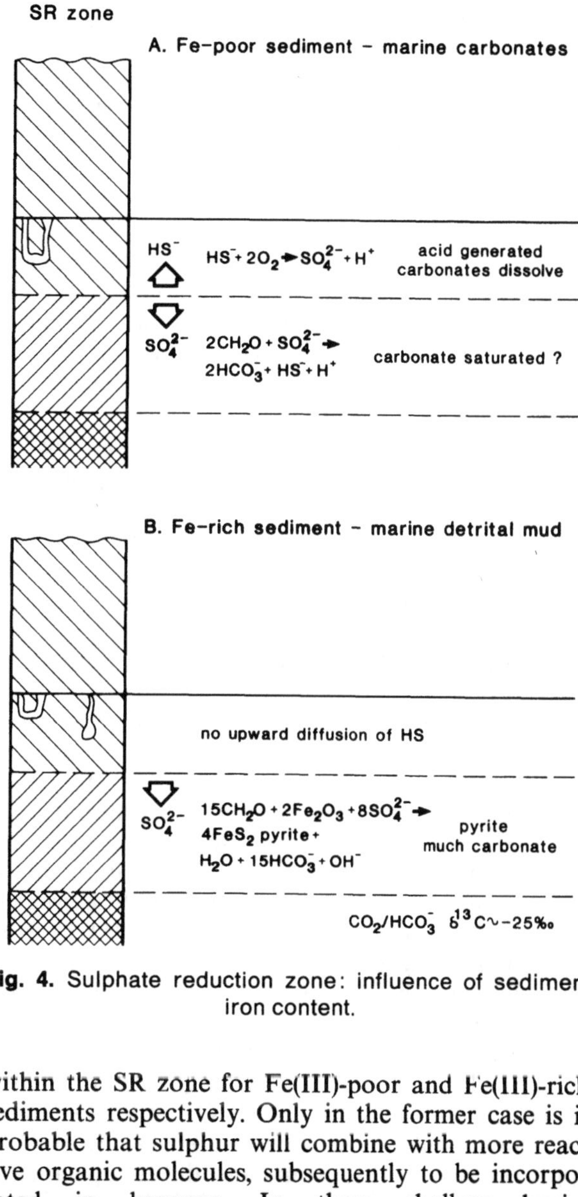

Fig. 4. Sulphate reduction zone: influence of sediment iron content.

Two common sedimentary situations are illustrated in Fig. 4: sulphate reduction in iron-poor sediments (such as carbonates) and sulphate reduction in iron-rich sediments (such as muds). In case A, vigorous microbial activity might lead to H_2S evolution and carbonate saturation. Any authigenic cement would be absolutely non-ferroan because of the excess H_2S and very low solubility of iron sulphides. In the Ox zone, dissolution of the least stable carbonates seems likely.

In case B (ample Fe(III) available) massive coprecipitation of carbonate and sulphide seems very likely. An excess of SO_4^{2-} reduction would give Fe-poor carbonate and possible loss of HS^- upwards to the Ox zone. Generation of small amounts of sulphuric acid might destabilize detrital carbonate. Excess of Fe(III) reduction would prevent upward loss of HS^- and make coprecipitation of pyrite with ferroan carbonates likely.

Bicarbonate produced by both SR and FeR is likely to retain the characteristic isotopic signature of organic matter ($\delta^{13}C \sim 25‰$). Fractionation as a result of selective oxidation of certain components of organic matter is possible, but has not been documented to my knowledge.

Mineral authigenesis in the Me zone

It seems useful again to consider two specific cases: iron-poor and iron-rich sediments beneath 'normal' marine waters. The burial sequence is Ox, SR, Me, as illustrated in Fig. 5 (iron-poor case). Carbon dioxide and CH_4 are both produced in the Me zone: the former with 'heavy' isotopic composition ($\delta^{13}C$ approximately $+15‰$), the latter with extremely 'light' values ($\delta^{13}C$ approximately $-75‰$). Both would tend to diffuse upwards into the SR zone and, possibly, into the Ox zone. It appears that CH_4, under certain circumstances, can be utilized as substrate by sulphate reducers, in which case values of $\delta^{13}C$ much lower than $-25‰$ might be anticipated for HCO_3^- (similarly also for HCO_3^- in the Ox zone). As pointed out above, however, HCO_3^- with $\delta^{13}C$ approximately $+15‰$ would also diffuse upwards. It is extremely difficult to make any kind of prediction for the isotopic composition of HCO_3^- which would be a mixture of unknown proportions of these very different components. One thing, however, does seem clear: authigenic carbonates are unlikely to precipitate in the Me zone of iron-poor sediments (equation (5)).

Iron-rich sediments present a very different picture. HCO_3^-($\delta^{13}C$ approximately $+15‰$) would tend to precipitate as siderite, $FeCO_3$. By analogy with Fig. 5, CH_4 would still tend to move upwards but not be accompanied by 'heavy' HCO_3^-. In this case, methane oxidation would cause HCO_3^- in the Ox and SR zones to be lighter ($\delta^{13}C < -25‰$) by some uncertain but possibly large amount.

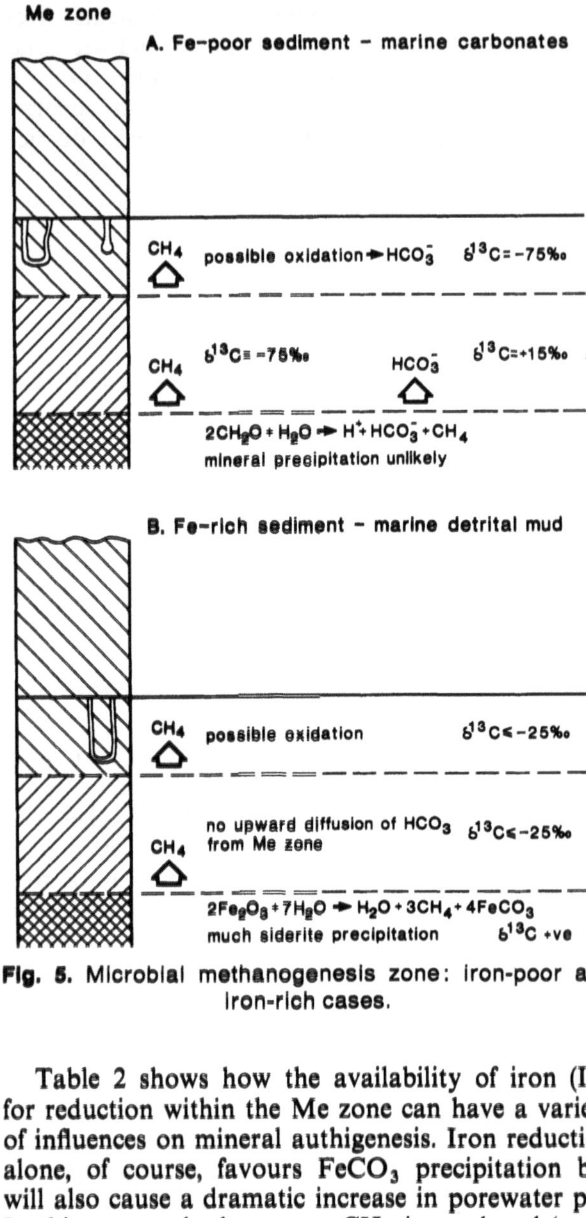

Fig. 5. Microbial methanogenesis zone: iron-poor and iron-rich cases.

Table 2 shows how the availability of iron (III) for reduction within the Me zone can have a variety of influences on mineral authigenesis. Iron reduction alone, of course, favours $FeCO_3$ precipitation but will also cause a dramatic increase in porewater pH. In this case and when some CH_4 is produced (cases A and B in Table 2), excess Fe^{2+} is introduced to the porewater system. From a chemical standpoint, these would be the conditions most likely to favour authigenesis of low-temperature iron silicates. Berthierine (a 0.7 nm phyllosilicate of approximate composition $Fe_5Al(Si_3AlO_{10})(OH)_8$—formerly known as chamosite) is a common constituent of iron-rich fine-grained sedimentary rocks.

At intermediate ratios of Me to FeR, copious quantities of $FeCO_3$ are likely to precipitate. When, for any reason, iron availability is somewhat restricted (case D, Table 2) Fe^{2+} would be stabilized in solution. Porewater migration would result in

TABLE 2. Methanogenesis and iron reduction. The balance between these is probably controlled by kinetic factors (Curtis, 1983b; Coleman, 1985).

A	$2Fe_2O_3 + CH_2O + 3H_2O \rightarrow 4Fe^{2+} + HCO_3^- + 7OH^-$	(7)B
B	$2Fe_2O_3 + 3CH_2O + 3H_2O \rightarrow 4Fe^{2+} + 2HCO_3^- + CH_4 + 6OH^-$	(9)
C	$2Fe_2O_3 + 7CH_2O + 3H_2O \rightarrow 4Fe^{2+} + 4HCO_3^- + 3CH_4 + 4OH^-$	(10)
D	$2Fe_2O_3 + 11CH_2O + 3H_2O \rightarrow 4Fe^{2+} + 6HCO_3^- + 5CH_4 + 2OH^-$	(11)
E	$12CH_2O + 6H_2O \rightarrow 6HCO_3^- + 6CH_4 + 6H^+$	(5)

Authigenic mineral precipitation

A	Reduction of iron alone	$FeCO_3$(siderite) + $3FE(OH)_2^*$
B	Minor methanogenesis	$2FeCO_3 + 2Fe(OH)_2^*$
C	Both important	$4FeCO_3$
D	Major methanogenesis	$2FeCO_3(+2Fe^{2+} + 4HCO_3^-)$**
E	Zero iron reduction	dissolution

* Conditions favouring Fe(II) precipitation as silicates such as berthierine (0.7 nm) or chamosite (1.4 nm)

** Conditions favouring iron mobility; for example, in 'bog iron' ores.

transport to some different environment where precipitation might occur. Upward migration to the SR zone would be one example (FeS_2 precipitation) or to the Ox zone (especially in fresh-water sediments where H_2S is unlikely to be present). An example here would be lateral migration, as in fresh-water swamps, until oxidation causes precipitation of hydrous oxides—the 'bog-iron' ores.

Microbially induced diagenesis—examples from the ancient record

Sediments of all ages and compositions reveal evidence of early diagenetic alteration. Leeder (1982) provides an excellent general summary and cites numerous publications dealing with different sediment compositions and depositional and diagenetic environments. For the several reasons cited above, iron-rich sediments are most likely to preserve the products of organic degradation as sulphides, carbonates, phosphates and silicates. Most mudrocks contain significant amounts of Fe(III), derived directly from the fine-fraction of soils. Petroleum source-rocks, more often than not, consist of organic-rich mudrocks.

Within these fine-grained clastic sediments, the effects of early diagenic mineralization are often strikingly obvious. Concretions (sub-spherical bodies cemented by a variety of minerals) can be demonstrated to have formed very early during sediment burial. Within them, fossil plants and animals are usually preserved without physical distortion, attesting to precompaction cementation (Weeks, 1953). The same conclusion can be reached by tracing sediment laminae through concretions (Tomkieff, 1927). In general, most authors follow Lippmann (1955) in the view that concretionary cements (carbonates, phosphates, sulphides) often precipitate within the pore spaces of muddy sediments to exclude porewater without distorting the sediment fabric. These issues are reviewed concisely by Raiswell (1971) in a study of Cambrian and Liassic concretions.

Irwin *et al.* (1977) and Irwin (1980) studied early diagenetic concretions in the organic-rich Kimmeridge Clay Formation, Dorset, England. Stable oxygen isotope ratios were used to infer temperature (depth) of precipitation and carbon isotope ratios to deduce source of CO_3^{2-}. Very clear evidence for precipitation within the SR zone (calcite/pyrite) and Me zone (ferroan dolomite/ankerite) was obtained. Non-marine sediments (the Kimmeridge sequence is entirely marine) are often characterized by siderite concretions, and coal measures, with their wide range of depositional environments, contain early diagenetic cements of just about all possible kinds (Matsumoto and Iijima, 1981, Pleistocene, Palaeogene and Upper Triassic Coalfields of Japan; Curtis and Coleman 1985, Carboniferous coalfields of Europe).

Early concretionary diagenetic cements are also found in carbonate sediments (Sass and Kolodny, 1972, Cretaceous Mishash Formation, Israel; Campos and Hallam, 1979, Jurassic limestones, England). They commonly occur in sandstones but it is then more difficult to infer precisely the time of formation.

In summary, there are numerous examples of very early cements which undoubtedly precipitated from porewater before the host sediments had been buried to depths of more than a few tens of metres. Isotopic evidence strongly points to specific microbial degradation pathways for the source of mineralizing solutes. The scale of precipitation may be enormous, early diagenetic siderite accounting for as much as 10% of some mudrock outcrops in the British coal measures (Pearson, 1979). These sediments were worked formerly as iron ores. Degradation of organic matter to water-soluble products by microbial mechanisms is thus quantitatively important in determining the mineralogical composition of many sedimentary rocks.

In the data of Irwin *et al.* (1977), precipitation temperatures in excess of 50 °C were indicated by some samples, although estimates were subject to considerable uncertainty. More significantly, a trend from positive $\delta^{13}C$ values (Me zone) at low precipitation temperature to negative $\delta^{13}C$ values at depth (highest temperatures) was recorded. This trend was attributed to generation of isotopically light HCO_3 at depth by kerogen decarboxylation, a process normally associated with catagenesis.

DEEPER BURIAL: ORGANIC MATTER CATAGENESIS

Thermal decarboxylation and D Zone carbonate minerals

Using very different experimental approaches, Tissot *et al.* (1974) and Laplante (1974) both showed that the earliest stages of thermal maturation of kerogen involved selective loss of oxygen as CO_2 and H_2O. This implies that the immature kerogen structure contains carboxylic acid and hydroxide groupings which are easily lost. Breakdown products would join the sediment porewaters. With enormous simplification and with considerable risk of insulting organic geochemists, the overall reaction can be written as an equation:

$$\text{kerogen-}CO_2H + H_2O \rightarrow \text{kerogen-}H + HCO_3^- + H^+ \quad (9)$$

On its own, this reaction would acidify porewaters and cause carbonates and silicates to dissolve. If, however, the decarboxylation reaction were accompanied by Fe(III) reduction, a range of possibilities exists which includes copious precipitation of siderite and enhanced mobility of iron as soluble bicarbonate. Once again the critical factor is the relative importance of FeR and decarboxylation, precisely as depicted for the Me zone in Table 2 (Curtis, 1983; Coleman, 1985).

Tissot *et al.* (1974) recognized this zone of elimination reactions as being one of four key stages of kerogen maturation. It is certainly a critical stage for mineral reactions and, following Coleman (1985), will be referred to subsequently as the D (decarboxylation) zone. The bicarbonate liberated in equation (9) would be expected to be isotopically light, although some fractionation from the $\delta^{13}C =$ 25‰ value is possible. Irwin *et al.* (1977), as mentioned above, demonstrated the increasing importance of D zone HCO_3^- in diagenetic carbonates precipitated at greater depth. Curtis *et al.* (1975) showed that, within a diagenetic siderite concentration, the early, central portion is enriched in ^{13}C (Me zone) whereas the later, outer parts are lighter. Evidence from concretions, therefore, suggests that precipitation of carbonates from the porewaters of mudrocks continues down into the stage of catagenesis. The rate of precipitation, however, seems to decline sharply. This is, perhaps, not unexpected since the supply of Fe(III) must become progressively exhausted with depth, and acid-generating reactions become relatively more important (Curtis, 1983b).

More recently, perhaps starting with the reports of Carothers and Kharaka (1978, 1980), it has been recognized that organic acids as well as H_2O and CO_2 are produced in the early stages of thermocatalytic cracking. Much of the most common acid has proved to be acetic and 95 formation-water samples from California and Texas proved to contain up to 5000 mg litre^{-1} acetate. At temperatures below 80 °C only small concentrations were documented, between 80 °C and about 130 °C high concentrations and, again, very little above 200 °C. It was argued that acetate is generated from kerogen within source rocks, migrates with expressed porewaters and, at increased burial temperatures, thermal decarboxylation produces methane:

$$CH_3COO^- + H_2O \rightarrow CH_4 + HCO_3^- \quad (10)$$

Surdam and Crossey (1985), from a series of experiments specifically linking organic acids with mineral reactions, have suggested that difunctional acids (oxalic, maleic, malonic) can be generated, along with the more abundant monocarboxylic acids (acetic, propionic) in the D zone. Another group of acids found in some formation waters is the phenols. Two conclusions reached in this interesting work are relevant here: quite a range of different organic acids appears to be generated from kerogen in the 80–130 °C temperature range and remarkably little work, as yet, has been undertaken to establish the magnitude and timing of their production.

It seems probable that different types of kerogen will produce different quantities and types of acid. Oxidation of kerogen by Fe(III) minerals could also be important since an external source of oxygen might be provided. Localized alkaline environments might influence reaction mechanisms.

In summary, the early stages of catagenesis (D zone) are characterized by production of CO_2 and H_2O, some of the former finding its way into carbonate minerals. Where organic acids are generated, mineral dissolution and metal transport are more likely.

Mineral dissolution and porosity enhancement

The reason why some sedimentologists are particularly interested in acid generation is that sandstones commonly reveal evidence of substantial grain dissolution with consequent increase in porosity and permeability. Changes of this kind are obviously of interest to the petroleum geologist who, from an understanding of chemical porosity enhancement, might locate commercial reserves in (by conventional reasoning) unlikely places. The reviews by Hayes (1979) and Schmidt and McDonald (1979) summarize the evidence and make the essential link with kerogen maturation (CO_2 production).

Whereas it is relatively easy to envisage dissolution of carbonate minerals in waters acidified by addition of CO_2, dissolution of feldspars ($KAlSi_3O_8$, $CaAl_2Si_2O_8$) is more difficult to understand because of the very low solubility of Al in water at anything near neutral (or even mildly acid) pH values.

Curtis (1983b) argued that acid waters capable of dissolving feldspars as well as carbonates *were* generated in mudrocks, citing the earlier work of Hower *et al.* (1976) as evidence. It was suggested that aqueous solutions expelled from the mudrocks during compaction might be capable of transporting Al in solution and also enhancing sandstone porosity by dissolution.

There are a number of objections to this picture. The first is that Al is really not very soluble in acetic acid solutions until quite low pH values are reached. The work of Surdam and Crossey (1985) is relevant here. If difunctional acids are generated in the D zone (oxalic, catechol), the solubility of Al is increased by something like three orders of magnitude. Experiments with minerals allowed to react with these acids produced textures indistinguishable from those in sandstones with enhanced porosity. Surdam and Crossey (1985) conclude that organic acids generated within the D zone probably are important in determining the properties of sandstone reservoirs and, certainly, that sedimentologists should not expect to understand mineral diagenesis without taking due notice of inorganic–organic reactions.

Another serious problem, however, has been delineated by Bjorlykke (1985). Calculations of the total amount of CO_2 that could have been liberated from kerogens accessible to the Frio Formation (Gulf Coast Tertiary) fall far short of that necessary to generate a significant proportion of the observed 'secondary' or leached porosity observed in the sandstones. It is also true that 'aggressive' pore solutions generated within mudrocks would lose much of their acidity by dissolution of carbonates and silicates within the mudrocks—this was the observation of Hower *et al.* (1976) which provides evidence for acid generation in the first place. There is no doubt, however, that mineral dissolution does take place in sandstones and that this enhances porosity. In fact the enhancement may be a relatively localized phenomenon which amounts rather more to redistribution than to net dissolution (Bjorlykke, 1985). On a basin scale, of course, this must be the case.

An attractive mechanism for mineral redistribution has been offered by Wood and Hewett (1982). Thermally driven porewater convection cells within reservoir systems would move solutes from regions of under-saturation to regions of over-saturation. The problem of solvent volume (i.e. in shale compaction models, very large volumes of water are apparently needed to effect mass transport because of low solubilities) is overcome because the porewater is effectively recycled.

Whichever mechanism of water movement is advocated to account for mineral dissolution (shale porewaters, meteoric water, convection cells), there is no doubt that the existence of difunctional acids such as those discussed by Surdam and Crossey would greatly increase transport capacity. It is imperative to learn more about the natural occurrence and stability of such acids and also something of the amounts that might be produced from different kerogen types. This is a problem first and foremost for the organic geochemist.

Organic influenced silicate precipitation reactions

Scanning electron microscopy (SEM) demonstrates that many of the clay minerals found in sandstones are unquestionably authigenic in origin. Their delicate structure and perfect form show that they precipitate from true solution. Beautiful examples of many different minerals are illustrated by Wilson and Pittman (1976). Kaolinite and chlorite commonly

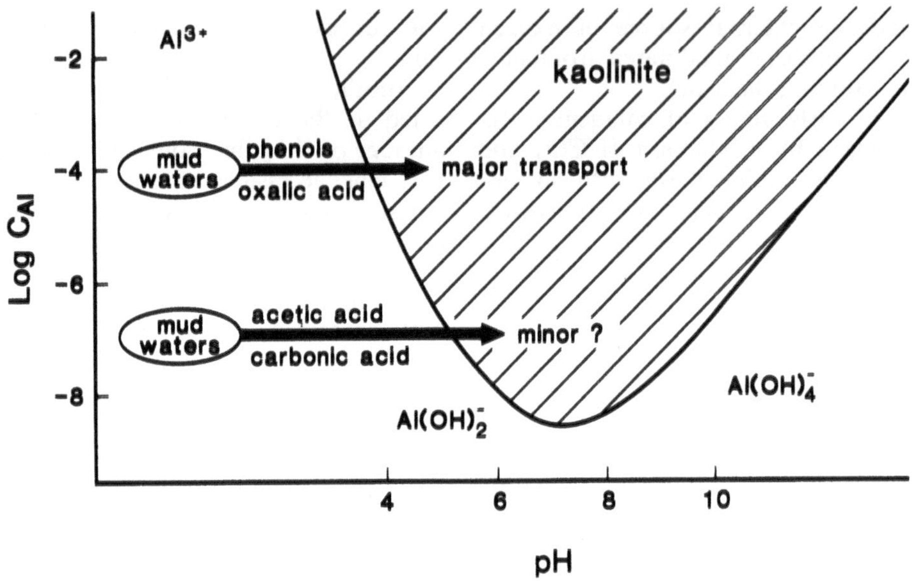

Fig. 6. Possible links between kaolinite precipitation and Al-transport via organic acids. For details see text.

develop in reservoir sandstones and exhibit almost perfect morphology. For these crystals, Al, Si and Fe must have existed for at least some time and migrated for certain minimal distances (cm to m?) in true aqueous solution.

In the previous section it was argued that kerogen undergoes elimination reactions within the D zone and acidifies porewater by addition of HCO_3^- and organic acid anions. Reaction with source-rock minerals takes place with dissolution of some aluminosilicate minerals. Compaction-induced porewater migration would tend to concentrate flow within more permeable sandstones where any residual acid would cause mineral dissolution and porosity enhancement. Such dissolution consumes acid and a rise in pH must follow. Curtis (1983a) argued that this would cause kaolinite precipitation, since the solubility of kaolinite decreases very sharply as neutral pH values are approached (Fig. 6). This explanation was developed to account for the not uncommon observation (V. Schmidt, personal communication, 1981) that kaolinite precipitation closely follows after porosity enhancement in sandstones and may even take place at the same time. Dissolution of calcite cements certainly would cause precipitation of kaolinite from Si- and Al-bearing acid solutions.

Another mechanism for precipitating kaolinite from solutions containing organic acid anions such as acetate is decarboxylation, as in equation (10) (Carothers and Kharaka, 1980). Carbonic is a weaker acid so pH would rise. The same mechanism would encourage carbonates to precipitate since the activity of HCO_3^- would also rise.

Authigenic chlorites commonly occur in sandstones as remarkably continuous and uniform grain coatings (Hayes, 1970). This very uniformity suggests significant solution transport of Al, Si and Fe; otherwise precipitation would be local and non-uniform in style. Again, acid solutions, most probably organic acid solutions, are likely to be involved. Although they are sometimes reported as being formed very early during diagenesis, most examples I have seen probably developed within the D zone.

A second, crucial, point about chlorites is that they are predominantly Fe(II) minerals. A necessary precondition for their development is reduction of Fe(III) by organic matter. For significant transport in solution, acid generation would have to outweigh considerably iron reduction (compare Case D in Table 2). This would not be unlikely, however, at depths typical of the D zone, bearing in mind earlier opportunities (at shallower depths) for iron reduction (i.e. the most reactive Fe(III) compounds would all have been reduced such that excess OH^- generation would be unlikely). Once again, the involvement of organic matter both as a reducing agent and in facilitating transport appears to be very important.

Recently, analytical transmission electron microscopy (AEM or ATEM) studies of authigenic chlorite coatings (Curtis *et al.*, 1985) from a variety of sandstones have revealed a certain uniformity of composi-

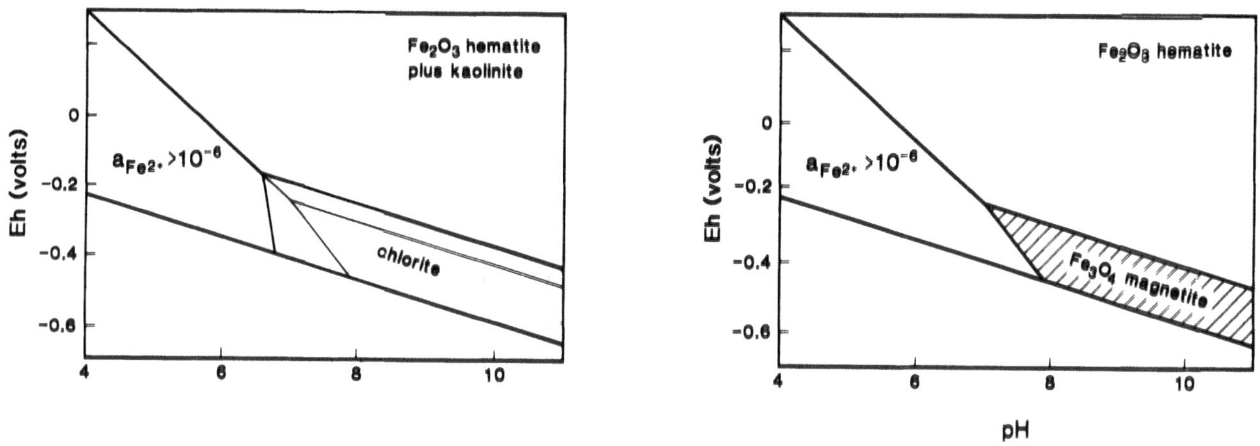

Fig. 7. Calculated stability fields for authigenic chlorite (magnetite field for comparison).

tion. All are relatively Al-rich and have a Si/Al ratio close to one. A typical composition might be:

$$Al_3Fe^{III}_{0.83}(MgFe^{II})_{7.5}(Si_{5.5}Al_{2.5}O_{20})(OH)_{16}$$

Using the method of Tardy and Garrels (1974) it is possible to estimate the formational free energy of phyllosilicates. A value of $\Delta G_f^0 = 13411$ kJ was obtained for the composition in which all Mg^{2+} is notionally replaced by Fe^{2+}. Figure 7 represents stability relations in the system Al-Si-Fe-O-H at 25 °C and one atmosphere total pressure in the form of a pH/Eh diagram. This suggests that chlorite (chamosite or berthierine) should be stable relative to hematite and kaolinite at the pH and Eh values likely to be encountered in sediments where reducing agents are available. The field of magnetite Fe_3O_4 is totally excluded. It is important to note that organic matter is the only reducing agent available and that development of chlorite is quantitatively dependent on oxidation of organic matter by Fe(III).

Development of chlorite early during diagenesis in organic-rich sediments is unlikely because either HS^- or HCO_3^- or both (equations (8) and (10)) would be plentifully available. Under these conditions pyrite FeS_2 or siderite $FeCO_3$ would precipitate in preference (Curtis and Spears, 1968). Chlorite minerals are therefore most likely to form in the Me and D zones—later diagenesis and catagenesis.

Several workers have drawn attention to the fact that chlorite minerals tend to replace kaolinite (Boles and Franks, 1979; Hutcheon *et al.*, 1980; Iijima and Matsumoto, 1982) but the role of organic matter as a reducing agent has not been stressed.

Fe(III) in sediments and ease of reduction to Fe(II)

The importance of organic matter oxidation by iron (FeR) has been stressed throughout. Iron in sediments at deposition is predominantly Fe(III); the mineral products of diagenesis and catagenesis are predominantly Fe(II)—pyrite FeS_2 or siderite $FeCO_3$—and chlorites. Solubility of Fe(III) at the pH values of sediment environments, however, is very low. This suggests that Fe(III) reduction proceeds via a heterogeneous mechanism at the surface of solids and is likely to be slow or, at least, very dependent on the chemical reactivity of the Fe(III) mineral.

In a recent paper (Curtis, 1985) I suggested that three essentially different forms of Fe(III) were likely to be present in fine-grained detrital sediments at deposition:

(a) soil 'sesquioxides': hematite Fe_2O_3, goethite FeOOH and amorphous hydroxides;
(b) interlayer Al-Fe(III) hydroxide sheets in soil clay–vermiculites;
(c) in the dioctahedral layer of beidellite–illite and nontronite–glauconite minerals.

It seems likely that this represents a sequence of increasing resistance to reduction. Soil sesquioxides probably disappear at the earliest stages of diagenesis whereas evidence was cited for believing that Fe(III) in clay minerals of the illite/smectite group continues to be reduced until temperatures as high as 130 or 140 °C are reached at depths of the order 4 km. Figure 8 is modified from Curtis (1985) and shows

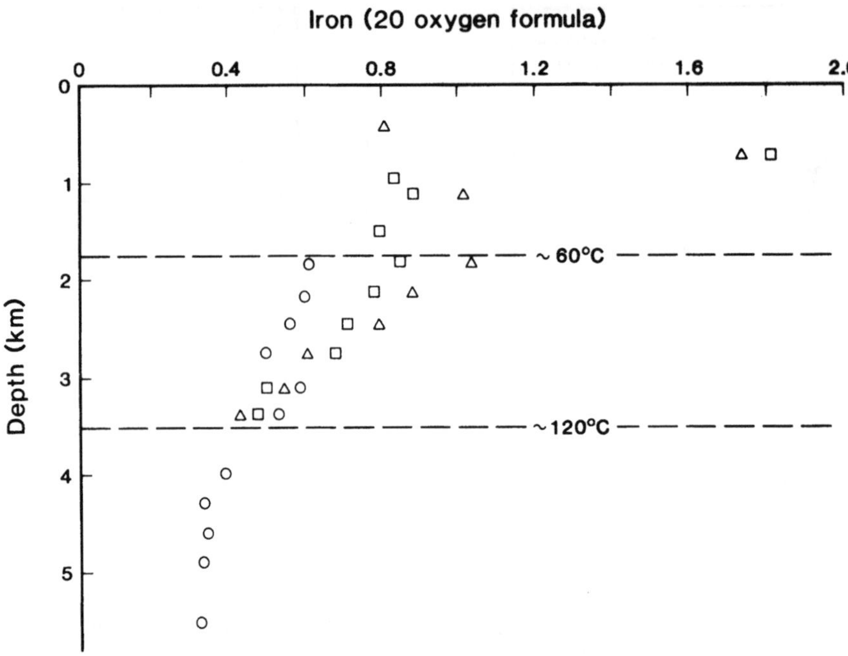

Fig. 8. Decrease in iron(III) content of the fine clay fraction of mudrocks with depth. After Curtis (1985) and references therein. Chemical data expressed as (I/S) mineral formulae.

the systematic loss of iron from the fine clay fraction of mudrocks in two different sedimentary basins.

The important conclusion to be drawn from this initial suggestion (very much more work needs to be done, data are few) is that organic matter is being oxidized by Fe(III) throughout much of the D zone. The maturation of kerogen may thus be significantly influenced by the mineral constituents of source-rocks in two ways: firstly, by catalysis (Johns, 1979) and secondly, by oxidation. This applies to the zone of liquid hydrocarbon generation as well as the earlier phase of CO_2 and H_2O elimination reactions.

Deep H_2S generation

Certain kerogens, particularly those affected by SR zone processes in the absence of Fe(III) reduction, are sulphur-rich and generate H_2S late in the 'dry gas' stage of catagenesis (H_2S also may be generated deep by microbial reduction of evaporite mineral/hydrocarbon assemblages). The very low solubility of metal sulphides means that further mineral transformations will take place as a direct consequence of organic matter degradation reactions. Precipitation of pyrite at the expense of iron in glauconite has been documented (Ireland *et al.*, 1982). Here it should be noted that H_2S also acts as a reducing agent since sulphur in pyrite is present in the form of a polysulphide:

$$Fe_2O_3 + 2H_2S = FeS_2 + H_2O + 2OH^- + Fe^{2+}$$

CONCLUSIONS

There are several ways in which organic matter maturation influences the mineralogy of sediments in chemical reactions which take place after burial and which continue to operate to depths of 4 or 5 km.

1. Low molecular-weight water-soluble products are incorporated directly into minerals. The most obvious examples are the carbonates (calcite, dolomite, siderite) precipitated at shallow (sulphate reduction, microbial methanogenesis zones) and intermediate (thermal decarboxylation zone) depths. Unambiguous evidence of direct incorporation is provided by stable isotope studies and the scale of precipitation, especially at shallow depths, is very considerable.

Phosphate minerals (apatite and francolite—a carbonate fluorapatite) are also formed from organophosphates. Although much less important volumetrically than carbonates, they are favoured in certain geochemical/sedimentary environments and achieve important concentrations. Diagenetic development of phosphate minerals is discussed in Coleman (1985).

Sulphur is eliminated from kerogen under conditions of deep burial and high temperature: direct precipitation of mineral sulphides (mostly pyrite) results.

2. The products of organic matter degradation reactions which do not themselves originate in the organic matter are incorporated into authigenic minerals. Hydrogen sulphide produced within the sulphate reduction zone is the most obvious example. Early pyrite is an abundant and characteristic mineral of organic-rich marine sediments.

Another important reaction is reduction of Fe(III). Oxidation of organic matter produces Fe^{2+} which is incorporated into many authigenic minerals: siderite, ferroan dolomite, ferroan calcite, ankerite, and chlorite minerals (berthierine and chamosite). These minerals form early concretions and later, pore-lining cements in almost every type of sediment.

3. Organic matter degradation reactions strongly influence and may constitute the principal control on the pH of sediment porewaters during diagenesis. Almost all authigenic mineral precipitation/dissolution reactions are pH sensitive (equations (1) and (2)). The influence can be positive or negative:

(a) pH rise; Fe(III) reduction—(equation (7))
(b) pH fall; acid generation (equations (3), (4), (5) plus organic acids)

Minerals such as kaolinite and albite (neither of which is a direct or indirect product of organic degradation as in (1) or (2) above) precipitate or dissolve in response to changes in porewater pH.

On a sedimentary basin scale, acid generation and mineral dissolution reactions probably have important consequences for petroleum migration—both primary and secondary. Porewater flow concentrates through more permeable pathways. Permeability enhancement by mineral dissolution would further focus flow and hence influence subsequent migration of liquid hydrocarbons.

The reverse effect, that of precipitation of minerals in pore space, obviously reduces porosity and permeability of reservoirs. These deleterious effects extend to hydrocarbon production since authigenic minerals, particularly clay silicates, react with chemical constituents of drilling, completion and production fluids (reservoir sensitivity).

4. Organic matter degradation products such as bicarbonate and, particularly, organic acid anions, facilitate the transport of metals in solution. Effective solubility is enhanced by provision of soluble anions, by complexation and chelation. The geochemistry of Fe and Mn are greatly influenced by organic–inorganic interactions during diagenesis; for this reason alone, many other metals could be expected to be equally affected.

5. The four points above emphasize the effects of organic matter diagenesis and catagenesis on the mineralogical composition of sedimentary rocks. The reverse emphasis is seen in the catalytic effects of clay minerals on kerogen maturation (Johns, 1979) and in direct oxidation by mineral oxidants. Iron Fe(III) appears to be the most important of these at depth, although surprisingly little work has been undertaken to study this oxidation.

References

Berner, R. A. (1964). Iron sulphides formed from aqueous solution at low temperatures and atmospheric pressure, *Journal of Geology* **72,** 293–306.

Berner, R. A. (1970). Sedimentary pyrite formation, *American Journal of Science* **268,** 1–23.

Berner, R. A. (1974). Kinetic models for the early diagenesis of nitrogen, sulphur, phosphorus and silicon in anoxic marine sediments, *In The Sea*, Vol. 5, Goldberg (Ed.), New York, Wiley-Interscience, pp. 427–450.

Bjorlykke, K. (1984). Formation of secondary porosity: how important is it? *In Clastic Diagenesis*, D. A. McDonald and R. C. Surdam (Eds), A.A.P.G. Memoir 37, pp. 277–286.

Boles, J. R. and Franks, S. G. (1979). Clay diagenesis in Wilcox sandstones of south-west Texas: implications of smectite diagenesis on sandstone cementation, *Journal of Sedimentary Petrology* **49,** 55–70.

Boudreau, B. P. and Westrich, J. T. (1984). The dependence of bacterial sulphate reduction on sulphate concentration in marine sediments, *Geochimica et Cosmochimica Acta* **48,** 2503–2516.

Campos, H. S. and Hallam, A. (1979). Diagenesis of English Lower Jurassic limestones as inferred from oxygen and carbon isotope analysis, *Earth and Planetary Science Letters* **45,** 25–31.

Carothers, W. W. and Kharaka, Y. K. (1978). Aliphatic acid anions in oil-field waters and their implications for the origin of natural gas, *Bulletin American Association Petroleum Geologists* **62,** 2441–2453.

Carothers, W. W. and Kharaka, Y. K. (1980). Stable carbon isotopes of HCO_3^- in oil-field waters—implications for the origin of CO_2. *Geochimica et Cosmochimica Acta*, **44,** 323–332.

Cerling, T. E. (1984). The stable isotope composition of modern soil carbonate and its relationship to climate, *Earth and Planetary Science Letters* **71,** 229–240.

Claypool, G. E. and Kaplan, I. R. (1974). The origin and distribution of methane in marine sediments, *In Natural Gases in Marine Sediments*, I. R. Kaplan (Ed.), New York, Plenum Press, pp. 99–139.

Coleman, M. (1985). Geochemistry of diagenetic non-silicate minerals: kinetic considerations, *Philosophical Transactions Royal Society, London Series A*, **315,** 39–56.

Curtis, C. D. (1977). Sedimentary geochemistry: environments and processes dominated by involvement of an aqueous phase, *Philosophical Transactions, Royal Society, London Series A*, **286,** 353–372.

Curtis, C. D. (1983a). The link between aluminium mobility and destruction of secondary porosity, *Bulletin American Association Petroleum Geologists* **67,** 380–384.

Curtis, C. D. (1983b). Geochemistry of porosity enhancement and reduction in clastic sediments, *In Petroleum Geochemistry and Exploration of Europe*, Brooks (Ed.), Blackwell Scientific/Geological Society London, pp. 113–125.

Curtis, C. D. (1985). Clay mineral precipitation and transformation reactions during burial diagenesis, *Philosophical Transactions Royal Society, London, Series A.*, **315,** 91–105.

Curtis, C. D. and Coleman, M. L. (1985). Controls on the precipitation of early diagenetic calcite, dolomite and siderite concretions in complex depositional sequences, *In The Relationship of Organic Matter and Mineral Diagenesis*, Gautier (Ed.), S.E.P.M. Special Publication, No. 38, 23–33.

Curtis, C. D., Hughes, C. R., Whiteman, J. A. and Whittle, C. K. (1985). Compositional variation within some sedimentary chlorites and some comments on their origin, *Mineralog. Mag.*, **49,** 375–386.

Curtis, C. D., Pearson, M. J. and Somogyi, V. (1975). Mineralogy, chemistry and origin of a concretionary siderite sheet (clay–ironstone band) in the Westphalian of Yorkshire, *Mineralogical Magazine* **40,** 385–393.

Curtis, C. D. and Spears, D. A. (1968). The formation of sedimentary iron minerals, *Economic Geology* **63,** 257–270.

Demaison, G. J. and Moore, G. T. (1980). Anoxic environments and oil source bed genesis, *Bulletin American Association Petroleum Geologists* **64,** 1179–1209.

Drever, J. I. (1971). Early diagenesis of clay mineral, Rio Ameca Basin, Mexico, *Journal of Sedimentary Petrology* **41,** 982–994.

Froelich, P. N., Klinkhammer, G. P., Bender, M. L., Luedtke, N. A., Heath, G. R., Cullen, D. and Dauphin, P. (1979). Early oxidation of organic matter in pelagic sediments of the eastern equatorial Atlantic: suboxic diagenesis, *Geochimica et Cosmochimica Acta* **43,** 1075–1090.

Garrels, R. M. and Christ, C. L. (1965). *Solutions, Minerals and Equilibria*, W. H. Freeman, San Francisco.

Hayes, J. B. (1970). Polytypism of chlorite in sedimentary rocks, *Clays and Clay Minerals* **18,** 285–306.

Hayes, J. B. (1979). Sandstone diagenesis—the hole truth, *In Aspects of Diagenesis*, P. A. Scholle and P. R. Schluyer (Eds), S.E.P.M. Special Publications no. 26, pp. 127–139.

Hower, J., Eslinger, E. V., Hower, M. E. and Perry, E. A. (1976). Mechanism of burial-metamorphism of argillaceous sediment: 1. Mineralogical and chemical evidence, *Geological Society of America Bulletin* **87,** 725–737.

Hunt, J. M. (1979). *Petroleum Geochemistry and Geology*, W. H. Freeman, San Francisco.

Hutcheon, I., Oldershaw, A. and Ghent, E. D. (1980). Diagenesis of Cretaceous sandstones of the Kootenay Formation at Elk Valley (south-eastern British Columbia) and Mt Allan (south-western Alberta), *Geochimica et Cosmochimica Acta* **44,** 1425–1435.

Iijima, A. and Matsumoto, S. R. (1982). Berthierine and chamosite in coal measures of Japan, *Clays and Clay Minerals* **30,** 264–274.

Ireland, B. J., Curtis, C. D. and Whiteman, J. A. (1983). Compositional variation within some glauconites and illites and implications for their stability and origins, *Sedimentology* **30,** 769–786.

Irwin, H. (1980). Early diagenetic carbonate precipitation and pore fluid migration in the Kimmeridge Clay of Dorset, England. *Sedimentology* **27,** 577–591.

Irwin, H., Curtis, C. D. and Coleman, M. (1977). Isotopic evidence for source of diagenetic carbonates formed during burial of organic-rich sediments. *Nature* **269,** 209–213.

Johns, W. D. (1979). Clay mineral catalysis and petroleum generation, *Annual Review Earth and Planetary Science* **7,** 183–198.

Laplante, R. E. (1974). Hydrocarbon generation in Gulf Coast Tertiary sediments, *Bulletin American Association Petroleum Geologists* **58,** 1281–1289.

Leeder, M. R. (1982). *Sedimentology: Process and Product*, George Allen and Unwin, London.

Lippmann, F. (1955). Ton, Geoden und Minerale des Barreme von Hoheneggelsen, *Geolograches Rundschau* **43,** 475–503.

Matsumoto, R. and Iijima, A. (1981). Origin and diagenetic evolution of Ca–Mg–Fe carbonates in some coalfields of Japan, *Sedimentology* **28,** 239–259.

Pearson, M. J. (1979). Geochemistry of the Hepworth Carboniferous sediment sequence and origin of the diagenetic iron minerals and concretions, *Geochimica et Cosmochimica Acta* **43,** 927–941.

Raiswell, R. (1971). The growth of Cambrian and Liassic concretions, *Sedimentology* **17,** 147–171.

Sass, E. and Kolodny, Y. (1972). Stable isotopes, chemistry and petrology of carbonate concretions (Mishash Formation, Israel), *Chemical Geology* **10,** 261–286.

Schmidt, V. and McDonald, D. A. (1979). The role of secondary porosity in the course of sandstone diagenesis, *In Aspects of Diagenesis*, P. A. Scholle and P. R. Schluyer (Eds), S.E.P.M. Special Publication no. 26, pp. 175–207.

Surdam, R. C. and Crossey, L. J. (1985). Organic–inorganic reactions during progressive burial: key to porosity and permeability enhancement and preservation, *Philosophical Transactions, Royal Society, London Ser. A*, in press.

Tardy, Y. and Garrels, R. M. (1974). A method of estimating the Gibbs energies of formation of layer silicates, *Geochimica et Cosmochimica Acta* **38,** 1101–1116.

Tissot, B., Durand, B., Espitalié, J. and Combaz, A. (1974). Influence of nature and diagenesis of organic matter in formation of petroleum, *Bulletin American Association Petroleum Geologists* **58,** 499–506.

Tissot, B. and Welte, D. H. (1978). *Petroleum Formation and Occurrence*, Springer-Verlag, Berlin.

Tomkeiff, S. (1927). On the occurrence and the mode of origin of certain kaolinite-bearing nodules in the coal measures, *Proceedings Geological Assocation* **38,** 518–547.

Weeks, L. G. (1953). Environment, mode of origin and facies relationships of carbonate concretions in shales, *Journal of Sedimentary Petrology* **23,** 162–173.

Wilson, M. D. and Pittman, E. D. (1977). Authigenic clays in sandstones: recognition and influence on reservoir properties and palaeo-environmental analysis, *Journal of Sedimentary Petrology* **47,** p. 3–31.

Wood, J. R. and Hewett, T. A. (1982). Fluid convection and mass transfer in porous sandstones—a theoretical model, *Geochimica et Cosmochimica Acta* **46,** 1707–1713.

Chapter 7

Evolution of Silurian Depositional Systems in the Southern Uplands, Scotland

Alan E. S. Kemp, Department of Oceanography, University of Southampton, Southampton SO9 5NH

ABSTRACT

Throughout the early Silurian, south-west-flowing palaeocurrents in the turbidite sequences of the Southern Uplands record major axial flow and sediment dispersal parallel to the margin consistent with the presence of a topographic trench. Mid-Silurian sequences, however, record a major change in the character of sedimentation at the north-west margin of the Iapetus ocean. Early Wenlock depositional trends follow those of the Upper Llandovery. In the central Southern Uplands thick sequences dominated by massive sandstones are characterized by south-east-flowing palaeocurrents indicating transverse input to the trench via high gradient sand-rich fans. In the south-west Southern Uplands, however, relatively thin sequences of thin-bedded turbidites were deposited by south-west-flowing palaeocurrents suggesting deviation of transverse currents to flow axially along the trench. Mid-Wenlock facies represent a profound change from the monotonous classical turbidites of the late Llandovery and early Wenlock sequences. Diverse facies associations, which can be correlated both across and along strike, record the development of extensive channel/overbank complexes and coeval channel-mouth, depositional lobe and fan-fringe facies. These sedimentary associations represent the outbuilding of large passive-margin style fans which probably extended at least 200 km from the inner trench slope and infilled the topographic trench. The existence of such large fan bodies suggests a pause or slowing in subduction during the mid-Wenlock prior to closure in the Southern Uplands sector in the late Wenlock/early Ludlow interval.

TECTONOSTRATIGRAPHIC FRAMEWORK

Lack of good correlation frequently poses major problems in the interpretation of ancient turbidite sequences. The Southern Belt of the Southern Uplands has both high-quality continuous coastal exposure of a variety of turbidite facies and abundant fossiliferous hemipelagic sediment. New stratigraphical data (Kemp, 1986; Kemp and White, 1985) resulting from collection of over two thousand graptolites permits detailed correlation and subdivision into units of *c.* 0.5 Ma. This has facilitated interpretation of the diverse turbidite facies associations of the Southern Belt and constrained reconstructions of the NW margin of Iapetus during the mid-Silurian. This paper reports the principal results of detailed sedimentological studies in the Southern Belt and presents models for the U. Llandovery and Wenlock development of depositional systems in the Southern Uplands.

The Southern Uplands has a distinctive tectonostratigraphic pattern of progressively younger fault-bounded, subvertical units or tracts encountered to

Marine Clastic Sedimentology © J. K. Leggett & G. G. Zuffa (Graham and Trotman, 1987) pp. 124–155

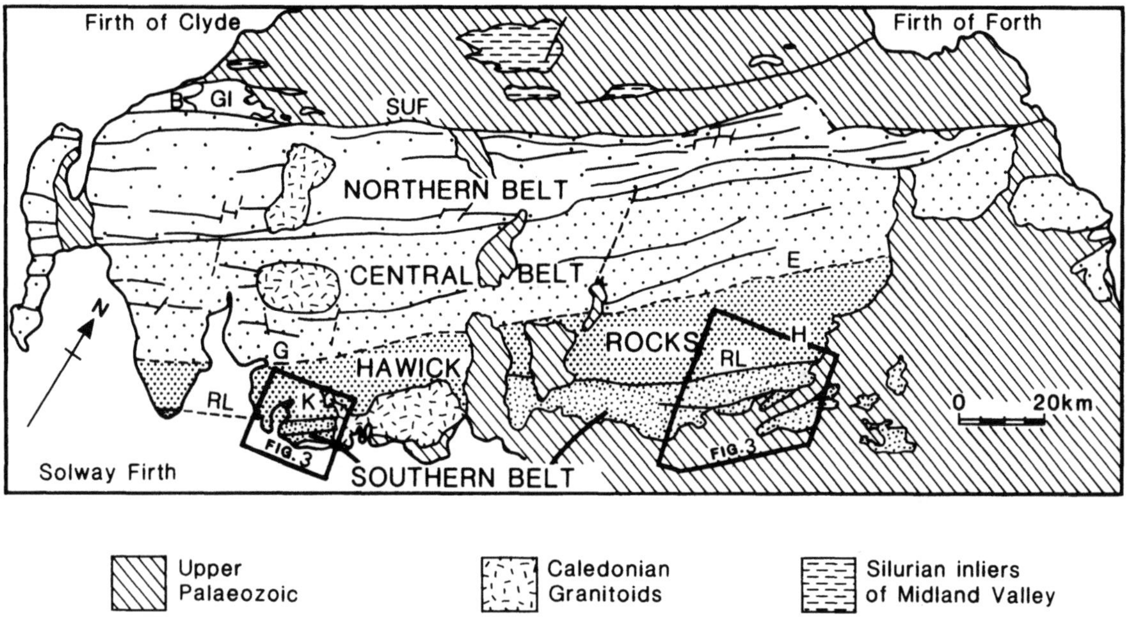

Fig. 1. Synoptic geological map of the Southern Uplands showing major subdivisions and faults. Northern Belt—Ordovician; Central Belt—Llandovery; Hawick Rocks (southern Central Belt)—uppermost Llandovery; Southern Belt—Wenlock; G1—Ordovician sequence of Girvan; B—Ballantrae ophiolite; E—Ettrickbridgend; G—Gatehouse of Fleet; H—Hawick; K—Kirkcudbright; L—Langholm; RL—Riccarton Line; SUF—Southern Upland Fault, Raeberry; (18)—Mullock Bay.

the south-east which individually contain north-west-younging sequences (Figs 1 and 2).

Following Peach and Horne (1899) the terrane is conventionally subdivided into a Northern Belt of Ordovician age, a Central Belt of Llandovery age and a Southern Belt of Wenlock age. A further subdivision may be made within the broad Central Belt by distinguishing a northern part in which the turbidite sequences are conformably underlain by pelagic sediments (the Moffat Shale) from a southern, largely unfossiliferous part known as the Hawick Rocks (Fig. 1).

THE PRE-SILURIAN CONTEXT

Episodic arc-type magmatism records north-west-directed subduction along the NW margin of Iapetus from early Ordovician to end-Silurian/early Devonian times (Fitton *et al.*, 1982; Pankhurst *et al.*, 1982; Thirlwall, 1981). The earliest accretion event is recorded along the margin in the Southern Uplands in the late Llanvirn/early Caradoc—the age of the first accreted packet. Subduction-accretion continued for some *c.* 35 Ma at least until the lundgreni zone of the Wenlock, resulting in the present tectonostratigraphic configuration of the Southern Uplands (Fig. 1).

Earlier suggestions by Piper (1972), interpreting previous studies, that the Southern Uplands sediments represented trench deposits, were made on the basis of the linear (NE–SW) palaeocurrent trends in the Southern Uplands, together with lack of evidence for south-east derivation. Subsequent work in the Northern and Central Belts of the Southern Uplands has shown that dominant axial flow (with localized NW-derivation) persisted throughout the Ordovician and into the Early Silurian (Casey, 1983; Hepworth *et al.*, 1982; Kassi, 1985; Leggett, 1980; Leggett *et al.*, 1982). These authors have interpreted the sediment dispersal patterns to indicate diversion and confinement of flow by the outer trench slope and hence infer the presence of a topographic trench throughout this period.

Limited evidence for south-east derivation of andesitic material occurs in some Upper Ordovician turbidites. This has been taken to infer a back-arc basin origin for the Northern Belt sequences by Stone *et al.* (in press), who invoke subsequent removal of the (now vanished) arc terrane by strike slip movement. A more likely provenance for this material is from minor volcanic edifices or rifted

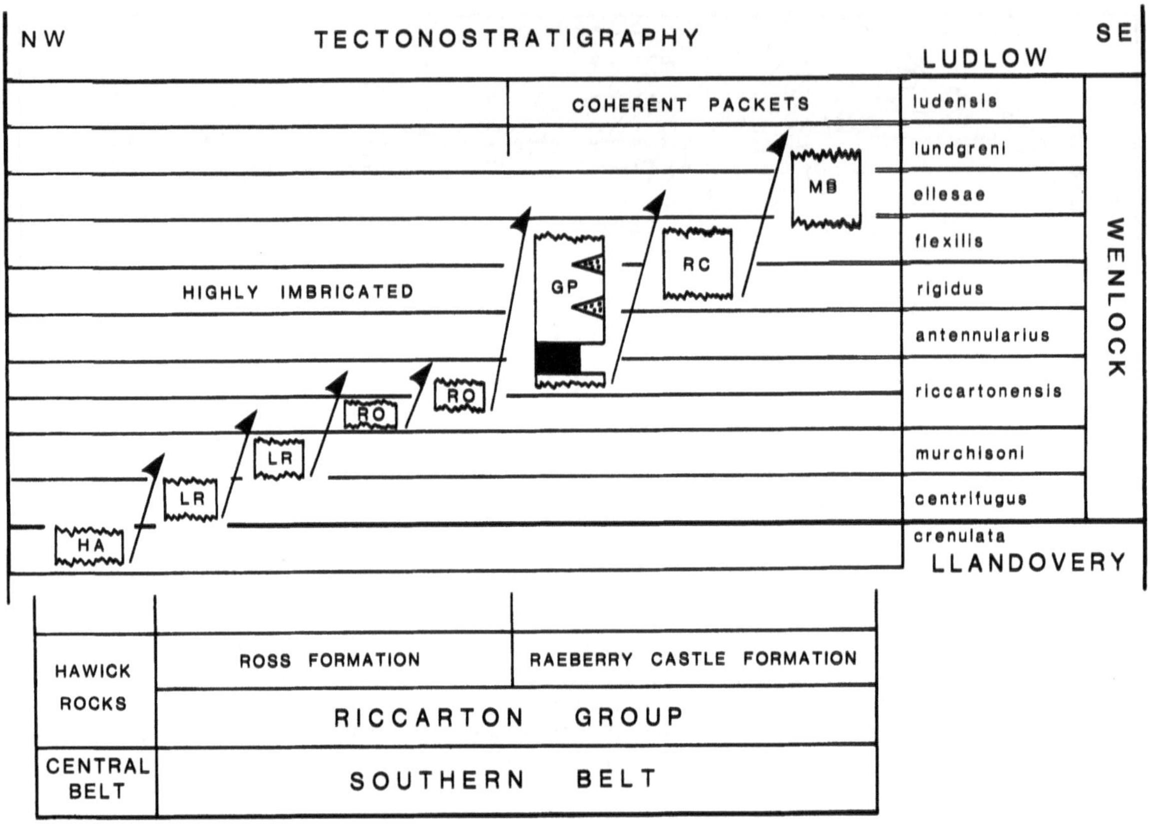

Fig. 2. Detailed stratigraphy of Southern Belt (bottom) with tectonostratigraphy of Kirkcudbright area (top). Abbreviations: HA—Hawick; LR—Long Robin; RO—Ross; GP—Gipsy Point; RC—Raeberry; MB—Mullock Bay.

remnant arc fragments on the subducting Iapetus plate (McMurtry, 1980).

LLANDOVERY

Leggett (1980) proposed the existence of an outer trench high separating a trench turbidite province from an ocean floor turbidite province to the south-east during early and middle Llandovery times before being swamped by laterally derived turbidites in late Llandovery times. Since the age of the Raeberry Castle Formation has been revised to Wenlock (Kemp, 1986; Kemp and White, 1985) there is now little evidence to support this.

Sedimentation patterns inferred from palaeocurrent directions from mid- to late Llandovery in the Southern Uplands and Longford-Down are remarkably consistent in indicating dominant south-westward flow with more minor but significant transverse input from the north-west (Cameron, 1979; Casey, 1983; Leggett, 1980; Walton, 1965). From these predominantly axial sediment dispersal patterns and by analogy with modern trench sedimentation, the pre-Wenlock sedimentary history of the Southern Uplands is entirely consistent with the existence of a topographic trench which restricted the transverse out-building of depositional systems.

Carbonate detritus which includes bioclastic material becomes an important component in the greywacke sandstones from the late Llandovery. This evidences the emergence of a trench-slope break and development of shallow marine carbonates which continued to provide sediment throughout the Wenlock.

TECTONOSTRATIGRAPHY OF THE SOUTHERN BELT

The Southern Belt or Riccarton Group is characterized by the presence throughout of a graptolitic, dark grey (N2–N3) laminated argillaceous siltstone which occurs as an interturbidite or hemipelagic sediment (Kemp, 1985). The Southern Belt is divided into an early Wenlock Ross Formation comprising highly

TABLE 1 Wenlock turbidite facies in the Southern Uplands.

Facies A	Coarse grained sandstones and conglomerates A1 Disorganized conglomerates A2 Organized conglomerates A3 Disorganized pebbly sandstones A4 Organized pebbly sandstones
Facies B	Fine to coarse grained sandstones B1 Massive sandstones with dish structures B2 Massive sandstones without dish structures
Facies C	Very fine to medium sandstones with interbedded mudstones—classical proximal turbidites beginning with Bouma 'a' division
Facies D	Very fine to fine grained sandstones, siltstones and mudstones—classical distal turbidites beginning with Bouma 'b' 'c' or 'd' division. Subdivided into: D_b—most beds beginning with Bouma 'b' D_c—most beds beginning with Bouma 'c' D_d—most beds beginning with Bouma 'd' D_m—graded mudstone units with occasional parallel laminae of siltstone at base D_cG, D_dG, D_mG } as above but including frequent hemipelagite interbeds
Facies E	Similar to $D_c/D_d/D_m$ but with common stacked ripples and irregular laminae, whispy and fading laminae; hemipelagite absent or as very rare thin laminae
Facies F	Chaotic deposits—slumps, etc.
Facies G	Hemipelagite. Subdivided into G1 continuous hemipelagite G2 hemipelagite with interbedded thin to very thin mudstone laminae G3 very thin discrete laminae with equally thin interbedded mudstones

imbricated, sedimentologically monotonous sequences of classical turbidites and massive sandstones and the mid- to late-Wenlock Raeberry Castle Formation comprising intact sequences containing diverse turbidite facies associations (Kemp, 1986; Kemp and White, 1985) (Fig. 2). The abundance of graptolitic hemipelagic bands and the resultant high quality of biostratigraphic control facilitates detailed correlation both across and along strike. Sedimentological analysis of the Raeberry Castle Formation in the Kirkcudbright Bay area is further aided by the local low metamorphic grade and lack of intense deformation (Kemp *et al.*, 1985; Kemp, in press). To accommodate the presence of abundant hemipelagic sediment facies, these are described with reference to a modified version of the Walker and Mutti (1973) facies scheme (Table 1).

SEDIMENTOLOGY OF THE EARLY WENLOCK ROSS FORMATION

The Ross Formation can be traced some 170 km between the southern tip of the Whithorn peninsula

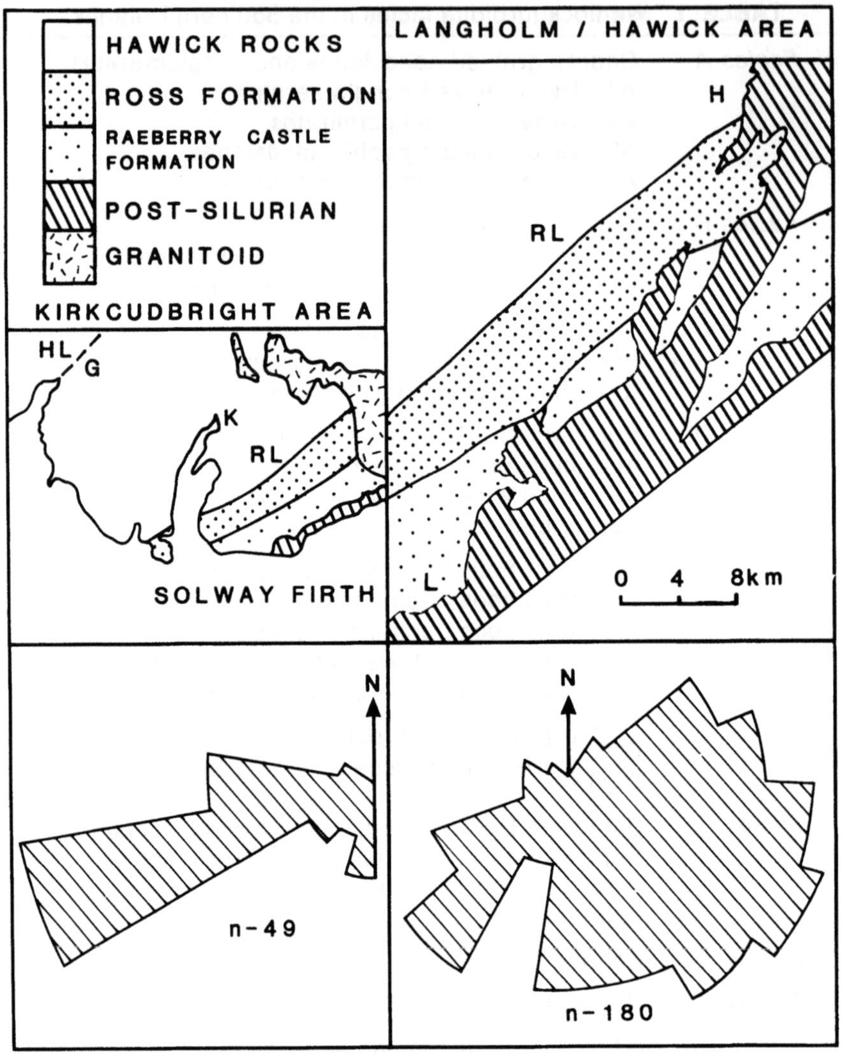

Fig. 3. Summary maps of the study areas showing differing cross-strike widths of the Ross Formation and directional palaeocurrent data. Abbreviations: H—Hawick; L—Langholm; K—Kirkcudbright; G—Gatehouse; HL—Hawick Line; RL—Riccarton Line. Palaeocurrents from the Langholm/Hawick area are from Warren (1963) and include data from the Hawick Rocks.

and the Cheviot inliers (Fig. 1). The study areas, where the Formation is best exposed, are separated by 100 km along strike (Figs 1, 3). The most striking feature of the Ross Formation is the relative lack of variation in lithology compared with younger Wenlock turbidites. It is dominated by thin- to medium-bedded Facies C and D turbidites (*sensu* Walker and Mutti, 1973) in the Kirkcudbright area and Facies B2, C and D turbidites in the Langholm area. Although Facies C and D alternate, the development of significant (>5 m) thicknesses of Facies D turbidites is rare. The hemipelagite most commonly occurs in isolated units (0.05–1.5 m thick) but also occurs as very thin beds or laminae below turbidite slits in Facies D sections. Vertical facies sequence analysis in the Ross Formation in both the Kirkcudbright and Langholm areas is severely hampered by intense faulting (Kemp, 1986; Kemp, in press). Consequently, one can only be confident of the continuity of the sequences of the order of 30 m.

Langholm area

The Langholm area contains the most sand-rich turbidites observed in the Southern Belt with massive sandstone (Facies B2) units frequently dominating the section. The rocks of *riccartonensis* Zone age contain several continuous cross-strike stream sections which were recently exposed by a flash-flood e.g. Fig. 4. Despite detailed bed by bed measurement and attempts to match hemipelagic bands, correlation between stream sections lying a few hundred metres apart along strike proved unsuccessful due to tectonic complexity. The detailed measurement was, however, useful in determining the extent of tectonic repetition in sequences (Kemp, 1985).

The most striking feature of these sequences is the apparent absence of either symmetrical or asymmetrical cycles. Where massive sandstone facies predominate, sandstone units are periodically punctuated by thin bundles (a metre or so thick) of turbidite silts and muds (Facies Dd) (Fig. 4). Individual massive sandstone units are characteristically capped by a normal Bouma sequence including a thick Tc (ripple laminated) division. This is a relatively common feature of massive sandstones in sand-rich fans and is typical, for example, of massive sandstones in the Jurassic Magnus Sandstone of the North Sea (S. Rainey, verbal communication, 1984). The association of Facies B and D with only restricted development of Facies C is more characteristic of slope-apron than of fan environments (Stow et al., 1985).

Limited palaeocurrent data from these outcrops indicate dominant derivation from the north-west. Warren (1964), working in an adjacent area south of Hawick, recorded dominant north-west derivation during deposition of the Hawick Rocks and the Ross Formation (Fig. 3).

Kirkcudbright area

The Ross Formation in the Kirkcudbright area (Fig. 6) is characterized by Facies C and D (classical proximal and classical distal) turbidites. The main differences in turbidite facies from those of the Langholm area are the predominance of thin- to medium-bedded turbidites, the almost total absence of massive sandstones, and the presence of thicker and more frequent hemipelagic interbeds. The Kirkcudbright sequences are also significantly more mud-rich than those of the Langholm area. Abundant palaeocurrent data indicate 'axial' transport from north-east to south-west (Fig. 3).

Sedimentation rates

Determination of reliable sedimentation rates is made difficult by the tectonically fragmented nature of the sequence and problems of identifying major repetitions. The approximate duration of Llandovery and Wenlock graptolite zones is of the order of 0.5–2.0 Ma (R. B. Rickards, personal communication, 1984). The time-scale of Harland *et al.* (1982) gives 7 Ma as the approximate duration of the Wenlock. (Since other recent time-scales give 5 Ma for the duration of the Wenlock—McKerrow *et al.*, 1985—the estimates below may be considered minimum rates.) Dividing the approximate 7 Ma duration by 9 (the number of zones present) gives 0.78 Ma as the mean zone time-span. The present structural thickness of the Ross Formation is about 2 km in the Kirkcudbright area and 6–8 km in the Langholm area. Taking the present structural thickness as an original sedimentary thickness gives rather high sedimentation rates for the facies present (up to 5 km/Ma). Sedimentation rates estimated in intact sequences of facies similar to the Ross Formation suggest values of 500–1000 m (max)/Ma (see below). This implied tectonic thickening of the Ross Formation supports evidence for small-scale structural thickening of the sequence (Kemp, 1985).

A depositional model for the Ross Formation

Reconstruction of Upper Llandovery and early Wenlock depositional environments is made difficult by the highly imbricated nature of these sequences (Kemp, in press). The original thickness of individual accreted packets is below palaeontological resolution and may have been as small as 100–200 m (Kemp, 1986). There is however, no evidence for a greater degree of stratigraphic repetition in the Langholm area to account for the contrast in outcrop width. Therefore, the structural thickness variation of the Ross Formation (Figs 1, 3), with the thicker section (5–8 km) in the Langholm area and the thinner section (*c.* 2 km) in the Kirkcudbright area, probably reflects an original sedimentary thickness variation (Fig. 7). This thickness variation is consistent with the contrasting facies developed in the two areas. The Ross Formation is particularly difficult to interpret because of the lack of long continuous sequences for facies analysis, but any interpretation must take the following features into account:

1. In the Langholm–Hawick area: sequences with abundant massive and thick-bedded sandstones derived from the north-west.
2. In the Kirkcudbright area: sequences dominated by thin- to medium-bedded turbidites derived from the north-east.
3. The presence of relatively thin, isolated units of hemipelagite.

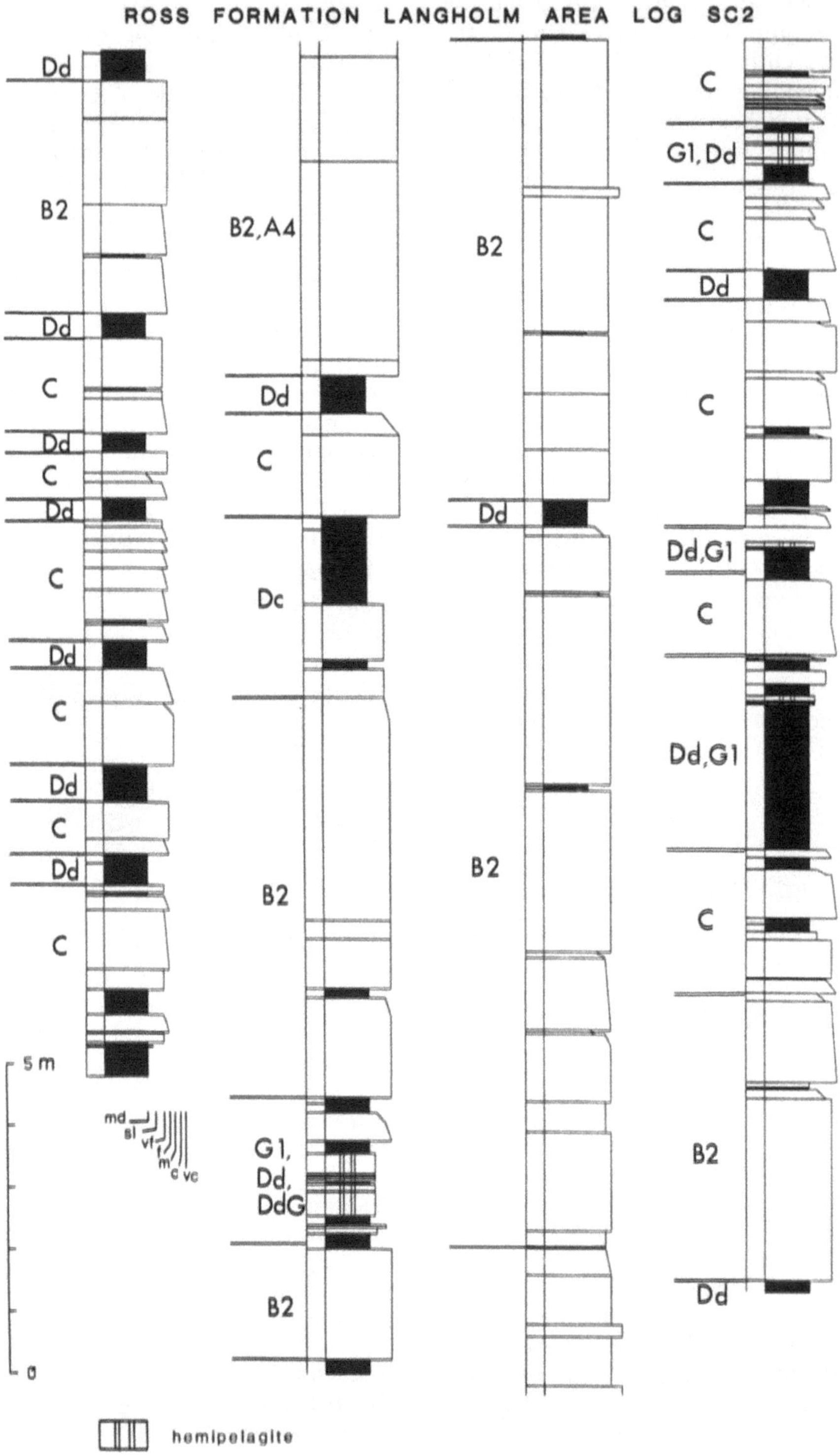

Fig. 4. Sedimentary log of Ross Formation (Log SC2) in *riccartonensis* Zone age rocks of the Langholm area showing facies distribution with dominance of massive sandstones (facies abbreviations given in Table 1).

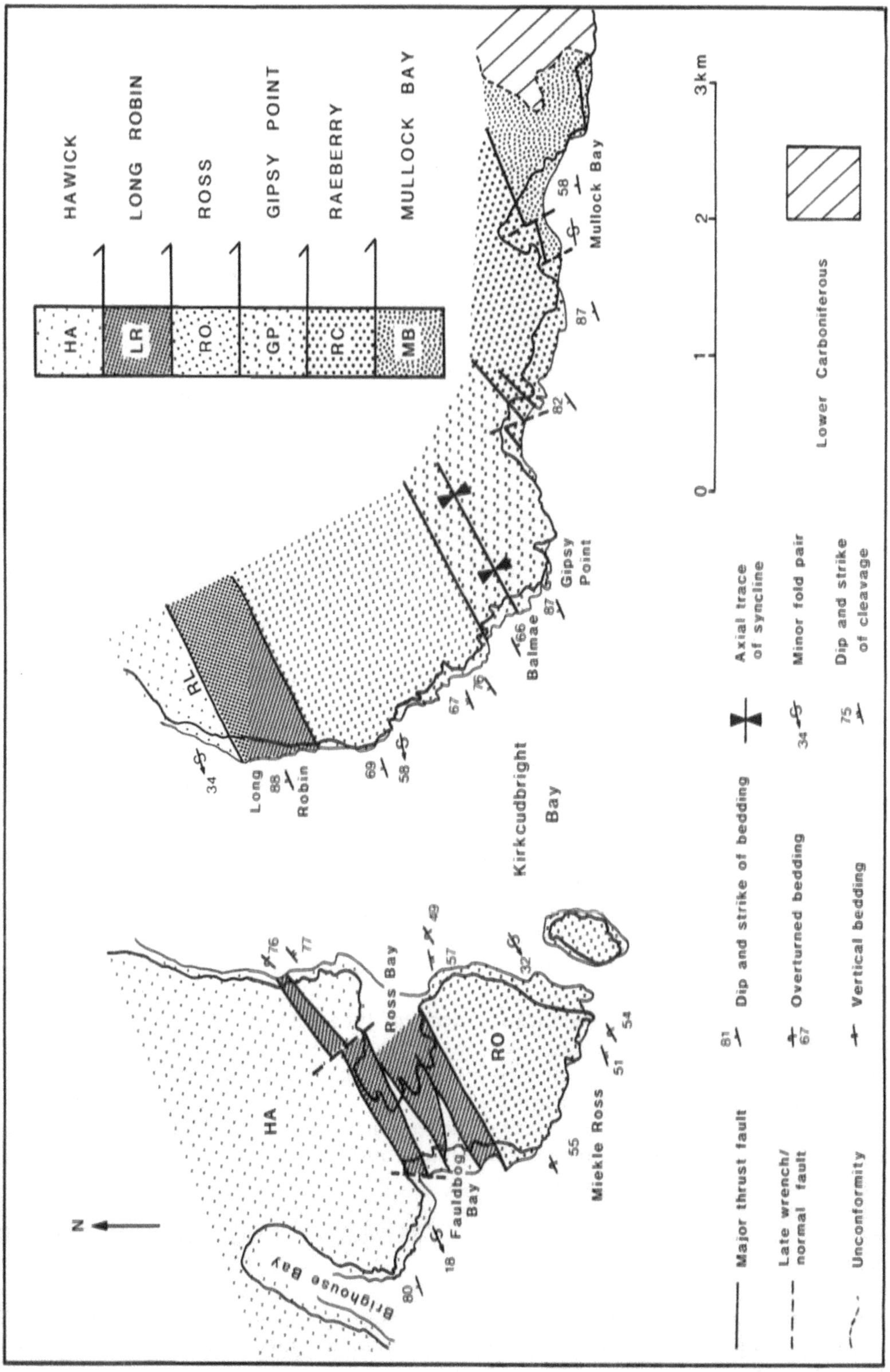

Fig. 5. Detailed geological map of the Kirkcudbright area showing distribution of tectonostratigraphic units.

Fig. 6. Typical Ross Formation of Kirkcudbright area showing dominance of thin- to medium-bedded classical turbidites.

4. No development of thin-bedded sequences, Facies D or hemipelagite thicker than about 5 m.

The apparent absence of significant thicknesses of hemipelagite in both study areas suggests either one or a combination of high turbidity current frequency or high sedimentation rates. The rarity of thick (>5–10 m) sequences of thinly bedded, mud-rich turbidite facies indicates the lack of significant sediment by-passing (which would have developed significant thicknesses of thin-bedded overbank facies—see Gipsy Point Unit, below). There is therefore no evidence for the existence of an 'inner fan' channel/overbank complex, which is an essential characteristic of most fan models (e.g. Mutti and Normark, this volume; Shanmugam and Moiola, 1985). The absence of an inner fan-type sequence due to selective subduction seems unlikely in the light of the extensive imbrication of the Ross Formation. This absence of inner fan-type sequences, together with the abundance of massive and thick-bedded sandstones, suggests rapid dumping of sediment and is consistent with the development of a sand-rich 'low efficiency' fan. However, the dominance of massive and thick-bedded sandstones and thinly bedded silty turbidites in the Langholm area may be more consistent with a slope-apron than a fan model (cf. Stow *et al.*, 1985). This may have been related to high uplift and erosion rates in the source area and/or a relatively narrow width of the depositional basin and associated high depositional gradients.

The axial palaeocurrent directions in the Kirkcudbright area suggest deviation of the north-west-derived turbidites in the Langholm area to flow axially to the south-west along the basin for at least 100 km (Fig. 7). This is also consistent with a narrow width for the basin. The most likely cause of the deflection would be the topographic confines of the trench outer slope.

SEDIMENTOLOGY OF THE RAEBERRY CASTLE FORMATION

The Raeberry Castle Formation comprises three discrete tectonic units, which, in contrast to the Ross Formation, possess varied and distinctive facies associations and are only lightly deformed. The aim

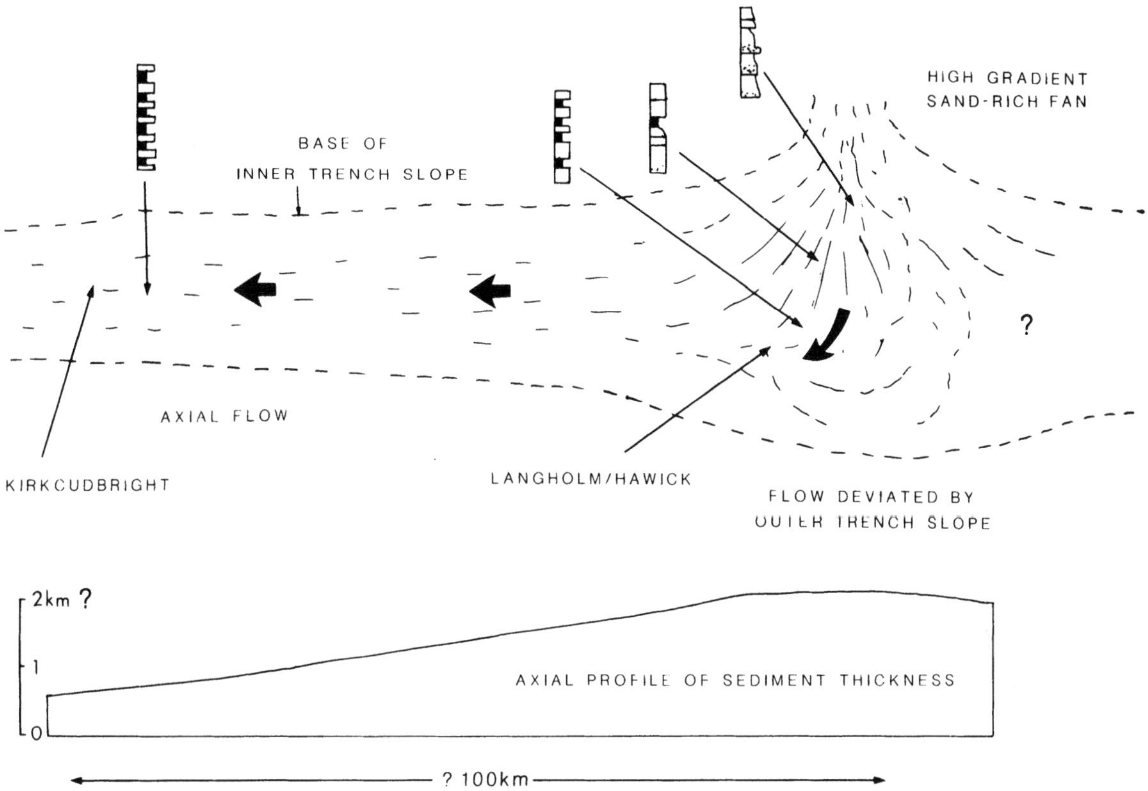

Fig. 7. Depositional model for Ross Formation.

of the following account is primarily to use the partly coeval sequences of the Gipsy Point and Raeberry and Mullock Bay Units to reconstruct mid–late Wenlock depositional systems. The sedimentology of the Gipsy Point Unit is discussed in detail elsewhere (Kemp, in preparation) and a synopsis is presented below.

Gipsy Point Unit—channel overbank sequence

The Gipsy Point Unit comprises a basal 'A' Member (Fig. 8) of similar facies to the Ross Formation showing derivation mainly from the north-east. However, the A Member also contains a thick unit of hemipelagite and displays some variability in palaeocurrent towards the top which heralds a dramatic change in the character of sedimentation.

The B Member of the Gipsy Point Unit (Fig. 8) comprises a sequence of channelized arenites, rudites and slumps alternating with overbank or levee deposits of thin-bedded very fine sandstones, siltstones and mudstones. A basal, thin-bedded sequence of ripple-laminated siltstones, silty mudstones and mudstones overlies the classical turbidites of the A member. This is succeeded by north-west-derived channelized thick-bedded arenites and rudites. Some beds at the top of the overbank sequence pass laterally into the thick-bedded arenites of the channel, defining its depositional margin (Figs 9, 10). The maximum observed down-cut of the arenite beds forming the channel margin is some 1.6 m. Marked lateral migration is manifested in a (minimum) 210 m eastward shift in the channel margin achieved in 15 m of vertical section and associated with a 15 degree anticlockwise shift in palaeocurrent direction (Fig. 10). The lower channel, which is a minimum of 325 m wide, is infilled by a slump unit (Fig. 8) showing a transport direction similar to that of the channel sediment. This slump unit, which is topped by a series of sand volcanoes, is overlain by a thin overbank sequence which signifies abandonment of the lower channel as a locus of coarse sediment transport and deposition. The arenites and rudites of the upper channel, which succeeds the overbank sequence, are derived from the north-east—a markedly different orientation to that of the lower

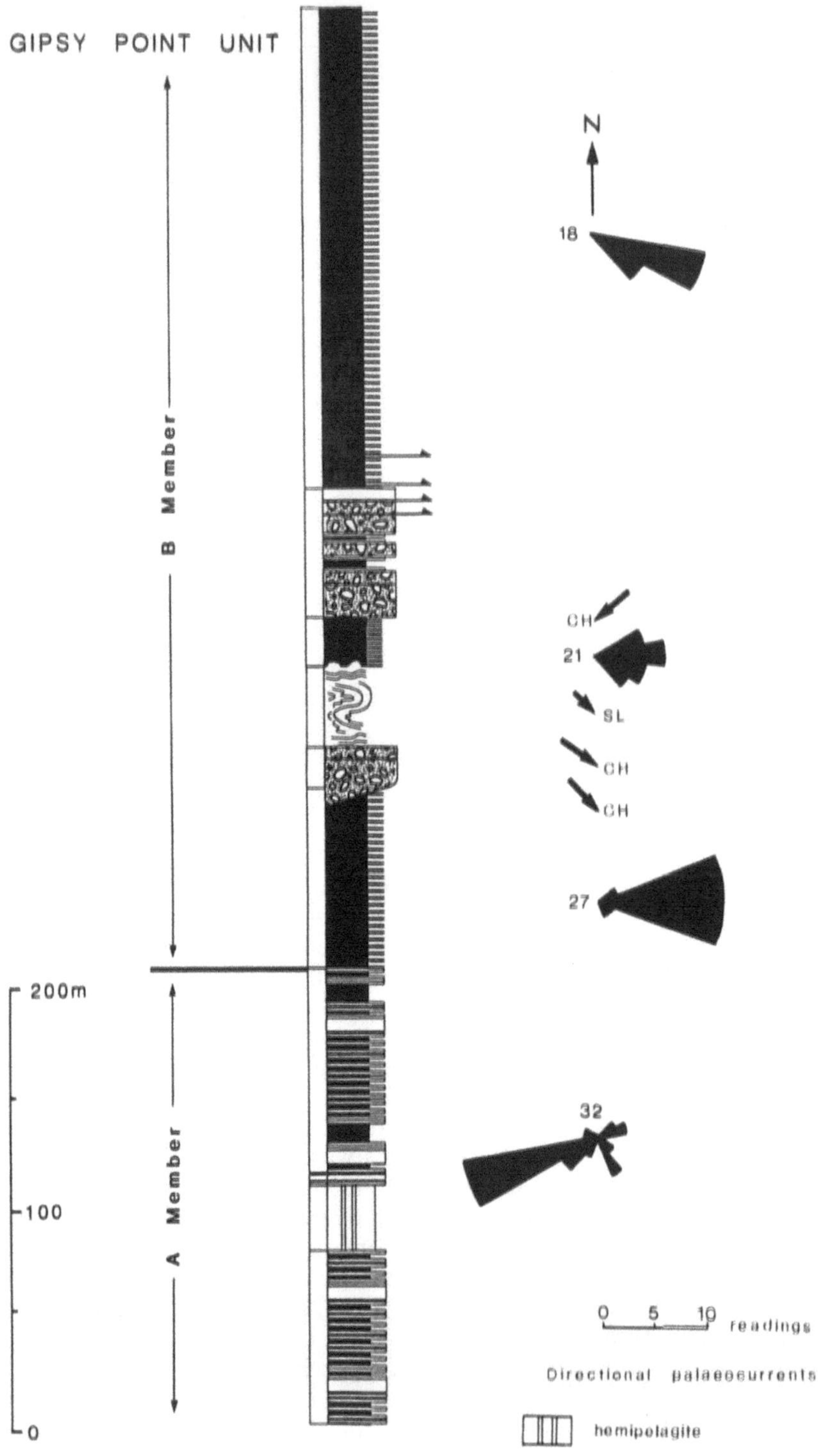

Fig. 8. Schematic sedimentary log of Gipsy Point Unit showing main facies units and palaeocurrent data.

Fig. 9. Low-level air photograph of base of Gipsy Point lower channel showing lateral migration up-section from left to right.

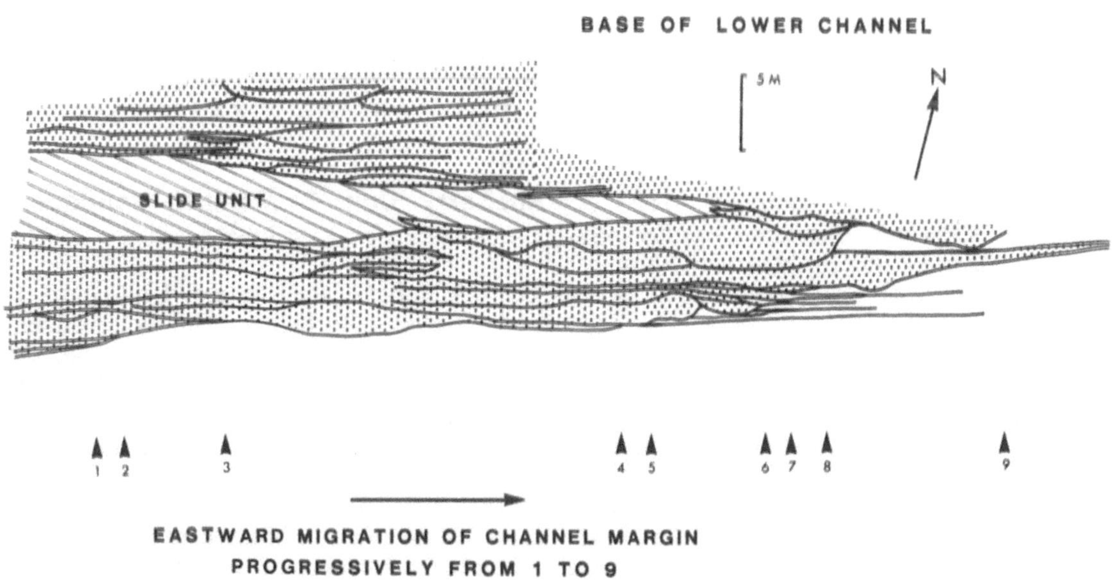

Fig. 10. Diagram of Fig. 9 showing lateral migration of channel through events 1–9.

channel. Several slide sheets of thinly bedded facies overlie the arenites and rudites of the upper channel and these are topped by a further *in situ* overbank sequence (Fig. 8).

The combined evidence of channel migration and markedly differing channel trends suggests the development of meandering channels. In the light of this evidence the Gipsy Point sequence is best interpreted as part of a meandering channel–levee complex (Kemp, 1985, and in preparation).

Channels over 1 km wide also occur in the Langholm area suggesting that the margin-parallel width of the mid-Silurian channel–overbank system was at least 100 km. Changes in the character of sedimentation in these channel–overbank sequences coincide with the emplacement of major slumps/slide sheets (Fig. 8). Estimates of sedimentation rates for the Gipsy Point Unit based on turbidite deposition for two graptolite zones give a value of 420 m/Ma (590 m/Ma for uncompacted sedimentation rate following Whitman and Davies, 1979).

The detailed biostratigraphic data obtained from the Southern Belt (Kemp, 1986) facilitate correlation of the Gipsy Point Unit with the structurally lower Raeberry Unit. Both these units pass through the *rigidus/flexilis* Zone boundary and therefore possess coeval sequences (Fig. 11).

SEDIMENTOLOGY OF THE RAEBERRY UNIT

The Raeberry Unit is divided into three members on the bases of facies type and palaeocurrent variation (Fig. 12). The A Member comprises some 256 m of Facies C and D turbidites. The B Member comprises 44 m of base-absent (Facies D) turbidites (generally commencing with the Tc division) and interbedded hemipelagite. The C Member comprises 418 m of medium-bedded base-absent turbidites commencing with Tb or Tc divisions interbedded with thin- to very thin-bedded Tc,d,e or Td,e siltstones and mudstones.

The Raeberry Unit ranges in age from the *rigidus* Zone to the early *flexilis* Zone (Kemp, 1986; see Fig. 2) spanning approximately one zone and comprising 720 m of sediment. This gives a sedimentation rate of 925 m/Ma (or 1300 m/Ma uncompacted). Since the C Member is entirely in the *flexilis* Zone (Fig. 11) its sedimentation rate is probably in excess of the above figure. Estimates for frequency of turbidity current frequency give minimum values of one flow per hundred years for the C Member.

Raeberry Unit A Member—depositional lobe facies

The basal A Member is similar in facies to the Ross Formation of the Kirkcudbright area (Facies C and D turbidites) (Figs 13, 14). There are only three substantial units of hemipelagite: two relatively thin units near the base and one ten-metre unit in the middle of the member. Other than these, only a few very thin laminae of hemipelagite occur below siltstones in Facies D sequences. Palaeocurrents are mainly from the west or south-west but there is locally significant variation within the sequence (Fig. 15).

Most of the medium- to thick-bedded sandstones comprise Ta,e divisions although occasional Ta,b,c or Ta,c beds occur. The sandstone beds are parallel sided and no significant erosion is indicated (Fig. 14). Only ripple-laminated beds show lateral thickness variation. The overall sandstone/shale ratio for the top 50 m is 4.5. The top bed of this member shows an abrupt change in palaeocurrent direction (Fig. 13), being derived from the north-west—the same direction as for erosive currents in the overlying B Member.

The Facies C and D turbidites and significant internal palaeocurrent variation are consistent with location on a depositional lobe where local thickness variations might cause local deviation in the path of successive currents (Hiscott, 1981; Mutti and Sonnino, 1981) or where more than one channel was responsible for deposition of the lobe. Certain sections of the A Member, which show markedly contrasting palaeocurrents (Fig. 15), may have formed in an area influenced by two distinct depositional lobes. Strongly contrasting palaeocurrents such as these may be a characteristic of areas of lobe interaction (Fig. 22).

Raeberry Unit B Member—fan/lobe fringe facies

The B Member of the Raeberry Unit comprises 44 m of base-absent turbidites with interbedded hemipelagite (Fig. 16). These regular, thin-bedded, base-absent (mainly Tc,d,e) turbidites (Facies DbG, DcG and DdG), represent an abrupt change in sedimentation from the medium-bedded Ta,e sandstones of the A Member.

The B Member has an overall sandstone/shale ratio of 1.24 and a turbidite/hemipelagite ratio of 5.7. Erosional palaeocurrents indicate north-west derivation (Fig. 17). However, many depositional current indicators (ripples) show derivation from west or

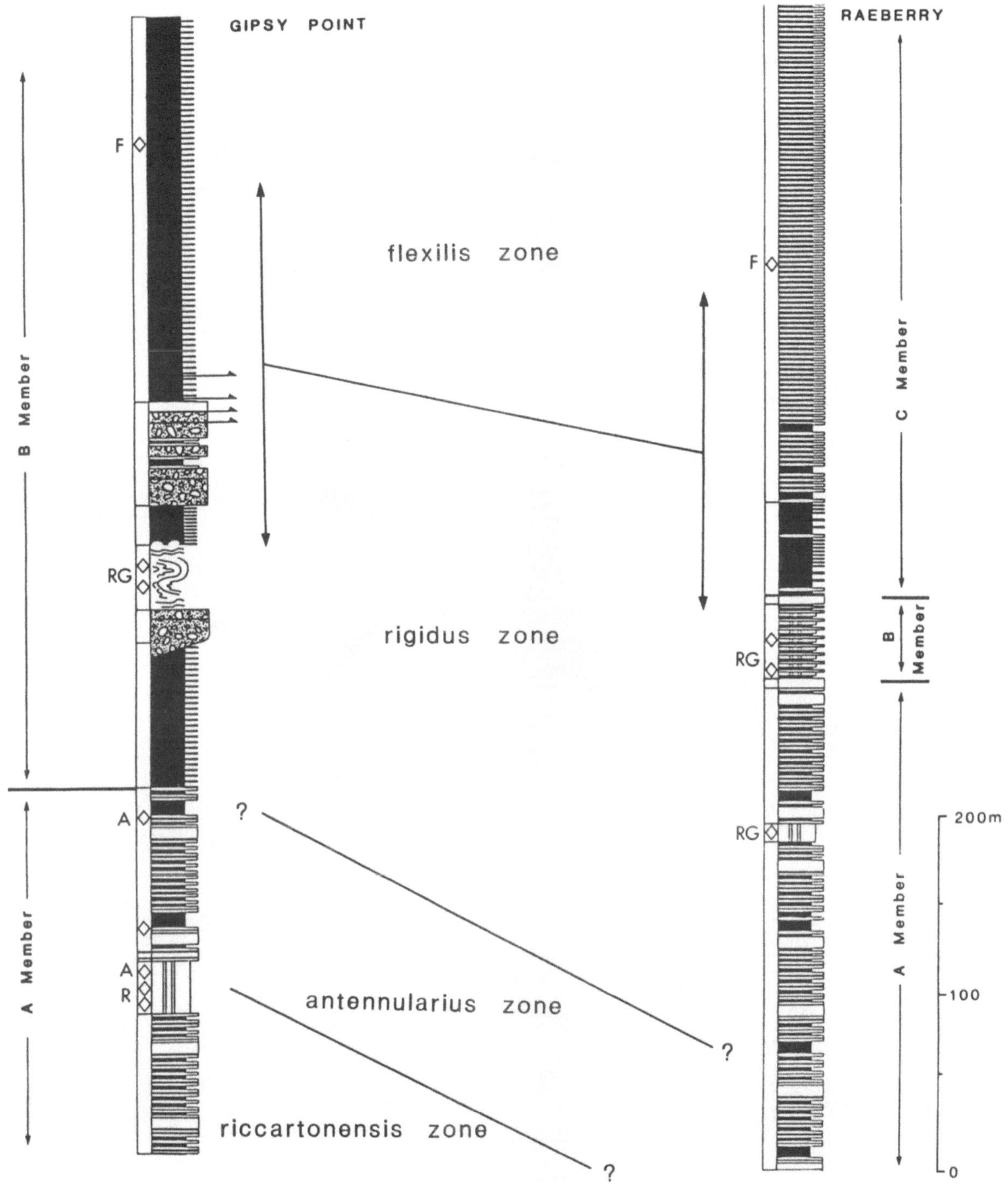

Fig. 11. Summary sedimentary/stratigraphic diagram showing correlation between Gipsy Point and Raeberry Units.

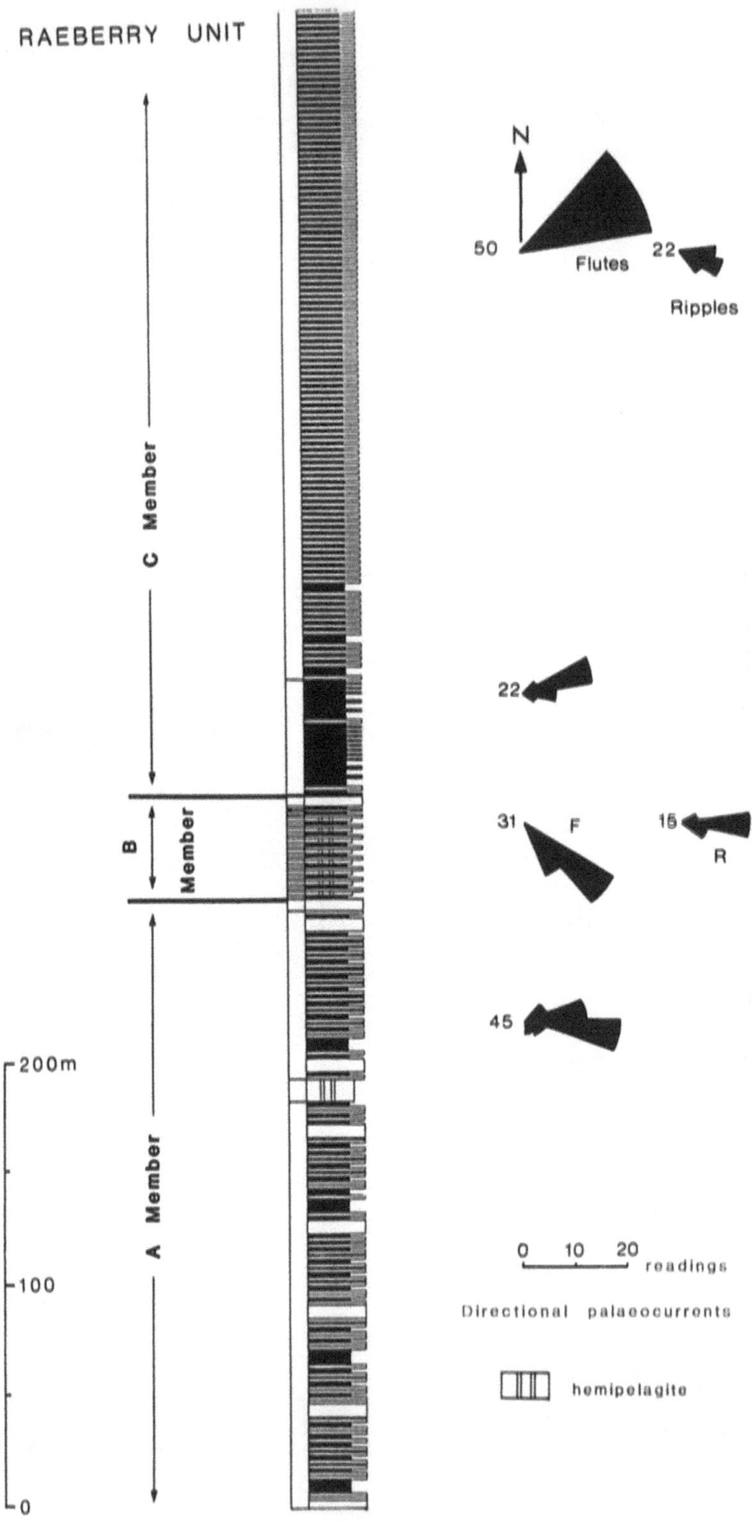

Fig. 12. Schematic sedimentary log of Raeberry Unit showing stratigraphic facies distribution and palaeocurrent data.

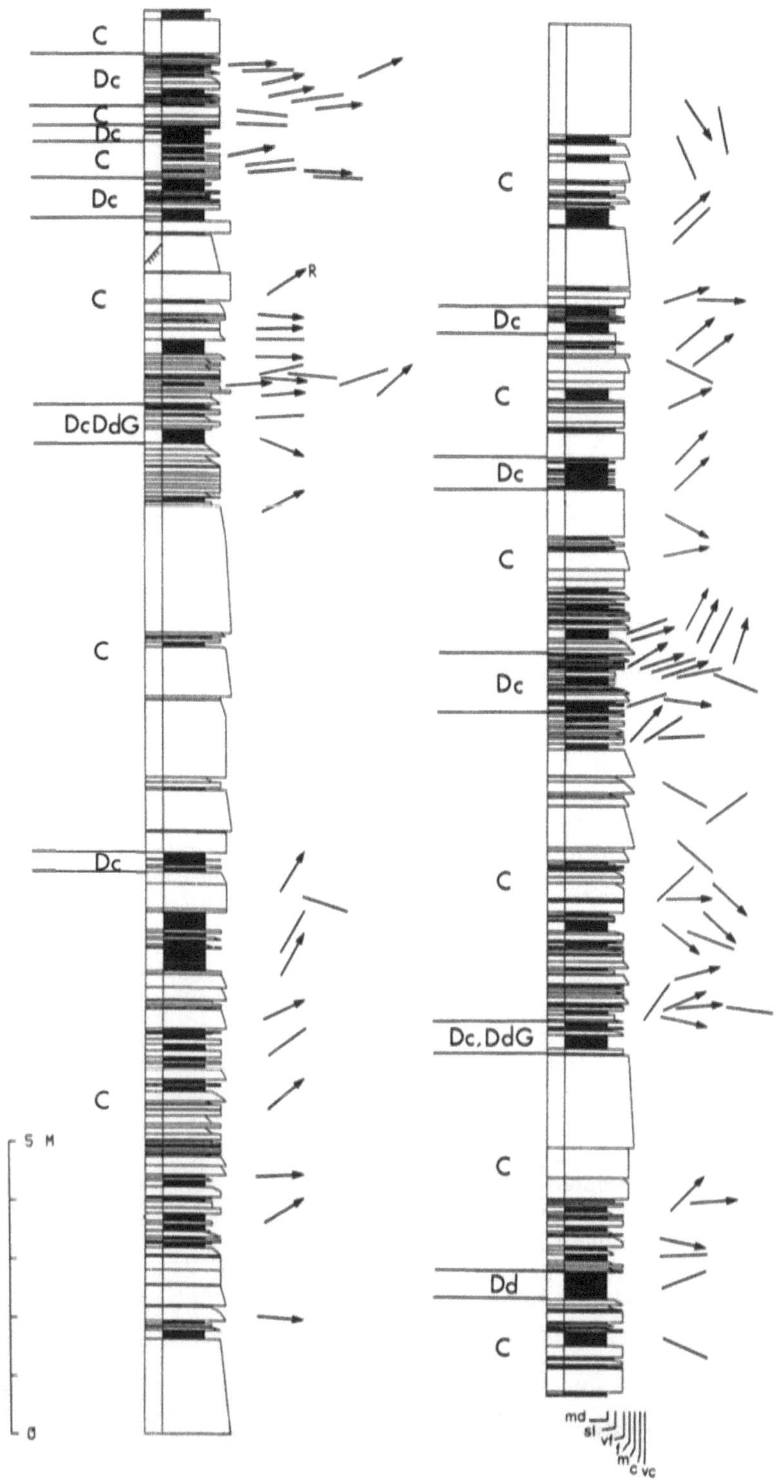

Fig. 13. Sedimentary log of top of Raeberry Unit A Member with bed-by-bed palaeocurrent analyses comprising mainly Facies C turbidites with subordinate intervals of Facies D. Arrows are flutes or directional L-ridges. Bars are grooves. Right-hand end of arrow or bar is opposite base of source bed for palaeocurrent determination.

Fig. 14. Parallel-sided Facies C and D turbidites from log of Fig. 13.

west-south-west (Fig. 12). The top of the member is marked by an unusual two-metre bed of structureless siltstone and very fine sandstone (Fig. 16).

The association of thin-bedded turbidite facies and common interbedded hemipelagic sediment suggests a marginal or distal position with respect to the main depositional area—perhaps a fan or lobe fringe setting. The abrupt change in palaeocurrent direction above the B Member (Fig. 12) suggests the occurrence of some large-scale event which profoundly affected the basin. The structureless siltstone which marks the top of the B Member has many affinities with the 'unifites' or 'homogenites' of the Eastern Mediterranean (Cita *et al.*, 1984; Hieke, 1984; Stanley, 1981) which are correlated with major events. In the light of the biostratigraphic data (Fig. 11) a correlation of this 'homogenite' bed at the top of the C Member with the Gipsy Point slump unit seems probable and the homogenite could therefore represent a more distal expression of the same event which produced the slump.

Raeberry Unit C Member—channel mouth facies

The C Member of the Raeberry Unit comprises a thick (418 m) sequence dominated by base-absent (Tb,c) turbidites (Figs 18, 20). There are several intriguing features about this sequence. The grain

Fig. 15. View of sole structures near top of Raeberry Unit A Member showing bed-by-bed palaeocurrent variation.

Fig. 16. General view of Raeberry Unit B Member. Thick bed at top of A Member marked A, and 'homogenite' bed at top of B Member marked H. Distance A–H is 44 m.

Fig. 17. Erosional palaeocurrents (flutes and grooves) from Raeberry Unit B Member showing transport to south-east.

size of the decimetre bedded Tb,c units is commonly medium and coarse sand grade, being markedly coarser than the fine to very fine sand grade material which dominates the Wenlock turbidite sequences of the rest of the Southern Belt. The Tc divisions, especially those occurring towards the top of the sequence, generally comprise climbing ripples, and in many cases these units are quite thick and steep angles of climb indicate rapid deposition (Fig. 20). In detail, the decimetre bedded base-absent units are interbedded with very thin-bedded very fine sands and silts, alternating with mudstones and very thin laminae of hemipelagite. This intercalation of medium-bedded sandstones with very thin-bedded siltstones and mudstones (Fig. 19) represents an intriguing, bimodal bed thickness pattern.

A puzzling feature of the medium-bedded sandstones is the almost total lack of Ta divisions. Structureless sand is generally only present below the Tb divisions in small infilled scours (Fig. 19).

An abrupt change in palaeocurrent direction occurs above the two-metre structureless, poorly sorted, silty mudstone/very fine sandstone bed. Ripples and flutes from the more thinly bedded lowermost 50 m of the C Member show current flow to the north-east. Locally, intrabed palaeocurrent variation (previously documented by Scott, 1967) is developed with erosional currents (flutes) indicating flow to the north-east, and depositional currents (ripples) commonly indicating currents flowing to the south-east at up to 90 degrees to the erosive currents (Fig. 12). Several minor slumped horizons near the top of the sequence and sparse subhorizontal fold hinges suggest a south-east-facing palaeoslope.

Towards the top of the sequence sandstone units display increasing lateral thickness variation, and locally complex erosion/amalgamation patterns occur. However, very thin laminae of hemipelagite can be traced for over 200 m laterally along the wave-cut platforms. Several dune structures also occur towards the top (Fig. 21) and increase in frequency, grain size and thickness up-section. These dune structures further emphasize the dominance of traction–deposition in sedimentation of the C Member. Dune structures are relatively common in coarse, conglomeratic, channelized and/or 'proximal' environments (Colella, 1979; Winn and Dott, 1977) but are very rare in classical turbidite sequences. The occurrence of dunes in the C Member may in part be explained by the coarser grain size, since the presence

Fig. 18. Alternation of medium- and thin-bedded base-absent sandstones in Raeberry Unit C Member (younging from right to left).

of at least medium sand grade seems necessary for dune formation (Allen, 1982). The formation of dune structures would also be promoted by rapid loss of flow power by the transporting current, allowing the current to pass through the field of dune formation (cf. Allen, 1970).

The combination of abundant traction deposition, unusually coarse grain size and high sedimentation rates suggests an abrupt loss of flow power by successive transporting currents and suggests that successive flows responsible for the deposition of this sequence underwent hydraulic jumps. Such effects can be caused by flows encountering a change in slope, fault scarp or emergence from the confines of a channel (Komar, 1971). In this case they can be related to the coeval channelized arenites and rudites of the Gipsy Point Unit (Fig. 11). Palinspastic restoration would leave the higher thrust sheet in the stack (Gipsy Point Unit) in a more proximal position than the Raeberry Unit (the lower thrust sheet: Kemp, in press) which would be consistent with the contrasting sedimentology of these two units. The south-west-derived palaeocurrents of the Raeberry Unit C Member appear to be inconsistent with the north-east or north-west-derived palaeocurrents of the Gipsy Point Unit channels. However, as these channels were probably meandering and certainly underwent lateral migration, apparent inconsistencies in orientation would be expected.

In this context, the C Member would represent a 'channel mouth' facies. In detail, the sedimentology of the C Member is comparable to other sequences interpreted as 'channel mouth' facies and displays many of the characteristics cited as typical of this facies by Mutti (1977): irregular bedding patterns with lensing and lateral discontinuity; presence of small-scale and large-scale cross-bedding; sandstone beds bounded by an upper rippled surface; beds split both up-current and down-current into a number of thinner beds separated by mudstone partings; general characteristics indicating deposition mainly by tractional processes. Emergence of the flow from the channel would result in expansion and readjustment of the flow concomitant with a loss of flow power and 'dumping' of part of its load. The almost complete absence of Ta divisions, except in localized scours, may be explained by 'dumping' of sediment from the head of a turbidity current followed by tractional reworking by the rest of the current (cf. Mutti, 1977).

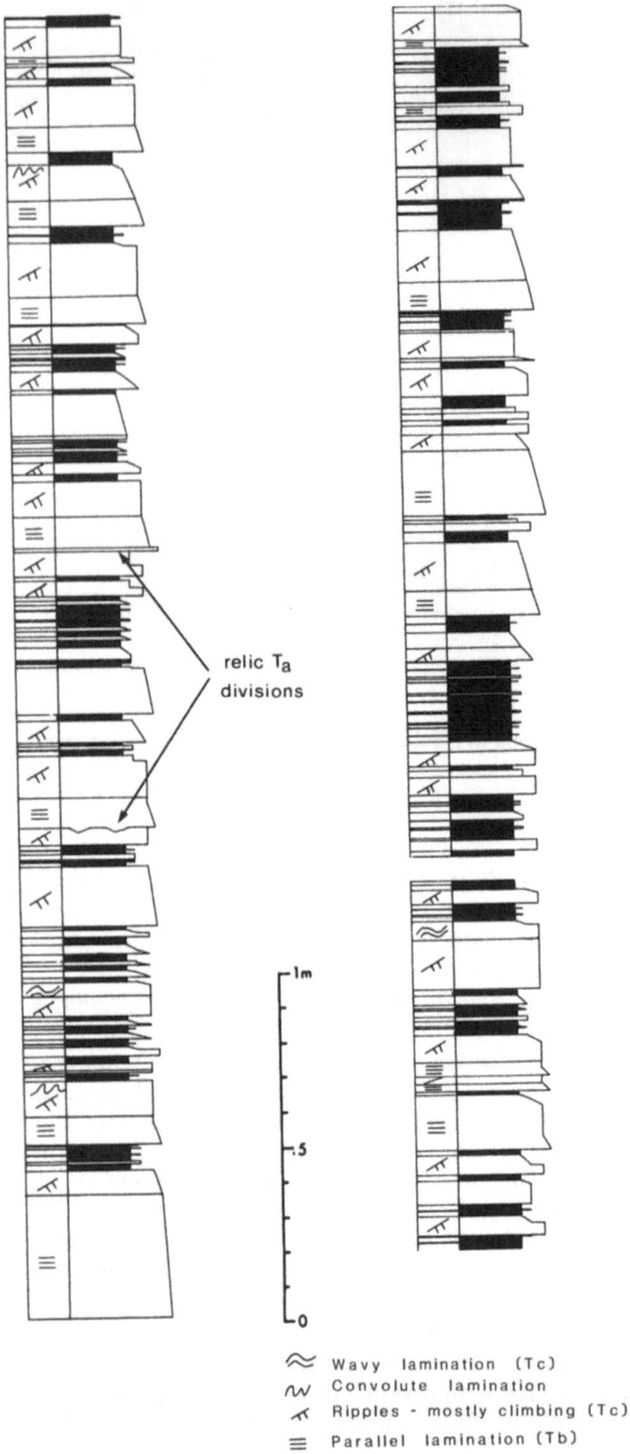

Fig. 19. Sedimentary log in middle part of Raeberry Unit C Member showing internal structures of medium-bedded turbidites and relative scarcity of Ta divisions.

The overall thickening and coarsening upwards progression through the C Member (Fig. 12) culminating in the abundant and coarse-grained dune structures at the observed top would be consistent with a progradation of a channel mouth towards the area of deposition of the Raeberry Unit.

SEDIMENTOLOGY OF THE MULLOCK BAY UNIT

The Mullock Bay Unit (574 m) is subdivided into an A Member (228 m) and a B Member (346 m) (Fig. 23). It is characterized by diverse facies associations, and more varied palaeocurrent trends, significantly lower sandstone/shale ratios, and more abundant hemipelagite than earlier Wenlock sediments. The unit has not yielded any zone fossils but assemblages recorded are characteristic of the *ellesae* and *lundgreni* zones although there may be limited overlap with the Raeberry Unit (Kemp, 1986).

Mullock Bay Unit A Member—channel and overbank sequence

The lower 139 m of the A Member comprises thin-bedded very fine sandstones and siltstones interbedded with mudstones and minor hemipelagite. Limited palaeocurrents indicate derivation from the north-east. This sequence is abruptly succeeded by 15 m of channelized rudite and arenite which contain four fluted horizons indicating derivation from the south-west. These rudites and arenites are overlain by 69 m of thin-bedded very fine sandstones, siltstones and mudstones. At the top of the sequence beds thin rapidly up-section over about 5 m and increasing amounts of hemipelagic sediment occur.

The facies associations of the A Member—thinly bedded very fine sandstones, siltstones and mudstones alternating with channelized arenites and rudites—are very similar to those of the B Member of the Gipsy Point Unit. These are consistent with development in a channel–levee system. From the stratigraphical evidence in the Gipsy Point Unit (Fig. 3) the channel–levee systems first developed in *rigidus* Zone times and continued into the *Linnarssoni* Zone. The age evidence available for the Mullock Bay Unit suggests that this sedimentation pattern may have persisted into *ellessae* Zone times.

Mullock Bay Unit B Member—prograding depositional lobes

The B Member comprises alternating intervals of thin- to medium-bedded Facies C and D turbidites

Fig. 20. Details of base-absent sandstones of Raeberry Unit B Member (top) showing high-angle climbing ripples and (bottom) Tb,c unit.

and thin-bedded Facies D and G, labelled MBS1–7 and MBP1–7 (Fig. 23). The first pair of packets (MBP1 and MBS1) take the form of an overall thickening and coarsening upwards sequence. The proportion of turbidite to hemipelagite progressively increases through MBP1 (Fig. 24). Near the base, very thin (mm–cm) interbeds of mud punctuate the hemipelagite which individually attains a maximum of 15 cm in thickness. These mud units become thicker up-section at the expense of the hemipelagite. They become detectably graded and some contain diffuse laminae of silt near their base (Piper's (1978) E1 division). Towards the top of MBP1, thin Tc divisions appear with increasing frequency at the base of turbidite units (Fig. 24). Most of these ripple-laminated divisions show transport from the east (Fig. 23). A two-metre zone at the top contains thin sand units of increasing thickness with some Tb divisions prior to the thicker, Ta,e sandstones which mark the start of MBS1.

The dominance of hemipelagic sediments in the lower part of MBP1 suggests a basin–plain environment and a marked shift in the locus of deposition from the A Member. The increase in turbidite/hemipelagite ratio and the increase in frequency of Tc divisions up-section in MBP1 is broadly compatible with progradation of a depositional system towards the area and transition to a fan or lobe–fringe environment.

The upper part of MBS1 contains prominent minor thickening and coarsening-upwards cycles (Figs 25 to 27). Sandstone/shale ratios and palaeocurrent variation reinforce definition of the asymmetrical nature of these cycles with occasionally abrupt palaeocurrent changes from one cycle to the next (see especially cycles marked A1, A2 and B of

Fig. 21. Dune bedding occurring towards top of Raeberry Unit C Member. Younging from right to left.

Fig. 25). Palaeocurrent directions in MBS1 indicate transport mainly to the north-east, except for one thickening and coarsening-upwards unit which shows opposing transport direction—to the south-west (Fig. 26).

The cycles comprise between thirty and seventy layers, which is significantly more than make up the suggested number in the compensation cycles of Mutti and Sonnino (1981), and although smaller cycles may be superimposed, there is no good evidence for this. The coherency of each individual cycle and the frequently marked palaeocurrent orientation difference between cycles would be consistent with the presence of overlapping fan lobes.

These cycles more probably represent the progradation of depositional lobes in a mid- to outer-fan setting. If this evidence is taken together with the overall thickening and coarsening-upwards 'megasequence' represented by MBP1 and MBS1, then the area would be consistent with several phases of fan progradation over the area, by analogy with the Italian models of the Apennine fans (e.g. Ricci Lucchi, 1975).

The top of MBS1 is marked by a prominent sandstone packet of which the top bed contains a granule grade material and is 2.3 m thick. Thus the overall thickening and coarsening-upward sequence formed by MBP1 and MBS1 has within it thinner subsidiary thickening and coarsening-upward sequences.

The rest of the B Member is characterized by alternations of pelitic and sandy intervals showing similar features to MBP1 and MBS1. The top of each sandy interval is characteristically marked by a thick, laterally extensive sandstone bed, containing up to granule grade material. Distinctive sandstone beds such as these commonly delimit individual sandstone–lobe sequences (Mutti and Normark, this vol.).

The repeated alternation of sandy intervals and pelitic intervals in the B Member is probably the result of frequent pulses of fan or lobe progradation over the area, punctuated by periods of quiescence.

The sequences of the B Member can be fitted into an overall model for the Raeberry Castle Formation. The relatively high sandstone/shale ratios (>4) and lack of hemipelagite in the Raeberry Unit A Member would be consistent with a mid-fan depositional lobe setting, whereas the sequences of the more thinly bedded and hemipelagic-rich Mullock Bay Unit B Member would fit a lower/outer fan setting (Fig. 22).

However, the abundant north-east and south-west-flowing palaeocurrents of the B Member do not fit the pattern of transverse outbuilding of the model of Fig. 22. Since there is no evidence either in the Kirkcudbright area or Langholm area for the presence of channel–overbank systems above the stratigraphic level of the Mullock Bay Unit A Member, an alternative interpretation of the Mullock Bay Unit B Member is possible. There may have been a reduction in sediment input by late Wenlock times and the relatively thin prograding sequences could have formed in response to the development of depositional lobes of relatively small fans. The opposed palaeocurrents of the B Member (Figs 23, 25) may therefore be due to the overlap of lobes from adjacent fans (cf. Hiscott, 1980).

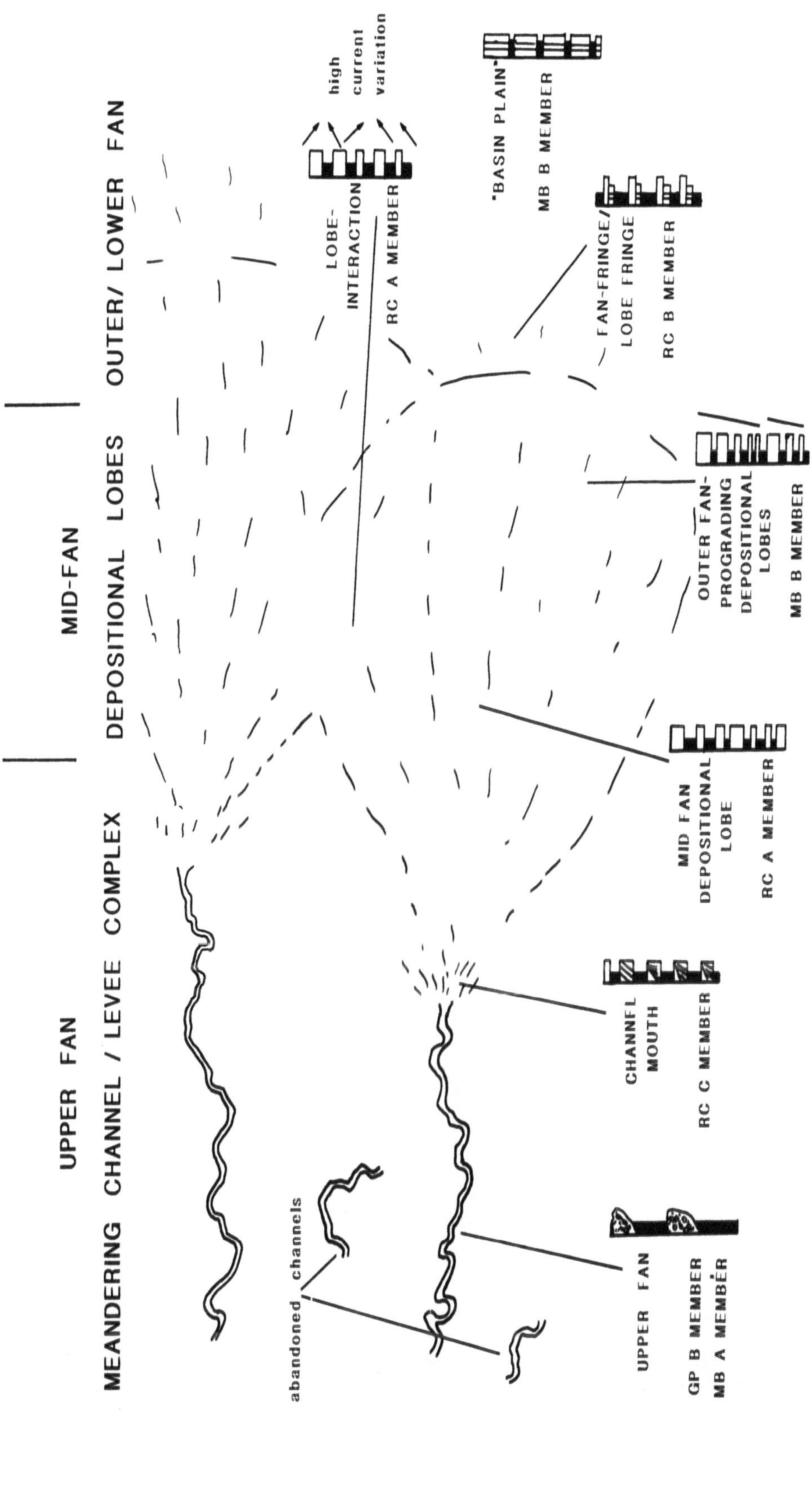

Fig. 22. Depositional model for Raeberry Castle Formation.

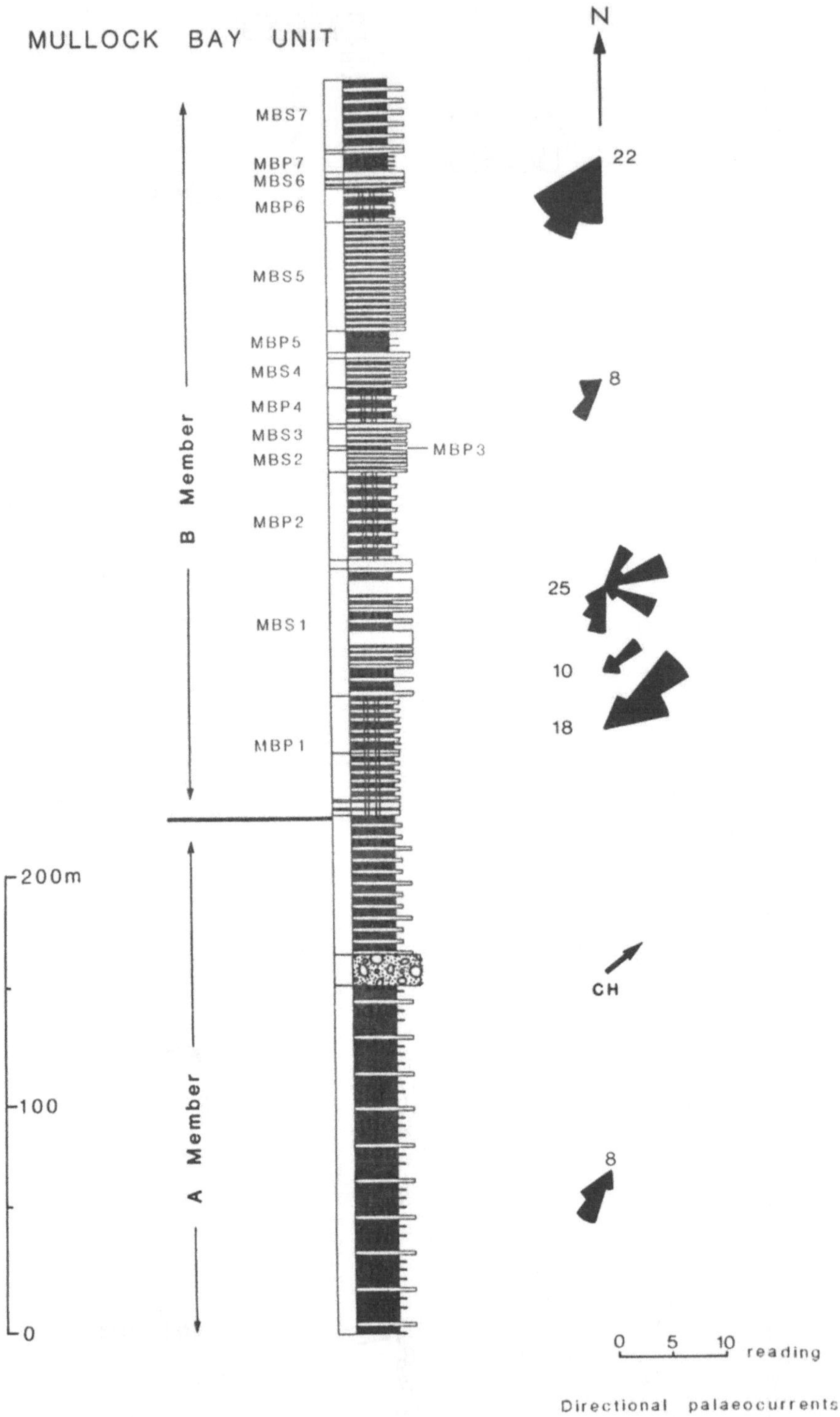

Fig. 23. Schematic sedimentary log of Mullock Bay Unit showing facies stratigraphy and palaeocurrent data. Intervals labelled MBS 1–7 comprise Facies C and D turbidites. Intervals labelled MBP 1–7 comprise mainly Facies G, DdG and DcG.

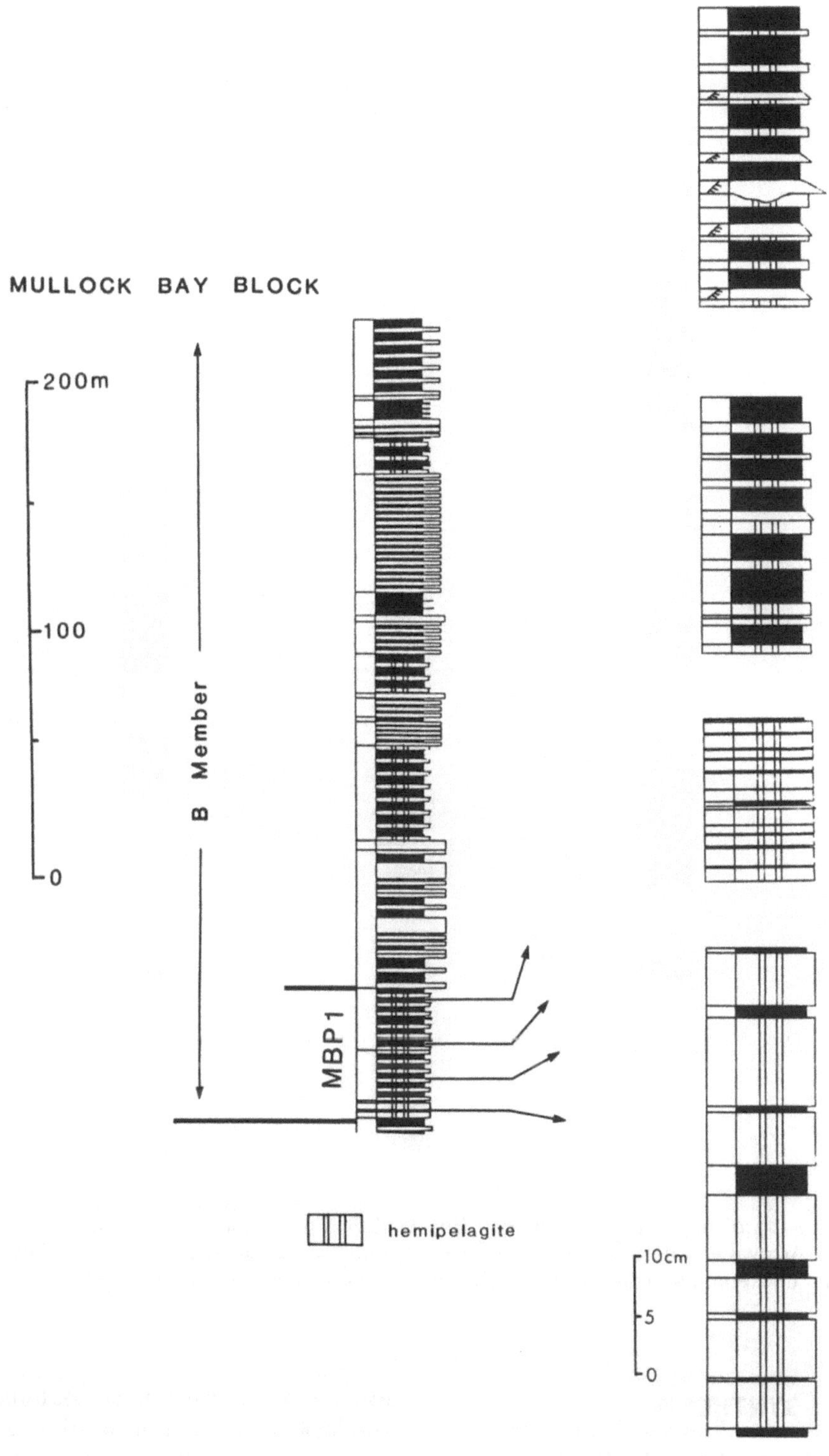

Fig. 24. Sedimentary logs in Mullock Bay Unit B Member pelitic interval 1 (MBP) showing increased influx of low-density turbidity currents.

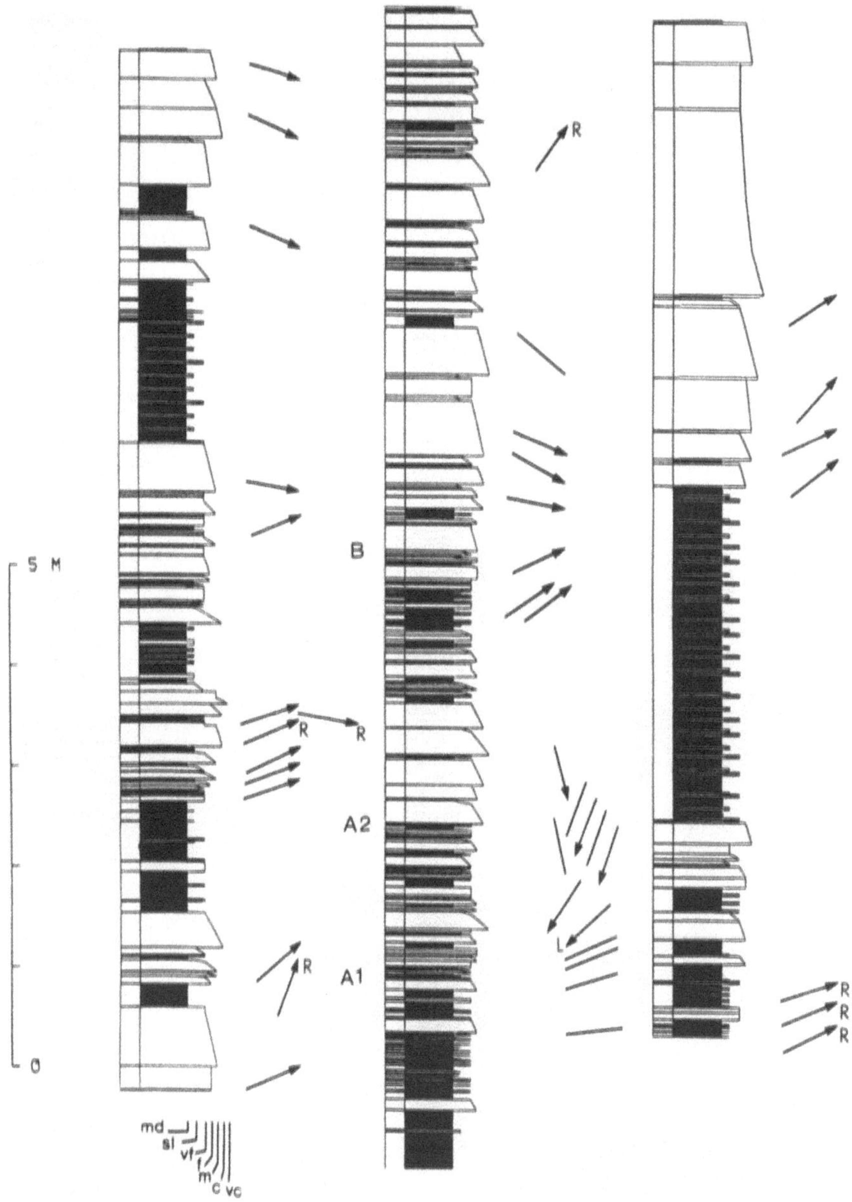

Fig. 25. Sedimentary log of upper part of Mullock Bay Unit B Member sandy interval 1 (MBS1) with bed-by-bed palaeocurrent data. Arrows and bars as for Fig. 16. R—ripples; L—non-directional L-ridge. This shows the development of apparent thickening and coarsening- upwards cycles each encompassing 30 to 70 depositional events.

SYNTHESIS

Late Llandovery to early Wenlock

The results of this study, together with those of Craig and Walton (1962), Warren (1963) and Rust (1965) show that a sediment dispersal pattern with dominant south-westward flow continued into the early Wenlock where major north-west-derived input is recorded in the Langholm area and axial flow to the south-west in the Kirkcudbright and Whithorn areas. This is consistent with models for the earlier development of the Southern Uplands which infer the existence of a topographic trench.

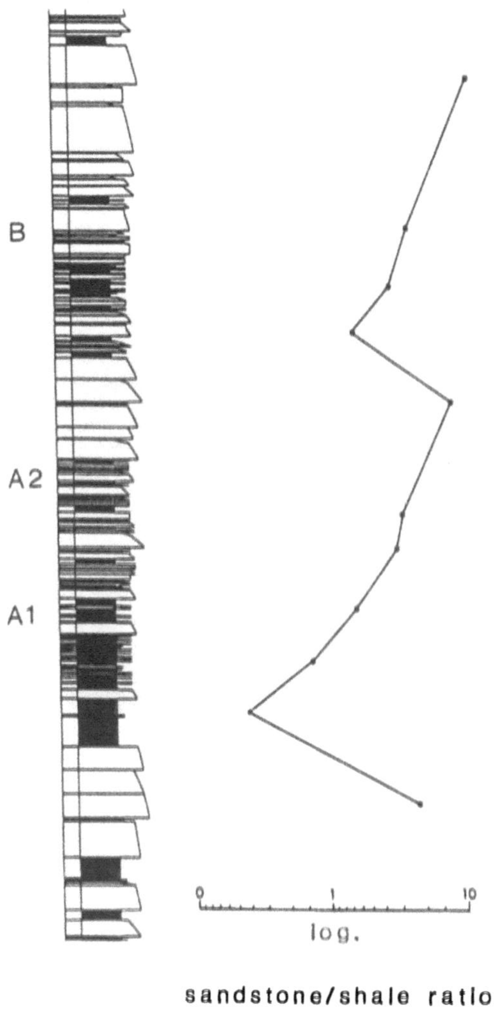

Fig. 26. Detail of cycles A1/A2 and B of Fig. 25 with sandstone/shale ratios.

The depositional model inferred for the Ross Formation (Fig. 7) proposes that transverse (north-west-derived) input to the trench in the Langholm area was deviated to produce axial flow in the Kirkcudbright area. Sedimentation in the Langholm area was probably by way of a sand-rich fan with high depositional gradients, which resulted in the dumping of large amounts of sandy detritus (the thick units of massive sandstones) and precluded sediment bypassing and the development of channel–levee systems.

The absence of thick sequences of thinly bedded turbidites and obvious channelization of sandstone bodies in the Kirkcudbright area suggests that an axial channel–levee system such as that of the Aleutian trench was not developed and that sediment dispersal along the trench may have been by 'sheet sands'.

The extent of south-westward flow may have been very large, as palaeocurrent data from the Upper Llandovery sequences of the Lecale Peninsula of Ireland also indicate dominant south-westward flow, with only minor transverse input (Cameron, 1981). Axial flow from north-east to south-west therefore persisted for at least 250 km along the trench in late Llandovery times.

The dominant south-westward flow recorded in many of the Southern Uplands sequences could relate to a source area to the north-east along strike, such as a collision zone, cf. Mitchell and McKerrow (1975). Since the evolution of the 'Tournquist line' branch of the Caledonides is poorly understood and much of the evidence is probably buried under the North Sea, the possible role of this is difficult to assess. However, during sedimentation of the late Llandovery Hawick Rocks and the early Wenlock Ross Formation, the major sedimentary source to the trench was transverse (north-west-derived) input probably via a canyon system on the inner trench slope, recorded in the sequences in the Langholm–Hawick area (Fig. 28).

Middle to late Wenlock

The middle and late Wenlock sediments of the Southern Uplands record a major change in the character of sedimentation. The first sign of this change is displayed in the relatively thick sections of hemipelagic sediment in the upper *riccartonensis* and *antennularius* zones which indicate several hiatuses in turbidite deposition. Middle Wenlock sediments record the development of extensive meandering channel–levee systems (Gipsy Point Unit B Member) which built out transversely from the north-west and passed laterally to the south-east into channel–mouth sequences and depositional lobes (Raeberry Unit—Figs 22, 29). Changes in sedimentation patterns were probably controlled by major tectonic events which generated slumps infilling and diverting the channel systems of the Gipsy Point Unit. These events resulted in abrupt facies and palaeocurrent changes and deposition of 'homogenites' in the more distal Raeberry Unit sequences.

Analogy with modern submarine fans with meandering channels indicates that these channel–levee depositional lobe systems must have built out at least 100 km and probably more than 200 km from the inner trench slope. Therefore, by mid-Wenlock times, the trench outer slope had ceased to confine the growth of depositional systems which built out over the adjacent sea floor (Fig. 29).

The high sedimentation rates inferred for Wenlock

Fig. 27. Photograph showing detail of cycle A2 of Figs 25 and 26.

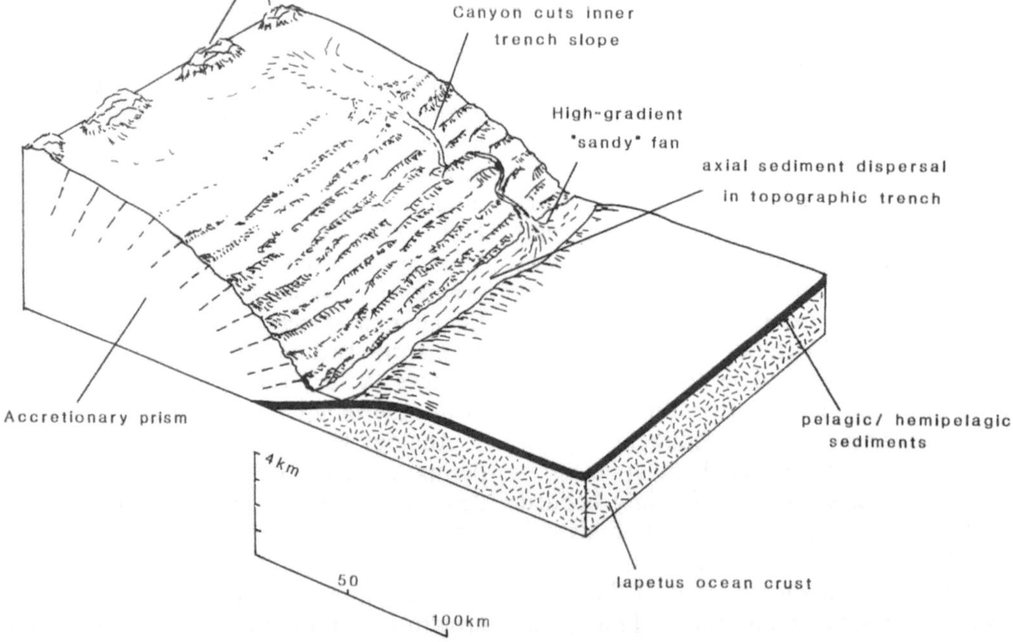

Fig. 28. Southern Uplands margin in the late Llandovery to early Wenlock period.

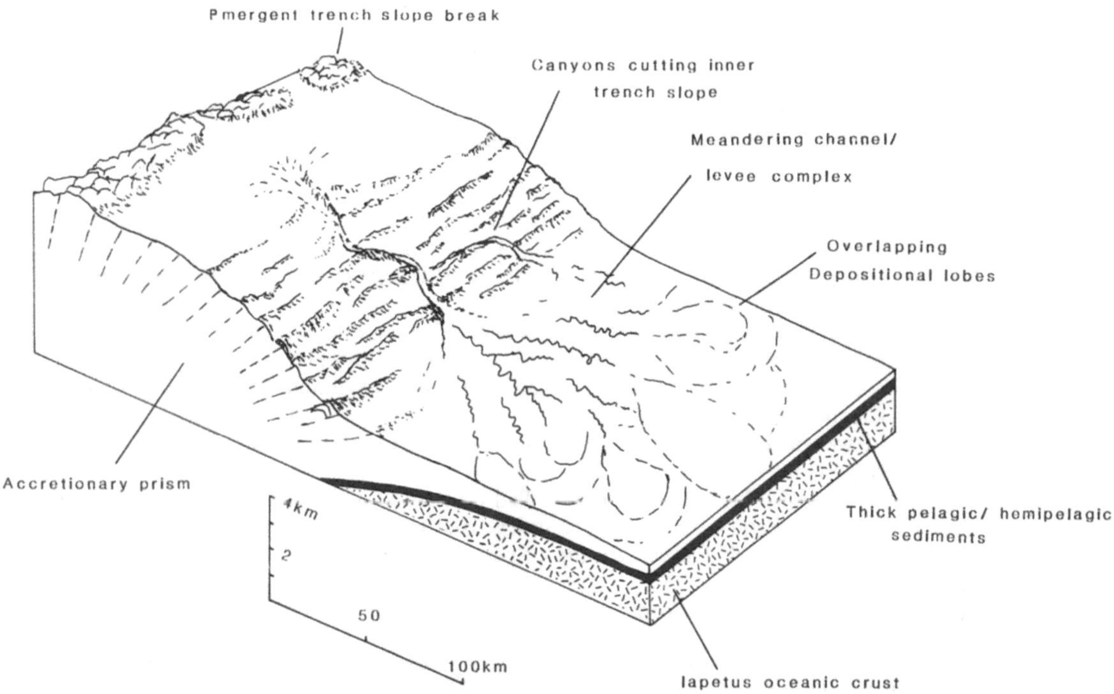

Fig. 29. Southern Uplands margin in the middle (and ?late) Wenlock period.

sediments (see above) must certainly have contributed to infill of the trench. However, there is no evidence to suppose that sedimentation rates prior to this were significantly lower (see above). It is perhaps more likely that infill of the trench was related to slowing subduction rate.

The most significant feature of the middle to later Wenlock sequences is the duration of sedimentation recorded in individual structurally intact accreted packets. In the Gipsy Point Block, for example, a 655-m sequence ranges from upper *riccartonensis* through *antennularius* and *rigidus* zones to the *flexilis* (*linnarssoni*) Zone—an interval spanning some 1.5–3 Ma. The existence of such a stratigraphically long-ranging and structurally intact sequence further suggests that subduction rates were greatly reduced and probably of the order of 1 mm/year or less. It was this slowing of subduction rates, or perhaps even a pause in subduction, which allowed mature, passive margin-style depositional systems to build up over periods of millions of years.

There are various possibilities for this inferred slowing of subduction rates or pause in subduction. Along the Washington–Oregon margin, slow subduction rates have allowed the building of relatively large fans 200 km or more across (Nitinat and Astoria fans: Kulm *et al.*, 1973; Nelson, 1985). The slowing of subduction rate was, in this case, due to subduction of a spreading ridge farther to the south and the subsequent formation of microplates, and change in convergence direction and rate and initiation of strike-slip motion along the margin (Atwater, 1970). Given the imminent closure of Iapetus in the late Silurian, a more likely cause in the case of the Southern Uplands would be collision along strike, perhaps in Ireland, related to oblique subduction.

A slowing or pause in subduction during the Wenlock could explain the paucity of igneous activity (no mid-Silurian granodiorites; rarity of tuff bands in the Southern Belt). The last phase of accretion and subduction in the late Wenlock/Ludlow could have been responsible for the renewed end Silurian/early Devonian magmatism.

References

Allen, J. R. L. (1970). The sequence of sedimentary structures in turbidites with special reference to dunes, *Scottish Journal of Geology* **6**, 146–161.

Allen, J. R. L. (1982). Sedimentary structures: their character and physical basis, *Developments in Sedimentology 30B*, Vol. II, Amsterdam, Elsevier.

Atwater, T. (1970). Implications of plate tectonics for the Cenozoic tectonic evolution of western North America, *Bulletin of the Geological Society of America* **81,** 3513–3535.

Cameron, T. D. J. (1979). The stratigraphy, sedimentology and structural geology of the Silurian rocks of East Lecale, County Down, Queen's University, Belfast, Ph.D. thesis (unpublished).

Casey, D. M. (1983). Geological studies in the Central Belt of the eastern Southern Uplands of Scotland, University of Oxford, D.Phil. thesis (unpublished).

Cita, M. B., Camerlenghi, A., Kastens, K. A. and McCoy, F. W. (1984). New findings of Bronze Age homogenites in the Ionian Sea: geodynamic implications for the Mediterranean, *Marine Geology* **55,** 47–62.

Colella, A. (1979). Medium-scale, tractive bedforms and structures in Gorgolione Flysch (Lower Miocene: Southern Apennines, Italy), *Boll. Soc. Geol. Ital.* **98,** 483–494.

Craig, G. Y. and Walton, E. K. (1962). Sedimentary structures and palaeocurrent directions from the Silurian rocks of Kirkcudbrightshire, *Transactions of the Geological Society, Edinburgh* **19,** 100–119.

Fitton, J. G., Thirlwall, M. F. and Hughes, D. J. (1982). Volcanism in the Caledonian orogenic belt of Britain, *In* R. S. Thorpe (Ed.), *Andesites*, Chichester, Wiley, pp. 611–636.

Harland, W. B. Cox, A. V., Llewellyn, P. G., Pickton, C. A. G., Smith, A. G. and Walters, R. (1982). *A Geologic Time-scale*, Cambridge, Cambridge University Press.

Hepworth, B. C., Oliver, G. J. H. and McMurtry, M. J. (1982). Sedimentology, volcanism, structure metamorphism of the northern margin of a lower Palaeozoic accretionary complex: Bail Hill–Abington area of the Southern Uplands of Scotland, *In* J. K. Leggett (Ed.), *Trench–Forearc Geology*, Special Publication of the Geological Society, London **10,** pp. 521–534.

Hieke, W. (1984). A thick holocene homogenite from the Ionian abyssal plain (Eastern Mediterranean), *Marine Geology* **55,** 63–78.

Hiscott, R. N. (1980). Depositional framework of sandy mid-fan complexes of Tourelle Formation, Ordovician, Quebec, *American Association of Petroleum Geologists Bulletin* **64,** 1052–1077.

Hiscott, R. N. (1981). Deep-sea fan deposits in the Macigno Formation (Middle–Upper Oligocene of the Gordana Valley, Northern Apennines, Italy. Discussion, *Journal of Sedimentary Petrology* **51,** 1015–1021.

Kemp, A. E. S. (1985). The later (Silurian) sedimentary and tectonic evolution of the Southern Uplands accretionary terrain. University of Edinburgh, Ph.D. thesis (unpublished).

Kemp, A. E. S. (1986). Tectonostratigraphy of the Southern Belt of the Southern Uplands, *Scottish Journal of Geology*.

Kemp, A. E. S. (1987). Tectonic development of the Southern Belt of the Southern Uplands, *Journal of the Geological Society, London*, (in press).

Kemp, A. E. S. and White, D. E. (1985). Silurian trench sedimentation in the Southern Uplands, Scotland: implications of new age data, *Geological Magazine* **122,** 275–77.

Kemp, A. E. S., Oliver, G. J. H. and Baldwin, J. R. (1985). Low-grade metamorphism and accretion tectonics: Southern Uplands terrain, Scotland, *Mineralogical Magazine* **49,** 335–344.

Komar, P. D. (1971). Hydraulic jumps in turbidity currents, *Bulletin of the Geological Society of America* **82,** 1477–1488.

Leggett, J. K. (1980). Sedimentological evolution of a Lower Palaeozoic accretionary forearc in the Southern Uplands of Scotland, *Sedimentology* **27,** 401–417.

Leggett, J. K., McKerrow, W. S. and Casey, D. M. (1982). The anatomy of a Lower Palaeozoic accretionary forearc: the Southern Uplands of Scotland, *In* J. K. Leggett (Ed.), *Trench–Forearc Geology*, Special Publication of the Geological Society, London **10,** 495–520.

McKerrow, W. S., Lambert, R. St. J. and Cocks, L. R. M. (1985). The Ordovician, Silurian and Devonian periods, *In* N. J. Snelling (Ed.), *The Chronology of the Geological Record*, Memoirs of the Geological Society, London **10,** 723–780.

McMurtry, M. J. (1980). Discussion of: evidence for Caledonian subduction from greywacke detritus in the Longford–Down inlier, *J. Earth Sci. R. Dubl. Soc.* **2,** 209–212.

Mitchell, A. H. G. and McKerrow, W. S. (1975). Analogous evolution of the Burma Orogen and the Scottish Caledonides, *Bulletin of the Geological Society of America* **86,** 305–315.

Mutti, E. (1977). Distinctive thin-bedded turbidite facies and related depositional environments in the Eocene Hecho Group (South-central Pyrenees, Spain), *Sedimentology* **24,** 107–131.

Mutti, E. and Sonnino, M. (1981). Compensation cycles: a diagnostic feature of turbidite sandstone lobes, Abstract, IAS, 2nd European Meeting, Bologna, pp. 120–123.

Mutti, E. and Normark, W. R. (1987). Comparing examples of modern and ancient turbidite systems, this volume.

Nelson, C. H. (1985). Astoria Fan, Pacific Ocean, *In* A. H. Bouma, W. R. Normark and N. E. Barnes (Eds), *Submarine Fans and Related Turbidite Systems*, New York, Springer-Verlag, pp. 45–59.

Pankhurst, R. J. (1982). Geochronological tables from British igneous rocks, *In* D. S. Sutherland (Ed.), *Igneous Rocks of the British Isles*, Chichester, Wiley, pp. 575–581.

Peach, B. N. and Horne, J. (1899). The Silurian rocks of Britain, 1. Scotland, *Memoirs of the Geological Survey, Scotland*

Piper, D. J. W. (1978). Turbidite muds and silts on deep-sea fans and abyssal plains, *In* D. J. Stanley and G. Kelling (Eds), *Sedimentation in Submarine Canyons, Fans and Trenches*, Stroudsburg, PA, Dowden, Hutchinson and Ross, pp. 163–176.

Ricci Lucchi, F. (1975). Depositional cycles in two turbidite formations of the northern Apennines (Italy), *Journal of Sedimentary Petrology* **45,** 3–43.

Rust, B. R. (1965). The sedimentology and diagenesis of Silurian turbidites in south-east Wigtownshire, Scotland, *Scottish Journal of Geology* **1,** 231–246.

Scott, K. M. (1967). Intra-bed palaeocurrent variations in a Silurian flysch sequence, Kirkcudbrightshire, Southern Uplands, Scotland, *Scottish Journal of Geology* **3,** 268–281.

Shanmugam, G. and Moiola, R. J. (1985). Submarine fan models: problems and solutions, *In* A. H. Bouma, W. R. Normark and N. E. Barnes (Eds), *Submarine Fans and Related Turbidite Systems*, New York, Springer-Verlag, pp. 29–34.

Stanley, D. J. (1981). Unifites: structureless muds of gravity-flow origin in the Mediterranean basin. *Geomarine Letters* **1,** 77–83.

Stone, P., Floyd, J. D., Barnes, R. P. and Lintern, B. C. (1987). A back-arc, thrust-duplex model for the Southern Uplands of Scotland, *Journal of the Geological Society, London* (in press).

Stow, D. A. V., Howell, D. G. and Nelson, C. H. (1985). Sedimentary, tectonic and sea-level controls, *In* A. H. Bouma, W. R. Normark and N. E. Barnes (Eds), *Submarine Fans and Related Turbidite Systems*, New York, Springer-Verlag, pp. 15–22.

Thirlwall, M. F. (1981). Implications for Caledonian plate tectonic models of chemical data from volcanic rocks of the British Old Red Sandstone, *Journal of the Geological Society, London* **138,** 123–138.

Walker, R. G. and Mutti, E. (1973). Turbidite facies and associations, Soc. econ. Palaeon. Mineral, Pacific Section, Short course at Anaheim, pp. 119–157.

Walton, E. K. (1965). Lower Palaeozoic rocks: stratigraphy, palaeogeography and structure, *In* G. Y. Craig (Ed.), *The Geology of Scotland*, Edinburgh, Oliver and Boyd, pp. 161–227.

Warren, P. T. (1963). The petrography, sedimentation and provenance of the Wenlock Rocks near Hawick, Roxburghshire, *Transactions of the Geological Society, Edinburgh* **19,** 225–255.

Whitman, J. M. and Davies, T. A. (1979). Cenozoic oceanic sedimentation rates: how good are they? *Marine Geology* **30,** 269–284.

Winn, R. D. and Dott, R. H. (1977). Large-scale traction produced structures in the deep-water fan channel conglomerates in Southern Chile, *Geology* **5,** 41–44.

Chapter 8

Palaeo-oceanography and Depositional Environment of the Scandinavian Alum Shales: Sedimentological and Geochemical Evidence

Andrew Thickpenny, Lomond Associates, 10 Craigendoran Avenue, Helensburgh, Glasgow G84 7AZ, UK

ABSTRACT

The Middle Cambrian to Tremadoc Scandinavian alum shales are highly condensed sediments, notably enriched in organic matter. Sedimentological studies alone suggest their deposition under dominantly anoxic or euxinic conditions in an epicontinental sea environment. Plots of carbon–sulphur data derived from drill cores from Västergötland and Östergötland in Sweden indicate deposition of the Middle and Upper Cambrian portions in a dominantly euxinic environment. The amount of pyrite formed was TOC-limited during most of the Middle Cambrian, but controlled by the amount of reactive iron (Fe_r) in Upper Cambrian sediments. By contrast, Tremadoc sediments were probably deposited under 'normal' oxygenated conditions.

Palaeofertility estimates, utilizing sedimentological, biostratigraphical and geochemical data, suggest that primary organic productivity within the alum shale sea was relatively low compared with the present day. This was probably a result of low global hydrospheric oxygen levels, compounded by local palaeogeographic restriction.

This integrated sedimentological and geochemical approach indicates that the alum shales are the product of a depositional and palaeo-oceanographic environment lacking modern analogues.

INTRODUCTION

The Middle and Upper Cambrian and locally Tremadoc alum shales of Scandinavia and the Baltic area (Fig. 1) are organic-rich mudstones occurring in a condensed early Palaeozoic sedimentary sequence (Fig. 2), for the most part overlying gneissic Precambrian basement of the Fennoscandian Shield (Martinsson, 1974; Thorslund, 1960). Over much of their outcrop they are effectively undeformed, flat-lying and immature. Within and along the Caledonian front, however, nappe emplacement buried the mudstones to a greater depth (Kisch, 1980), while normal faulting and igneous activity in south-west Sweden (Scania or Skåne district, Fig. 1) has had a similar effect on the sediment's maturity. If the shortening effect of Caledonian tectonism is removed, it is apparent that the original depositional area of the mudstones was probably 2000 km N–S by 800 km E–W (Thickpenny, 1984).

Owing to their economic importance as a source for both hydrocarbons and trace elements such as

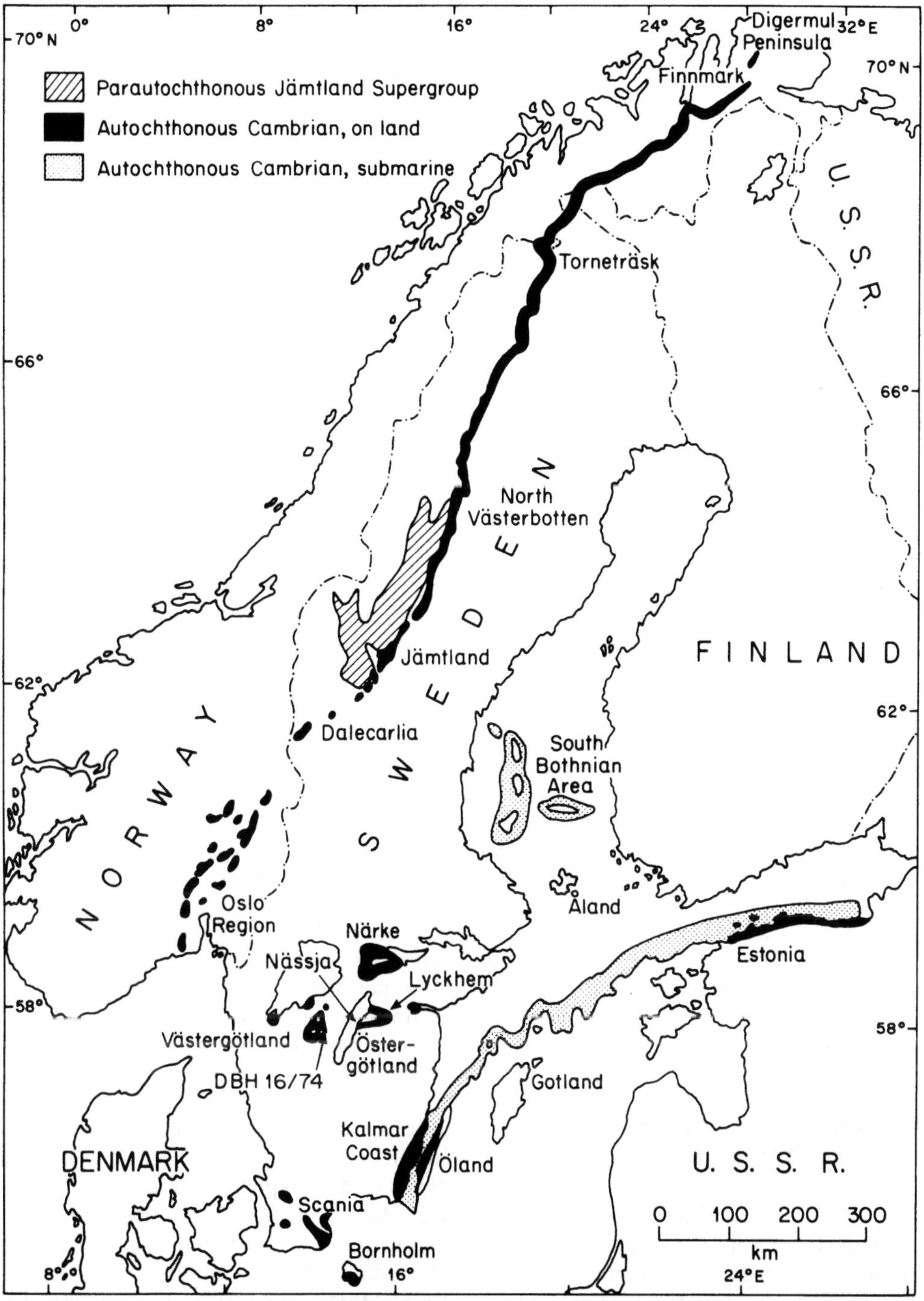

Fig. 1. Cambrian outcrop in Scandinavia (after Martinsson, 1974), showing location of boreholes discussed in text.

uranium, the palaeontology and inorganic geochemistry of the shales have been extensively studied. The abundant published and unpublished literature on these subjects is extensively reviewed by Andersson *et al.* (1985). Detailed sedimentological studies—essential in any attempt to understand the genesis of a deposit with such unusually high contents of organic carbon and trace elements—have only recently been undertaken (e.g. Thickpenny, 1984).

In this paper, new geochemical data are allied

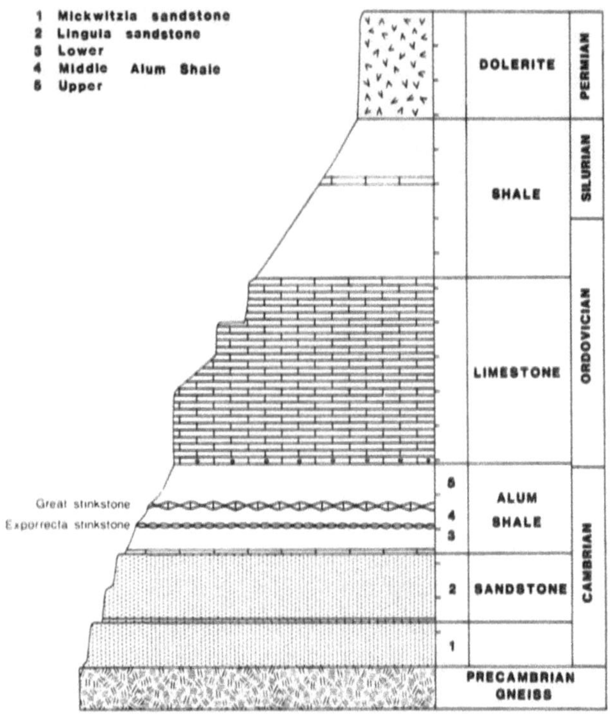

Fig. 2. Generalized Lower Palaeozoic stratigraphy of Billingen–Falbygden (after Dahlman and Gee, 1977).

with such sedimentological work to provide a more complete picture of the palaeo-oceanographic environment during deposition of the alum shales. Carbon–sulphur data are used as a palaeoenvironmental indicator, and TOC data are combined with sedimentological data to enable estimation of palaeofertility. The results underscore the importance of a multidisciplinary approach in the study of fine-grained sediments.

SEDIMENTOLOGICAL DEVELOPMENT

Thickpenny (1984) has discussed and interpreted the lithologies developed within the alum shales of southern Sweden, mainly with reference to two cores through the entire formation (DBH 16/74, drilled by the Swedish Alum Shale mining company (ASA) and Lyckhem No. 3, drilled by the Swedish Geological Survey (SGU)). Summary sedimentological logs of these cores are illustrated in Fig. 3. Since Andersson *et al.* (1985) fully review stratigraphical and compositional variations of the deposit, these topics are here touched on only briefly.

Typically, the mudstones are parallel laminated and contain high (5–20%) organic carbon contents (Andersson *et al.*, 1983, 1985 and references therein). Bioturbation and benthos are almost invariably absent (except in Scania, where sparse trilobites occur). Mudstone sedimentation is, however, punctuated to varying degrees by trilobite-rich limestone concretions. Thickest mudstone sequences occur in Scania (*c.* 90 m), while concretionary limestones are most abundant (and locally dominant) in Öland. Locally, sporadic redeposited silt and fine sand laminae may occur; these contain glauconite and phosphate clasts and fragmented bioclasts, which provide direct evidence of a probable sediment-starved, shallow-marine source. In addition, thin, pale grey mudstone laminae containing less than 2% total organic carbon (TOC) are also present in some sections, and may be finely interlaminated with more organic-rich horizons.

The lithological development of the deposit varies distinctly with time. Organic-poor mudstones, and more arenaceous laminae, are generally much more common within the Middle Cambrian—conditions favourable to organic matter preservation were apparently more geographically confined during this interval. Concretionary limestones are most abundant in Upper Cambrian sediments where coarser laminae are very rare. Tremadoc deposits are, however, typified by greatly reduced limestone contents, and a generally siltier character.

This lithological association points strongly towards dominantly very low-energy deposition, below wave base. The perfect parallel lamination and lack of benthic activity indicate that conditions were probably anoxic at the sediment–water interface for long periods of time, and abundant submillimetre laminae of framboidal pyrite hint at the possibility of truly euxinic conditions in the water column, with free H_2S. Sedimentary horizons are laterally persistent and vary only gradually in thickness, implying little or no tectonic control on sedimentation. These factors, taken in tandem with the areal extent of the deposit, seem typical of epicontinental sea conditions (e.g. Hallam, 1981 and references therein). Moreover, sediments over- and underlying the alum shales are of shallow marine affinities (e.g. Andersson *et al.*, 1985; Martinsson, 1974; Thickpenny, 1984), suggesting that prevailing water depths during alum shale deposition were less than 200 m. No modern analogues exist for such an environment, particularly where large amounts of organic carbon are preserved.

In summary, sedimentological considerations alone would suggest deposition of the alum shales in an epicontinental sea, which was anoxic and possibly

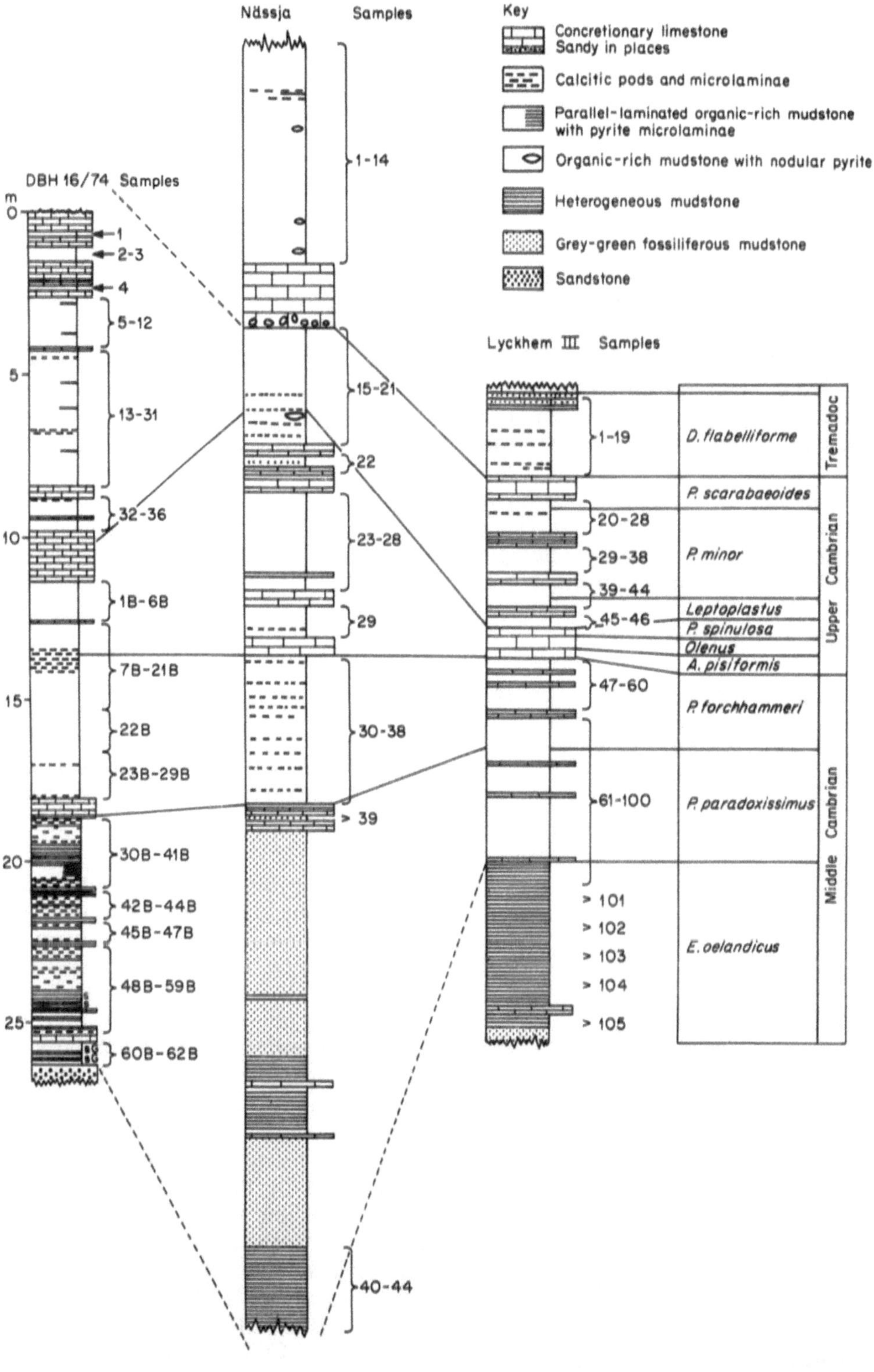

Fig. 3. Simplified sedimentological logs and sample locations for drill cores mentioned in text.

euxinic for long periods of time. Moreover, lithological variations indicate varying palaeo-oceanographic conditions with time. The remainder of this paper allies geochemical data to the sedimentology in an effort to embellish this model.

CARBON–SULPHUR RELATIONSHIPS

Recent work on carbon–sulphur relationships in marine and freshwater sediments illustrates the potential of using variations in this ratio as a palaeoenvironmental indicator (e.g. Berner and Raiswell, 1984; Leventhal, 1983a, b; Raiswell and Berner, 1985). Scatter plots of TOC against pyrite sulphur (or, less effectively, total sulphur) exhibit characteristic but differing ratios for samples from normal marine, euxinic and freshwater environments (Fig. 4). Since this technique can elegantly distinguish between recent environments, it is here applied to data obtained from the alum shales in an attempt to characterize their depositional environment.

Experimental procedure

Access to and sampling of core from DBH 16/74 borehole from the Ranstad area of Billingen was kindly allowed by ASA. Mudstones from the core were continuously sampled. Sample boundaries correspond to visible lithological boundaries, where present (Fig. 3); where the sediment is apparently homogeneous, each sample represents *c.* 10 cm of core. The Swedish Geological Survey (SGU) provided geochemical data and samples from the Nässja core from Östergötland and allowed access to the core for logging purposes. However, in this case the individual samples cover much larger depth ranges, and therefore obscure subtle small-scale lithological variations (Fig. 3).

Data obtained from the two cores are summarized in Tables 1 and 2. Analyses of DBH 16/74 were carried out by the author on powder splits. Total organic carbon was measured on a Perkin–Elmer elemental analyser at Imperial College. Reduced inorganic sulphur (equivalent to pyrite sulphur) was measured at Leeds University following the method of Canfield *et al.* (1986). Total Fe data were obtained at Imperial College using an Inductively Coupled Plasma (ICP) technique. Carbon and total sulphur data for the Nässja core were supplied by David Gee of SGU.

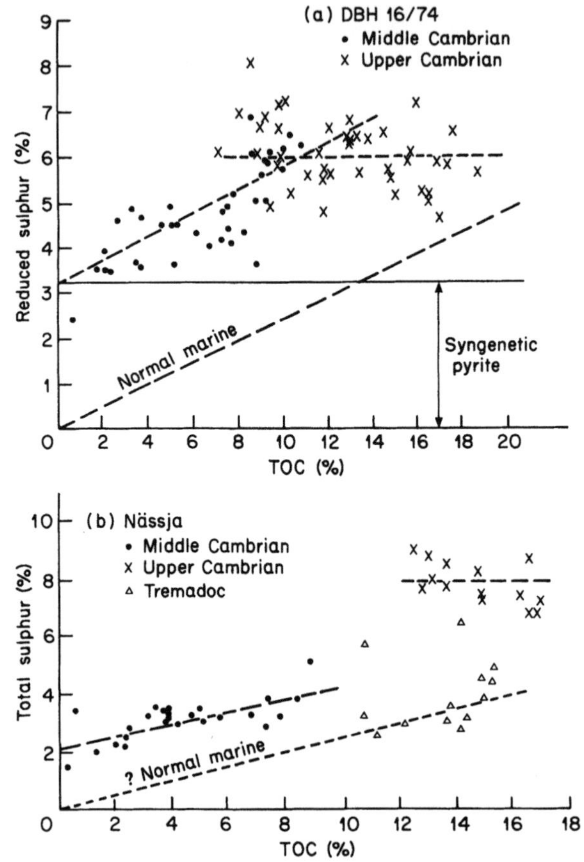

Fig. 5. C/S plots for the alum shales in (a) borehole DBH 16/74 (Billingen) and (b) Nässja borehole (Östergötland).

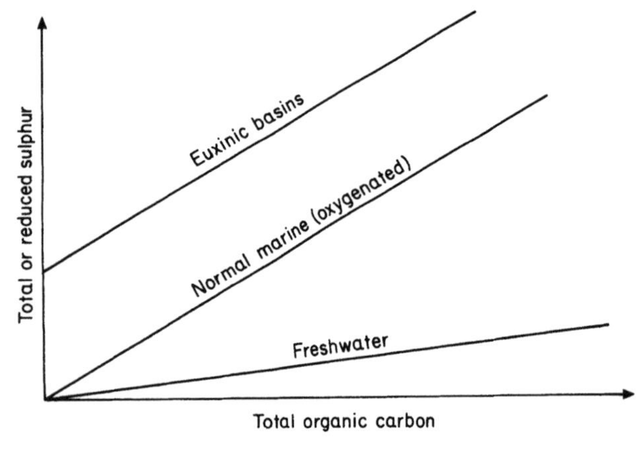

Fig. 4. Idealized C/S relationships of modern sediments (after Raiswell and Berner, 1985). Note that the euxinic basin curve has a positive intercept on the S-axis; it has a positive slope only if: (a) DOP is constant and Fe correlates positively with TOC, or (b) DOP increases with increasing TOC and Fe and TOC are not positively correlated (see also Fig. 6).

TABLE 1. Sulphur–carbon–iron data, DBH 16/74.

Sample	% *SPy*	% *Py*	% *Py Fe*	% *Fe*	*DOP*	% *TOC*
16/74/1	4.86	9.11	4.25	5.00	0.85	9.74
2	6.35	11.90	5.55	6.60	0.84	12.93
3	6.41	12.02	5.61	6.70	0.84	12.92
4	5.70	10.69	4.99	6.00	0.83	14.64
5	7.41	13.89	6.48	7.00	0.93	
6	5.65	10.59	4.94	6.10	0.81	18.67
7	5.82	10.91	5.09	6.00	0.85	17.34
8	5.14	9.63	4.49	5.50	0.82	14.96
9	5.86	10.99	5.13	5.70	0.90	16.98
10	4.40	8.25	3.85	5.40	0.71	15.97
11	6.36	11.92	5.56	6.50	0.86	13.72
12	5.22	9.79	4.57	5.40	0.85	16.11
13	6.06	11.36	5.30	6.10	0.87	15.68
14	5.01	9.39	4.38	4.90	0.89	16.51
15	7.15	13.41	6.26	7.00	0.89	15.97
16	5.15	9.66	4.51	5.10	0.88	16.50
17	6.46	12.11	5.65	6.30	0.90	17.56
18	5.54	10.39	4.85	5.40	0.90	11.04
19	5.51	10.33	4.82	5.40	0.89	13.28
20	4.64	8.70	4.06	5.10	0.80	17.01
21	4.75	8.91	4.16	5.10	0.82	11.77
22	6.74	12.64	5.90	7.30	0.81	12.90
23	5.47	10.26	4.79	5.60	0.86	11.78
24	6.06	11.36	5.30	6.10	0.87	11.56
25	6.36	11.92	5.56	5.70	0.98	12.86
26	5.89	11.04	5.15	5.70	0.90	15.47
27	6.42	12.04	5.62	6.70	0.84	13.20
28	4.51	8.46	3.95	6.20	0.64	14.78
29	5.52	10.35	4.83	5.60	0.86	14.76
30	6.50	12.19	5.69	7.20	0.79	14.46
31	5.61	10.52	4.91	7.20	0.68	13.30
32	5.18	9.71	4.53	5.60	0.81	10.31
33	5.71	10.71	5.00	6.20	0.81	11.76
34	5.59	10.48	4.89	6.20	0.79	12.00
35	6.58	12.34	5.76	6.70	0.86	12.01
36	6.56	12.30	5.74	7.50	0.77	9.76
1B	5.75	10.78	5.03	7.80	0.64	9.70
2B	6.08	11.40	5.32	7.70	0.69	7.10
3B	6.64	12.45	5.81	7.50	0.77	8.94
4B	8.04	15.08	7.04	8.50	0.83	8.52
5B	6.94	13.01	6.07	8.20	0.74	7.98
6B	7.20	13.50	6.30	8.20	0.77	10.03
7B	7.12	13.35	6.23	7.50	0.83	9.74
8B	6.85	12.84	5.99	7.60	0.79	9.19
9B	5.98	11.21	5.23	6.70	0.78	9.90
10B	6.04	11.33	5.29	6.80	0.78	8.74
11B	4.87	9.13	4.26	6.40	0.67	7.51
12B	5.85	10.97	5.12	6.80	0.75	9.30
13B	6.04	11.33	5.29	6.50	0.81	8.76
14B	5.87	11.01	5.14	6.30	0.82	9.24
15B	6.21	11.64	5.43	8.10	0.67	10.80
16B	6.86	12.86	6.00	7.50	0.80	8.47
17B	6.45	12.09	5.64	7.60	0.74	10.28
18B	5.72	10.73	5.01	7.20	0.70	9.86
19B	6.16	11.55	5.39	7.00	0.77	10.03
20B	6.92	12.98	6.06	7.60	0.80	9.11

TABLE 1 *(continued)*

Sample	*% SPy*	*% Py*	*% Py Fe*	*% Fe*	*DOP*	*% TOC*
16/74/21B	5.60	10.50	4.90	6.40	0.77	9.35
22B	5.00	9.38	4.38	5.90	0.74	9.19
23B	6.09	11.42	5.33	6.70	0.80	9.39
24B	5.57	10.39	4.82	6.30	0.77	8.97
25B	5.01	9.39	4.38	6.10	0.72	8.78
26B	4.32	8.10	3.78	5.50	0.69	8.24
27B	5.17	9.69	4.52	6.00	0.75	7.72
28B	3.61	6.77	3.16	4.80	0.66	8.71
29B	4.00	7.50	3.50	5.20	0.67	
30Ba	3.81	7.14	3.33	5.30	0.63	
30Bb	2.30	4.31	2.01	4.30	0.47	2.00
31B	4.77	8.94	4.17	6.30	0.66	
32B	4.89	9.17	4.28	6.80	0.63	5.00
33B	3.46	6.49	3.03	5.00	0.61	2.36
34B	2.40	4.50	2.10	3.20	0.66	0.67
35B	3.52	6.60	3.08	4.90	0.63	1.77
36B	3.65	6.85	3.20	5.20	0.62	3.47
37B	3.56	6.68	3.12	5.20	0.60	3.71
38B	4.58	8.60	4.02	6.10	0.66	2.70
39B	4.48	8.40	3.92	6.40	0.61	5.10
40B	4.80	9.00	4.20	7.00	0.60	4.90
41B	3.91	7.33	3.42	5.10	0.67	2.12
42B	4.66	8.74	4.08	6.20	0.66	3.67
43B	4.84	9.07	4.23	6.20	0.68	3.30
44B	3.48	6.53	3.05	5.10	0.60	2.17
45B	4.46	8.36	3.90	6.40	0.61	4.67
46B	3.59	6.73	3.14	6.00	0.52	5.21
47B	4.28	8.04	3.76	5.60	0.67	7.31
48B	4.76	8.92	4.16	6.00	0.69	7.68
49B	4.07	7.63	3.56	5.50	0.65	7.53
50B	4.39	8.23	3.84	6.00	0.64	7.52
51B	4.27	8.01	3.44	5.10	0.67	6.27
52B	4.02	7.54	3.52	5.50	0.64	6.74
53B	4.50	8.44	3.94	5.90	0.67	5.30
54B				5.30		5.54
55B				5.60		6.19
56B				5.00		4.75
57B				3.60		0.84
58B				5.80		3.01
59B				5.50		4.40
60B				5.20		8.78
61B				8.30		4.21
62B				9.30		1.88

Interpretation

Data for the two cores are presented in Figs 5 and 6. Data from core DBH 16/74 show a distinct trend from a positive slope in the Middle Cambrian to horizontality in the Upper Cambrian, although the change does not correspond precisely to the Middle–Upper Cambrian boundary. This pattern is very similar to that found by Armands (1972) from drill-core material from DBH-1A-65, also from the Ranstad area of Billingen. Where the slope is positive, the curve roughly parallels that of the present-day normal marine curve (Fig. 4) but lies above it with an intercept on the sulphur axis at *c.* 3% S (*c.* 5.6% pyrite). Such an intercept characterizes euxinic basins such as the Black Sea (Leventhal, 1983a and references therein). Free H_2S within the water column promotes the formation of syngenetic pyrite,

TABLE 2. Carbon–sulphur data, Nässja.

Sample	*TOC*	*Total S*	*Sample*	*TOC*	*Total S*
NS 1	11.20	2.45	NS 28	14.75	7.35
2	10.70	3.15	29	12.60	9.00
3	12.20	2.90	30	8.85	5.05
4	14.20	2.65	31	8.45	3.80
5	13.20	3.25	32	7.40	3.80
6	13.70	3.00	33	7.75	3.20
7	15.40	4.90	34	7.25	2.85
8	14.40	3.05	35	6.80	3.15
9	14.80	4.50	36	5.70	3.10
10	10.80	5.75	37	5.15	2.95
11	14.95	3.80	38	4.20	2.95
12	14.20	6.35	39	0.60	3.35
13	13.80	3.55	40	3.90	3.25
14	15.29	4.35	41	3.70	3.40
15	16.75	6.80	42	3.85	3.40
16	15.00	7.20	43	3.15	3.20
17	13.75	7.70	44	3.80	3.00
18	13.00	7.90	45	3.50	3.55
19	13.10	8.80	46	2.50	2.85
20	12.80	7.60	47	5.00	3.35
21			48	4.70	3.20
22	13.70	8.45	49	3.90	3.20
23	16.55	8.70	50	2.40	2.45
24	16.65	6.80	51	1.95	2.25
25	14.90	7.40	52	2.25	2.05
26	16.30	7.40	53	1.30	1.90
27	14.70	8.30	54	0.35	1.40

within the water column or at the sediment–water interface, and therefore not necessarily associated with preserved organic carbon.

The remainder of the pyrite sulphur presumably formed during early diagenesis where organic matter is necessary for *in situ* formation of pyrite (e.g. Berner, 1984 and references therein). Hence, organic carbon values increase in tandem with pyrite sulphur contents, generating the positive slope observed. Raiswell and Berner (1985) have suggested an alternative explanation for such a curve based on Fe-limited syngenetic pyrite formation, where Fe correlates with TOC. This produces a spurious correlation between pyrite sulphur and TOC. Figure 6 demonstrates that there is little or no Fe-TOC coupling for the analysed samples, so this explanation for the curve for DBH 16/74 is unlikely.

The best explanation of the sloping portion of the curve is therefore that it is produced by a combination of a syngenetic pyrite component (*c.* 5.6%) and a TOC-limited diagenetic component.

The horizontal sector of the curve comprising Upper Cambrian and uppermost Middle Cambrian sediments requires further explanation. Here, TOC contents increase, while pyrite sulphur concentrations remain roughly constant but very high. Levels of dissolved sulphate are high in marine waters and are unlikely to be a limiting factor for pyrite formation. Hence, it is most likely that the amount of reactive Fe associated with detrital clays and Fe-oxides has become the limiting factor in these sediments. Generally, the higher the concentration of H_2S and the longer its contact time with Fe-minerals, then the greater is the potential reactivity of these minerals (Berner, 1984; Raiswell and Berner, 1985). A measure of this reactivity is given by the degree of pyritization (DOP):

$$\mathrm{DOP} = \frac{\text{Pyrite Fe}}{\text{Total Fe}}$$

A plot of DOP versus TOC is shown in Fig. 6 and again shows a flattening trend at high TOC levels. Here, DOP values are *c.* 0.85, similar to those derived from the Jet Rock and Posidonia Shales (Raiswell and Berner, 1985), both interpreted to have been deposited under euxinic conditions. It is likely that DOP values for the alum shales are actually sig-

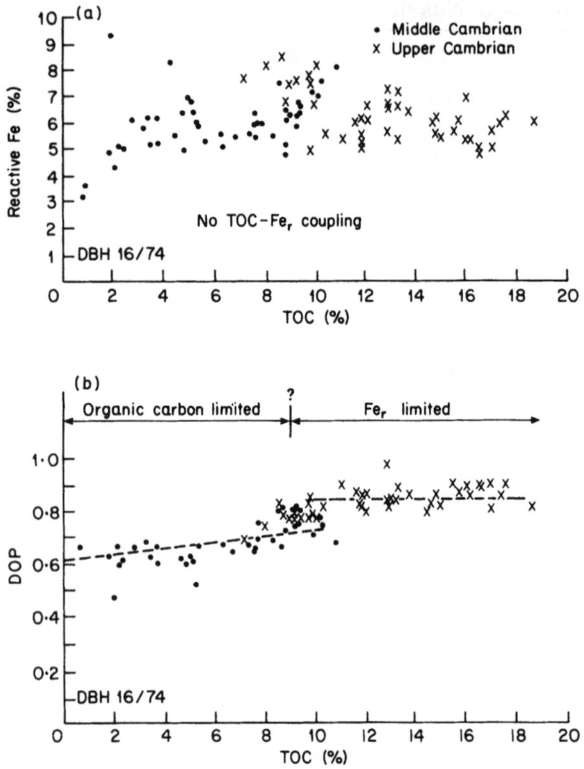

Fig. 6. (a) Fe versus TOC and (b) DOP versus TOC plots for the alum shales in borehole DBH 16/74.

nificantly higher than those presented by Raiswell and Berner (1985) since total Fe rather than (Fe_r + HCl-soluble Fe) has been used in the equation. This plot, therefore, indicates that during the Upper Cambrian, when detrital input was low, the amount of pyrite forming was Fe-limited, presumably within the diagenetic component.

Data from the Middle and Upper Cambrian of the Nässja Borehole (Fig. 5) essentially support those from DBH 16/74. It is interesting that geochemical conditions were apparently very similar in Västergötland and Östergötland during this interval. However, Tremadoc strata, which are absent in DBH 16/74, exhibit an important change in C/S ratios. These strongly suggest that all pyrite is of a diagenetic origin and therefore that the water column was no longer euxinic—conditions similar to present-day 'normal' marine conditions are implied. The two data points showing high S levels coincide with samples containing later diagenetic pyrite nodules, thus giving anomalously high concentrations.

The Upper Cambrian/Tremadoc boundary occurs at the base of a conglomeratic concretionary limestone and, as previously noted, in addition to their markedly differing C/S signatures, the mudstones above and below also show lithological differences. Accordingly, this boundary may be the most important within the alum shale succession since it may mark a major change in oceanographic conditions. The relatively abrupt changes in hydrospheric oxygen content indicated by these data may be related to changes in oceanic circulation and sea level, which in turn may be responsible for major changes in the trace element geochemistry of the mudstones (e.g. Andersson *et al.*, 1985).

Discussion

Several important conclusions may be drawn from the above interpretation. First, compared with other published data (Berner and Raiswell, 1984; Leventhal, 1983a, b; Raiswell and Berner, 1985), the amounts of syngenetic pyrite are extremely high. This is probably due to a combination of high H_2S levels in the water column and slow sedimentation rates, allowing long periods of contact between detrital Fe-minerals and H_2S. It is possible that the alum shales represent deposition in the most extreme marine euxinic conditions yet studied in the Phanerozoic. Secondly, even samples with low TOC contents, which appear pale grey in hand specimen, plot on the same curve, signifying the formation of syngenetic pyrite. It seems, therefore, that even during periods when organic production and preservation were relatively low, the water column apparently remained euxinic. This implies continued stable stratification and poor or non-existent oceanic circulation.

The variations in amounts of pyrite formed and the DOP agree well with the depositional environments determined purely from sedimentological studies. Where sedimentation rates are more rapid, sulphate reduction by bacteria is less efficient (Berner, 1984 and references therein). Moreover, the increase in detrital mineral content, with concomitant higher Fe content, helps explain the lower DOP. The DOP values at low TOC contents are, however, still high compared with other publised data. Thickpenny (1984) has attributed submillimetre-scale pyritic laminae to syngenetic pyrite formation on the basis of their sedimentology, and evidence presented here strongly supports this argument. Sedimentary pyrite may have formed within the water column as is the case, for example, in the Black Sea (Degens and Ross, 1974). The transition to extremely organic-rich laminated mudstones in the Upper Cambrian reflects the decreased detrital Fe input and decreased sedimentation rates; conditions during this period (when several trace elements are extremely enriched) must have been remarkably reducing. The implications of these results are that waters during alum shale deposition were (a) low in oxygen content and (b) stratified.

The results presented here have implications in a wider context. Several workers have recently discussed the relationships between the burial of organic carbon and pyrite sulphur over Phanerozoic time (Berner, 1984; Berner and Raiswell, 1983; Garrels and Lerman, 1981). On the basis of secular changes in the isotopic records of carbon in carbonates and sulphur in sulphates, these authors have demonstrated a coupling of the sedimentological carbon and sulphur cycles. Roughly reciprocal variations of the values of carbon and sulphur isotopes necessitate equal and opposite variations in the rate of burial of organic carbon and pyrite sulphur in order to maintain roughly constant atmospheric oxygen. As a result, a low C/S ratio is predicted globally for the early Palaeozoic (Berner and Raiswell 1983), i.e. relative to today, either increased pyrite or less organic carbon (or a combination) was buried. Limited data from normal (oxygenated) marine sediments from the Cambrian lend support to this hypothesis (Raiswell and Berner, 1986).

Berner and Raiswell suggest a possible explanation: a higher percentage of sediments may have been deposited under euxinic conditions, thus preserving pyrite sulphur. Coevally, reduced circulation decreased the degree of nutrient replenishment by watermass overturning, and therefore decreased primary organic productivity. In addition, refractory organic compounds were probably rare (since no terrestrial plants existed), and bioturbation may have been reduced, both factors contributing to lower C/S ratios.

The data and arguments presented here for the depositional environment of the alum shales provide support for Berner and Raiswell's theory. Euxinic conditions apparently prevailed over large areas of Scandinavia. Organic productivity may have been low (but preservation high) and stable stratification of the water column was widespread. This implies decreased oceanic circulation. If organic productivity above an oxic water column was also relatively low, then global organic carbon burial levels would also have been low, as predicted by the C/S values of Berner and Raiswell (1983).

PRIMARY ORGANIC PRODUCTIVITY

Bralower and Thierstein (1984) present a quantitative method for estimating primary organic productivity in ancient oceans, based on productivity and preservation factors measured from a variety of Holocene environments. They compare the sedimentary organic carbon accumulation rates with the measured primary production rates of overlying surface waters. The results can conveniently be represented as a log–log graphical plot (Fig. 7). Only euxinic environments (e.g. the Black Sea) combine high organic carbon accumulation rates with low primary productivity.

Outlined below is an iterative treatment of the alum shales, using this log–log graphical plot. These sediments satisfy the necessary sedimentological prerequisites, i.e. they are non-turbiditic, bituminous and dominantly parallel laminated. In order to estimate primary productivity it is necessary to calculate the accumulation rate of organic matter using the formula:

$$\text{accumulation rate} = \sigma * \text{TOC} * w$$

where σ = density, TOC = total organic carbon (as a fraction of unity), and w = sedimentation rate (cm/10^3 years).

The density is here estimated to be 2.3 g/cm^3, corresponding to *c.* 12% TOC (or *c.* 25% original organic matter), which is a rough mean for the alum shales. Clearly, higher TOC contents would lower the density and hence reduce calculated accumula-

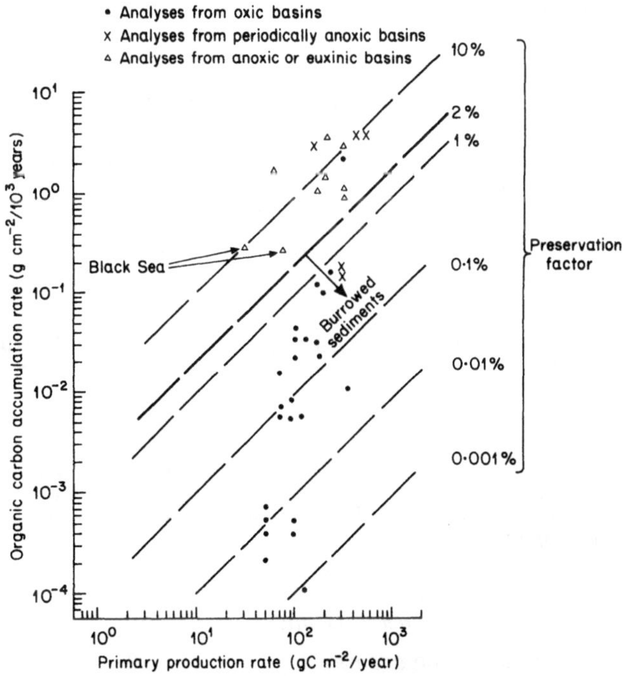

Fig. 7. Log–log plot of primary production rates versus organic carbon accumulation rates for Holocene sites (from Bralower and Thierstein, 1984).

tion rates. Increased errors in calculation are inherent in dealing with Cambrian rather than Mesozoic sediments, notably with respect to sedimentation rates and preservation factors. The effect of these variables is discussed below.

Sedimentation rates

Sedimentation rates are a vital prerequisite of any palaeofertility calculation, and depend strongly on biostratigraphic resolution. Typically, Cambrian biostratigraphic resolution is of the order of 1–5 Ma, much poorer than for the Mesozoic. Hence, the sedimentation rates calculated by Thickpenny (1984) for the alum shales are necessarily less precise, since they are averages over longer periods of time—the effect of subtle hiatuses and erosional events remain undetected.

Nonetheless, the biostratigraphy allows good order-of-magnitude calculations (Table 3). The results are clearly very low for a shelf environment, probably being comparable to present-day deep-sea pelagic oozes.

Preservation factors

Figure 8 illustrates the importance of preservation factors in estimating palaeofertility. Bralower and Thierstein (1984) regard the preservation of 2% of primary organic production as a likely conservative estimate for sediments deposited under an anoxic (or euxinic) water column. Importantly, their data from the Black Sea, a demonstrably euxinic basin, show significantly higher (*c.* 4.3%) preservation factors at low sedimentation rates. Evidence from the sedimentology and C/S relationships suggests that preservation factors within the alum shale sea would have been at least as high as for the Black Sea. Figure 8 clearly shows that the preservation factor has an inverse relationship with the primary productivity estimate.

Results

Total organic carbon data obtained from cores from borehole DBH 16/74 and Lyckhem No. 3 (a stratigraphic test well drilled by SGU in Östergötland) and biostratigraphic data, kindly provided by ASA and SGU, have been used to calculate both sedimentation rates and primary productivities (Table 3 and Fig. 8). Preservation factors of both 2% and 5% have been used in the palaeofertility calculations. Even with the lower preservation factor, the results suggest that primary productivity was of the same order as or less than the lowest measured Holocene values of Bralower and Thierstein. Although the apparent productivities are higher than the more extreme Cretaceous examples quoted by these authors, the results deserve further discussion, since the depositional environment is very different.

Discussion and interpretation

The palaeocontinent of Baltica probably occupied medial southerly latitudes (*c.* 50–60 °S) during alum shale deposition (Cocks and Fortey, 1982; Scotese *et al.*, 1979; Smith *et al.*, 1981). The present outcrop of the alum shales is confined to western and central areas of the palaeocontinent, and the deposit thickens westwards. Analogy with present-day conditions might suggest that this depositional setting satisfies criteria favourable to high organic productivity (e.g. Demaison and Moore, 1980; Müller and Suess, 1979; Summerhayes, 1983). Yet primary productivity has been shown to have been relatively

TABLE 3. Sedimentation rates and organic productivity, South Central Sweden.

Borehole	Age	Sedimentation rate ($mm/10^3$ years)	TOC (%)	Accumulation rate ($g\ cm^{-2}/10^3$ years)	Primary organic productivity ($g\ cm^{-2}$/year): 2% Preservation	5% Preservation
DBH 16/74	*Peltura*	7.5	14	0.24	125	48
	A. pisiformis	3	9	0.062	32	12.5
	P. forchhammeri	16	9	0.33	170	65
	P. paradoxissimus	10	4.5	0.103	56	22
Lyckhem 3	*Dictyonema*	3	11.5	0.08	35	16
	U. Cambrian	3.4	16.2	0.13	65	26
	M. Cambrian	5.3	7.4	0.09	45	18

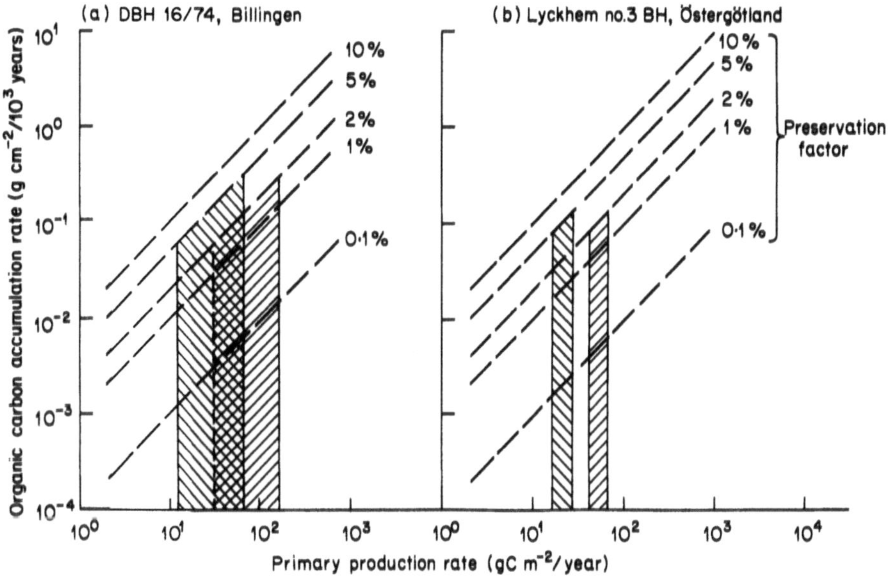

Fig. 8. Primary organic productivity estimates for the alum shales in (a) borehole DBH 16/74 and (b) Lyckhem No. 3 borehole (Östergötland) using the log–log plot of Bralower and Thierstein. Each diagram shows the range of calculated organic carbon accumulation rates for each borehole (Table 3). Two preservation factors are used to estimate palaeofertility: (i) the conservative 2% preservation factor used by Bralower and Thierstein; (ii) a probably more realistic 5% preservation factor, similar to modern euxinic basins. Note that there is clearly a reciprocal relationship between preservation factor and primary production rate.

low. Two alternative explanations exist for this apparent contradiction. Either (a) the alum shales typify early Palaeozoic shelf productivity, implying even lower productivity elsewhere in the ocean (an 'open system'), or (b) the deposit is atypical of shelves and was preserved due to unusual local oceanographic conditions (a 'closed system'). These possibilities are considered below, together with their palaeogeographic implications.

'Open system' deposition

If the alum shales represent typical shelf primary organic productivity in a site which in modern seas would be favourable to high productivity, then by analogy with the present-day distribution of productivity, contemporary global productivity must have been very low. Secular variations in the $^{13}C/^{12}C$ isotope ratio can be explained by low global rates of organic carbon burial during the early Palaeozoic (e.g. Berner and Raiswell, 1983; Garrels and Lerman, 1984). This could in turn be related to initial low organic productivity. If free contact between the alum shale sea and the open ocean is assumed, then by Bralower and Thierstein's reasoning, rates of oceanic circulation and consequent oxygen replenishment must have been greatly reduced.

Berry and Wilde (1978) argue that hydrospheric oxygen contents may have been markedly lower during the early Palaeozoic compared to the present day. Therefore, oxygen consumption by organic matter degradation might have generated and maintained anoxic or euxinic conditions within the water column, despite low primary organic productivity, due to the combined effects of low absolute hydrospheric oxygen content and reduced oxygen replenishment by circulation.

'Closed system' deposition

Since the epicontinental sea environment lacks modern analogues, likely palaeo-oceanographic conditions are difficult to model. Models such as that of Hallam (1981) suggest that these environments may be prone to anoxia since frictional loss of tidal and ocean current energy at the outer shelf margin might reduce watermass circulation and therefore oxygen replenishment. If oxygen replenishment were reduced below wave base, decomposition of organic matter might rapidly lead to the establishment of anoxic and subsequently euxinic conditions.

Moreover, sedimentological evidence (Thickpenny, 1984) indicates the possibility, particularly during the late Middle Cambrian, of local palaeo-

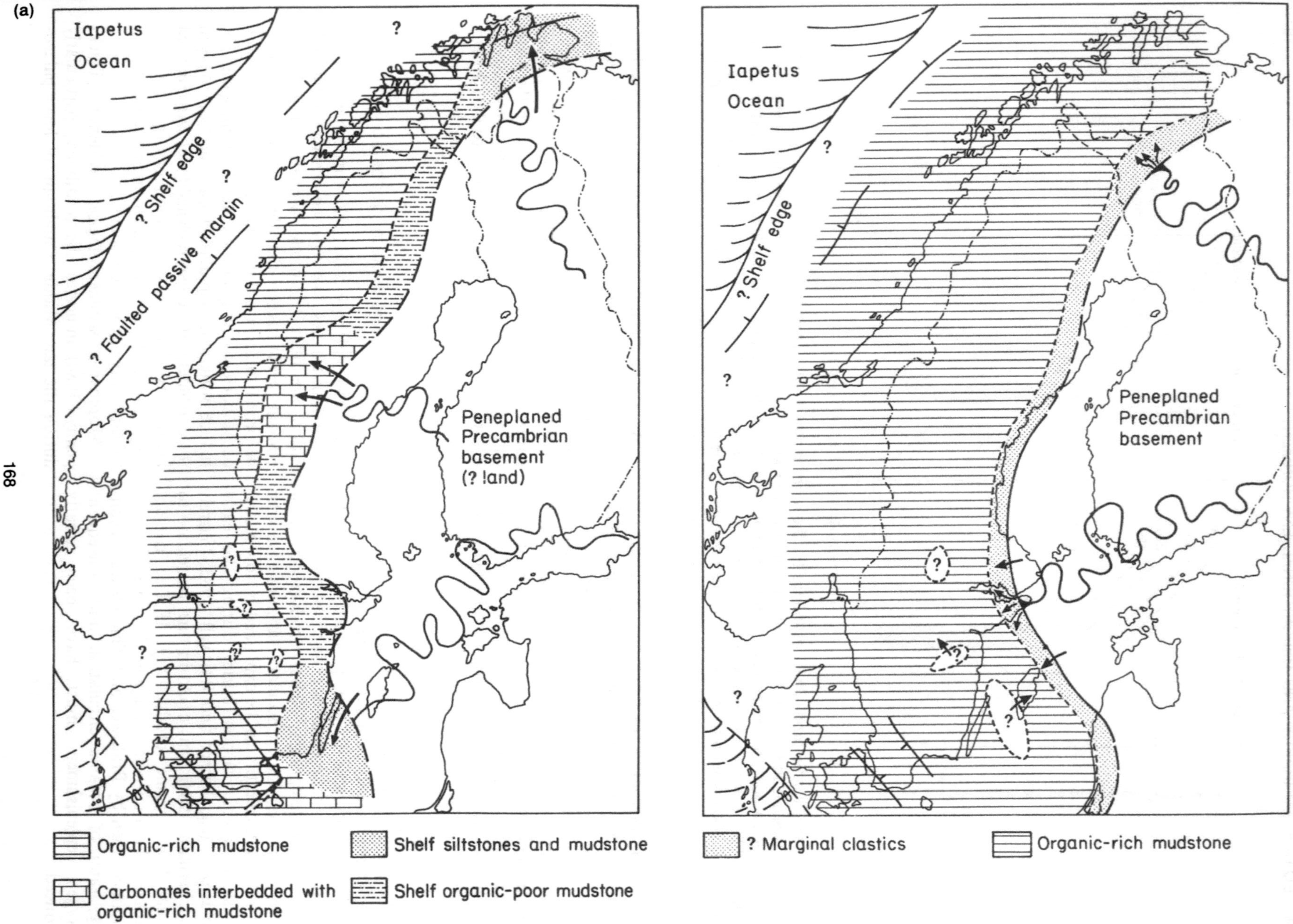

... during (a) the late Middle Cambrian and (b) the Upper Cambrian geographic restriction due to the presence of submarine highs. These may locally have been subaerially exposed, forming a very low-relief archipelago (e.g. Fig. 9). Such conditions may have been a contributory factor in initiating anoxic conditions. This situation of essentially local factors controlling the development of anoxic conditions has no global palaeo-oceanographic implications.

The models described above are end members of an environmental spectrum. It is likely that the actual palaeo-environment of the alum shales combines elements of both. In support of an 'open system' hypothesis, stratigraphic studies show that enhanced organic matter preservation was not a localized phenomenon within the Cambrian period. For example, coeval with alum shale deposition, organic-rich sediments are extensively developed in the British Isles (Leggett, 1980 and references therein), Poland (Sokolowski, 1970), Estonia (Nalivkin, 1973), the Ardennes (Walter, 1980) and south-east Canada (Hutchinson, 1952, 1962). At several intervals organic-rich mudstones blanket the south-east margin of the Iapetus Ocean (Gee *et al.*, 1974; Thickpenny and Leggett, 1987).

However, none of these deposits preserve such high levels of organic carbon as the alum shales, nor do they represent such a broad stratigraphic interval. Hence, in order to maintain anoxic or euxinic conditions in the Baltic area over long periods of time it is necessary to appeal to localized 'closed system' palaeo-oceanographic regimes.

CONCLUSIONS

Sedimentological studies alone indicate deposition of the Scandinavian alum shales in a broad, shallow epicontinental sea environment. The regional palinspastic geometry of this setting suggests potential restriction from ocean currents. Lithologies developed indicate that waters were deficient in oxygen, and this may have been due, in part, to reduced circulation and replenishment. Intrabasinal highs, possibly periodically subaerially exposed, are also inferred from the sedimentology, and these may have caused some palaeogeographic restriction.

Carbon–sulphur relationships provide corroborative evidence of anoxic conditions within the water column. Moreover, they show that euxinic conditions, only hinted at by the lithological development, were apparently dominant during the Upper Cambrian. These conditions coincided with reduced sedimentation rates and detrital input. The area was effectively starved of sediment, probably associated with a relative rise in sea level. Carbon–sulphur ratios also indicate a more abrupt and potentially profound palaeo-oceanographic change marking the Cambrian/Tremadoc boundary. In addition to lithological changes, Tremadoc sediments show a C/S signature more characteristic of oxygenated waters. The trace-element signatures of Upper Cambrian and Tremadoc sediments also differ markedly (Andersson *et al.*, 1985 and references therein), and may well be related to these palaeo-oceanographic variations and concomitant differences in early diagenesis.

Organic productivity calculations give clues as to the likely causes of such extreme reducing conditions during the Upper Cambrian. Since primary productivity was apparently low, and since coeval organic-rich deposits were widespread in north-west Europe, it is difficult to avoid the conclusion that global hydrospheric oxygen contents were low. This effect was compounded by the unusual palaeogeographic setting of the alum shales, to produce euxinic conditions over prolonged periods. Such a hypothesis fits well with the published stable carbon isotope data for the Cambrian (e.g. Berner, 1984; Berner and Raiswell, 1983; Garrels and Lerman, 1981).

The integration of sedimentological and geochemical approaches to the problem of the genesis of the alum shales indicates that they are likely to have been the product of a combination of circumstances which created a depositional and palaeo-oceanographic environment lacking any modern analogue.

ACKNOWLEDGEMENTS

I would like to acknowledge the help of Dr D. G. Gee and his colleagues of SGU and Astrid Andersson of ASA in making available core material and biostratigraphical and geochemical data. The manuscript benefited from comments by Dr J. K. Leggett and Dr R. Raiswell, who also provided analytical facilities at Leeds University. The work was supported by a Royal Dutch Shell scholarship at Imperial College and a research assistant post at Chalmers Tekniska Högskola, Gothenburg.

References

Andersson, A., Dahlman, B. and Gee, D. G. (1983). Kerogen and Uranium resources in the Cambrian alum shales of the Billingen–Falbygden and Närke areas, Sweden, *GFF* **104**, 197–209.

Andersson, A., Dahlman, B., Gee, D. G. and Snäll, S. (1985). The Scandinavian alum shales, *Sver. Geol. Unders., Ser. Ca*, No. 56.

Armands, G. (1972). Geochemical study of Uranium, Molybdenum and Vanadium in a Swedish Alum shale. *Stockh. Contr. Geol.* **27**, 1–148.

Berner, R. A. (1984). Sedimentary pyrite formation: an update. *Geochimica et Cosmochimica Acta* **48**, 605–615.

Berner, R. A. and Raiswell, R. (1983). Burial of organic carbon and pyrite sulfur in sediments over Phanerozoic time: a new theory, *Geochimica et Cosmochimica Acta* **47**, 855–862.

Berner, R. A. and Raiswell, R. (1984). C/S method for distinguishing freshwater from marine sedimentary rocks, *Geology* **12**, 365–368.

Berry, W. B. N. and Wilde, P. (1978). Progressive ventilation of the oceans—an explanation for the distribution of Lower Palaeozoic black shales, *American Journal of Science* **278**, 257–275.

Bralower, T. J. and Thierstein, H. R. (1984). Low productivity and slow deep-water circulation in mid-Cretaceous oceans, *Geology* **12**, 614–618.

Canfield, D. E., Raiswell, R., Westrich, J. T., Reaves, C. M. and Berner, R. A. (1986). The use of chromium reduction in the analysis of reduced inorganic sulphur in sediments and shales, *Chemical Geology* **54**, 149–155.

Cocks, L. R. M. and Fortey, R. A. (1982). Faunal evidence for oceanic separation in the Palaeozoic of Britain, *Journal of the Geological Society London* **139**, 465–478.

Dahlman, B. and Gee, D. G. (1977). Rversikt över billingen-Falbygdens geologi. In *Betäkande av Billingen-tredningen, Billingen 4, exampel.* Statens offentliga utredningar 1977; 47, Industri departmentet, Stockholm.

Degens, E. T. and Ross, D. A. (Eds), (1974). The Black Sea—Geology, Chemistry and Biology, *American Association of Petroleum Geologists* Mem. 20.

Demaison, G. J. and Moore, G. T. (1980). Anoxic environments and oil source-bed genesis, *Bulletin of the American Association of Petroleum Geologists* **64**, 1179–1209.

Garrels, R. M. and Lerman, A. (1981). Phanerozoic cycles of sedimentary carbon and sulfur, *Proc. Natl. Acad. Sci.* **78**, 4652–4656.

Garrels, R. M. and Lerman, A. (1984). Coupling of the sedimentary sulphur and carbon cycles—an improved model, *American Journal of Science* **284**, 989–1007.

Gee, D. G., Karis, L., Kumpalainen, R. and Thelander, T. (1974). A summary of Caledonian front stratigraphy, northern Jämtland/southern Västerbotten, central Swedish Caledonides, *G.F.F.* **96**, 389–397.

Hallam, A. (1981). *Facies Interpretation and the Stratigraphic Record*, Freeman, Oxford.

Hutchinson, R. D. (1952). The stratigraphy and trilobite faunas of the Cambrian sedimentary rocks of Cape Breton Island, Nova Scotia, *Geological Survey of Canada Memoir* **263**, 1–124.

Hutchinson, R. D. (1962). Cambrian stratigraphy and trilobite faunas of south-eastern Newfoundland, *Geological Survey of Canada Bulletin* **88**, 1–56.

Kisch, H. J. (1980). Incipient metamorphism of Cambro-Silurian clastic rocks from the Jämtland Supergroup, central Scandinavian Caledonides, western Sweden: illite crystallinity and 'vitrinite reflectance', *Journal of the Geological Society London* **137**, 271–288.

Leggett, J. K. (1980). British Lower Palaeozoic black shales and their palaeo-oceanographic significance, *Journal of the Geological Society London* **137**, 139–156.

Leventhal, J. S. (1983a). An interpretation of carbon and sulphur data in Black Sea sediments as indicators of environments of deposition, *Geochimica et Cosmochimica Acta* **47**, 133–138.

Leventhal, J. S. (1983b). Organic carbon, sulfur and iron relationships in ancient shales as indicators of environment of deposition, *EOS* **64**, 739.

Martinsson, A. (1974). The Cambrian of Norden, *In* C. H. Hollan, (Ed.), *Cambrian of the British Isles, Norden and Spitsbergen*, Wiley, Chichester.

Müller, P. J. and Suess, E. (1979). Productivity, sedimentation rate and sedimentary organic matter in the oceans—I. Organic carbon preservation, *Deep Sea Research* **26A**, 1347–1362.

Nalivkin, D. V. (1973). *The Geology of the USSR*, Oliver and Boyd, Edinburgh.

Raiswell, R. and Berner, R. A. (1985). Pyrite formation in euxinic and semi-euxinic sediments, *American Journal of Science* **285**, 710–724.

Raiswell, R. and Berner, R. A. (1986). Pyrite and organic matter in Phanerozoic normal marine shales, *Geochimica et Cosmochimica Acta* **50**, 1967–1976.

Scotese, C. R., Bambach, R. K., Barton, C., Vander Voo, R. and Ziegler, A. M. (1979). Palaeozoic base maps, *Journal of Geology* **87**, 217–278.

Smith, A. G., Hurley, A. M. and Briden, J. C. (1981). *Phanerozoic Palaeocontinental World Maps*, Cambridge University Press, Cambridge.

Sokolowski, S. (Ed.) (1970). *Geology of Poland*, Volume 1: *Stratigraphy*. Part 1: *Pre-Cambrian and Palaeozoic* Publishing House Wydawnictwa Geologiczne, Warsaw.

Summerhayes, C. P. (1983). Sedimentation of organic matter in upwelling regimes, *In* J. Thiede and E. Suess (Eds), *Coastal Upwelling*, Part B, Plenum, New York.

Thickpenny, A. (1984). The sedimentology of the Swedish alum shales, *In* D. A. V. Stow and D. J. W. Piper (Eds), *Fine-grained Sediments: Deepwater Processes and Facies*, Spec. Publ. Geol. Soc. London, No. 15, pp. 516–527.

Thickpenny, A. and Leggett, J. K. (In press). Stratigraphic distribution and palaeo-oceanographic significance of European early Palaeozoic organic-rich sediments, *In* J. Brooks and A. J. Fleet, (Eds). *Marine Petroleum Source Rocks*, Spec. Publ. geol. Soc. London, No. 24, pp. 231–247.

Thorslund, P. (1960). The Cambro-Silurian, *In* N. H. Magnusson, P. Thorslund, F. Brotzen, B. Asklund and O. Kulling, (Eds), Description to accompany the map of the pre-Quaternary rocks of Sweden. *Sver. Geol. Unders.* Ser. Ba **16,** 69–110.

Walter, R. (1980). Lower Palaeozoic Palaeogeography of the Brabant Massif and its southern adjoining areas, *Meded. Rijks. Geol. Dienst.* **32,** 14–25.

Chapter 9

Understanding Jurassic Organic-rich Mudrocks—New Concepts using Gamma-ray Spectrometry and Palaeoecology: Examples from the Kimmeridge Clay of Dorset and the Jet Rock of Yorkshire

Keith J. Myers, Department of Geology, Imperial College, London, and *Paul B. Wignall*, Department of Geological Sciences, University of Birmingham

ABSTRACT

Palaeoecology and gamma-ray spectrometry have been combined to study the controls on deposition of two well-known Jurassic organic-rich mudrocks. Portable gamma-ray spectrometry allows the rapid measurement of K, U and Th in the field. Th and K are contained mainly in the detrital clay fraction of the mudrocks and the Th/K ratio is found to decrease slightly with increasing water depth and distance from shoreline. Measured values of U have a detrital and an 'authigenic' component. Sediments deposited under anoxic conditions are consistently enriched in 'authigenic' U. Non-detrital 'authigenic' U has been estimated by the relative enrichment of U over Th, which is assumed to be immobile and present entirely in the detrital fraction.

The Kimmeridge Clay represents a several-million-year period of organic-rich mudrock deposition in which climate-induced stratification of the water column was the predominant control on the formation of the different lithologies. Sea level controlled the position of storm wave-base relative to the substrate and thus the frequency of oxygenation events during shale deposition. Coccolith limestones contain up to 10 ppm 'authigenic' U and represent the most anoxic sediments.

The lower Toarcian 'anoxic event' coincides with a period of rising sea level and transgression. The maximum 'authigenic' U value of 8.2 ppm was recorded 4.5 m above the base of the Jet Rock member and represents the most intense period of anoxia.

Organic carbon correlates well with 'authigenic' U in the lower Toarcian where the degree of bottom water anoxia appears to be the main control on organic carbon preservation. The organic carbon/'authigenic' U correlation is poor in the Kimmeridge Clay, where sedimentation rate and surface productivity are additional influences on organic carbon preservation. U/organic carbon ratios in the Jurassic case-study sediments are much lower than for Palaeozoic black shales.

INTRODUCTION

Thick organic-rich mudstone sequences are common in the British Jurassic and are of economic importance, being widely recognized as the main source for

Marine Clastic Sedimentology © J. K. Leggett & G. G. Zuffa (Graham and Trotman, 1987) pp. 172–189

both North Sea and southern England onshore hydrocarbon deposits (Glennie, 1984). Previous studies of the depositional environments of these sediments have been mainly concerned with sedimentological and palaeoecological analysis of benthic conditions (e.g. Morris, 1979). The benthic oxygen level at the time of deposition is one of the most important controls on the amount of bioturbation and fauna (Demaison and Moore, 1980), and on the preservation of organic matter (Curtis, this volume). The evaluation of palaeo-oxygen levels is thus of great importance in determining the location of the best hydrocarbon source rocks within a basin. We intend to show how gamma-ray spectrometry, when used in conjunction with palaeoecological studies, can be a powerful tool for predicting benthic palaeo-oxygen levels. We have combined detailed palaeoecology and gamma-ray spectrometry studies to elucidate depositional models and controls on organic carbon preservation for two well-known examples of organic-rich mudrocks from the British Jurassic.

The first example, the Kimmeridge Clay of Dorset, represents a prolonged period of organic-rich mudrock deposition (Cox and Gallois, 1981). The Jet Rock of Yorkshire, in contrast, manifests a relatively brief 'anoxic event' in the lower Toarcian (Jenkyns, 1985; Raiswell and Berner, 1985).

A portable gamma-ray spectrometer has been used in the field to measure the potassium (K), uranium (U) and thorium (Th) contents, and total radioactivity of both sequences. As this is a new application of a tool first developed for uranium exploration (Adams and Gasperini, 1970), we have included a review of the theory and methodology of portable gamma-ray spectrometry and the geochemistry of K, U and Th in sedimentary rocks.

In the two case-studies discussed, sedimentological and faunal evidence are first reviewed before the results of the gamma-ray spectrometry are presented.

In this work K. Myers is responsible for the gamma-ray spectrometry and P. Wignall for the palaeoecology, and the depositional models represent our combined views.

GAMMA-RAY SPECTROMETRY

Theory

Almost all natural gamma radiation originates from the decay of the three radionuclides, ^{40}K, ^{238}U and ^{232}Th. During radioactive disintegration these elements emit high-energy electromagnetic radiation (gamma rays), whose energy is characteristic of the emitting isotope. By measuring the intensity and energy of the gamma radiation at a particular location the abundance of the K, U and Th can be determined. Potassium only emits a single gamma ray, whereas the ^{232}U and ^{238}Th series each emit a number of gamma rays of different energies, of which only one from each series is normally measured by the detector. The Geometrics 410A Gamma-Ray Spectrometer used in this study is able to detect gamma rays emitted from rock surfaces and divide them into four channels: (1) total gamma rays of all energies; (2) ^{40}K gamma rays; (3) ^{214}Bi gamma rays from the uranium series; (4) ^{208}Tl gamma rays from the thorium series. Before the gamma-ray spectrometer can be used in the field it must be calibrated so that the detected gamma rays can be converted to known concentrations of the radioelements. In this case the spectrometer was calibrated over a series of five concrete pads doped with known concentrations of K, U and Th, using a method described by Lovborg (1984) (Fig. 1).

Field procedure

The following points are important in choosing sections suitable for study.

1. Sections should be as unweathered as possible since leaching of K and U may occur in deeply weathered rocks (Levinson and Coetzee, 1978).
2. The rock must be in a state of radioactive equilibrium, otherwise gamma rays emitted by ^{214}Bi and ^{208}Tl cannot be used to measure the concentration of the U and Th parents. Equilibrium can always be assumed for the thorium series and almost always for the uranium series, provided that the rocks are more than 1 Ma old and are not deeply weathered (Levinson and Coetzee, 1978).
3. As a rule, the geometry of the field location should closely approximate that of the calibration pads, i.e. a flat area of at least 3 m diameter (Fig. 2).
4. Variable moisture content can alter the emanating power of the rock and may add small errors to the U values—so sections of constant moisture content are preferable.

Wave-cut platforms are ideal sites since they satisfy all of these requirements, being flat, relatively fresh and permanently saturated. Cliff sections have also been found to give satisfactory results, though care must be taken with geometry and weathering. The two case-studies presented here are both from broad wave-cut platforms.

Fig. 1. Gamma-ray spectrometer on one of the calibration pads at the Risø National Laboratory, Denmark. Following primary calibration the spectrometer is monitored locally on control stations to check on its performance. The spectrometer used in this study has not required recalibration for over 20 months.

Fig. 2. Typical field exposure showing excellent measurement geometry.

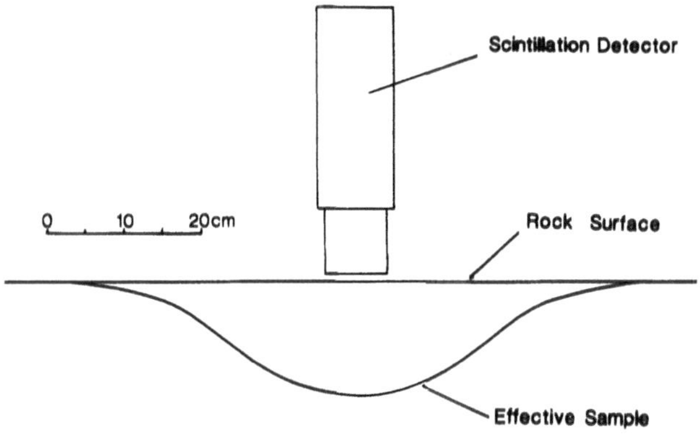

Fig. 3. This illustrates the effective sample of a field measurement as computed by Løvborg (1971).

When the detector is placed in contact with the rock surface the effective sample is a bowl-shaped volume of thickness 14 cm, surface diameter 80 cm and mass 49 kg (Løvborg, 1971; Fig. 3). Where possible the detector is placed on top of bedding surfaces in order to achieve maximum resolution (Fig. 4). Measurement spacing depends on the style of exposure, time available and lithological variation. In this work measurements were taken at least every one metre, and more frequently where lithological variations are less than one metre thick. The detector is left to accumulate counts for between 4 and 10 minutes depending on the radioactivity of the rock. This was found to maintain a precision of better than ±10% for all three radioelements except in rocks of lowest radioactivity, e.g. limestones. The reader is referred to Killeen (1979) and Løvborg (1984) for fuller technical and procedural discussions.

K, U and Th geochemistry in sedimentary rocks—general

On weathering of a typical granitoid source rock, potassium will largely be contained in detrital feldspar and mica grains, and as a component of soil clays such as illite. Some K will also go into solution. Clays continue to adsorb K during transport and upon deposition in the marine environment. Potassium does not substitute in the carbonate lattice but may be concentrated in late-stage K-rich evaporites (Heier and Billings, 1972).

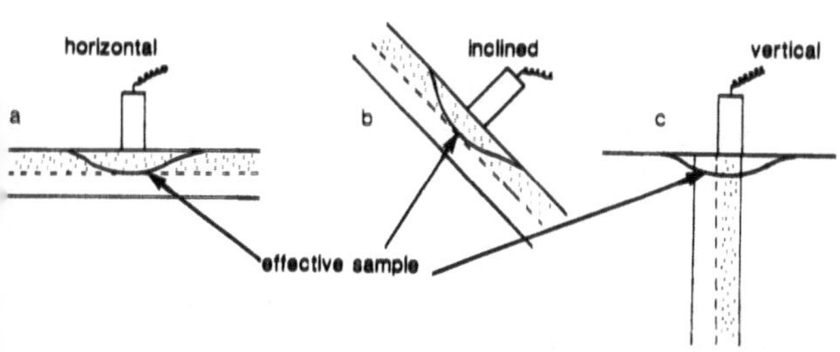

Fig. 4. This illustrates the relationship between the style of bedding and resolution. A and B give the finest resolution.

The geochemistry of thorium and uranium in the surface environment is closely linked to their mode of occurrence in granitoid source rocks where they are present both in accessory minerals, either as separate grains or as inclusions in major minerals, and loosely bound to mineral surfaces (Silver *et al.*, 1982; Guthrie and Kleeman, 1986). Uranium and Th contained in insoluble accessory minerals are leached to a very limited extent and transported and deposited as clastic material (Dongarra *et al.*, 1984). There is a strong tendency for Th released by weathering to be adsorbed and concentrated in soil clays such as illite (Tieh *et al.*, 1980). Thorium is effectively insoluble in water (Langmuir and Herman, 1980) and has a low residence time in the oceans of less than 100 years (Cochraine *et al.*, 1986). Uranium, like Th, is present in accessory minerals and clay grade material. In oxidizing environments it is unstable as U^{4+} and forms the highly soluble UO_2^{2+} uranyl ion. Unlike Th it is also carried partly in solution, mainly as uranyl carbonate complexes (Langmuir, 1978). The residence time of U in the oceans is therefore much greater than Th, being about 400 000 years (Cochraine *et al.*, 1986). Table 1 shows the range in the K, U and Th content of some common minerals found in sedimentary rocks.

K, U and Th in marine mudrocks

Most argillaceous mudrocks are composed of mixtures of clay minerals and silt-grade quartz with minor feldspar, carbonate, organic matter and iron oxide (Potter *et al.*, 1980 and Table 2). Potassium and Th will be contained mainly in the clay fraction, as the other K and Th carriers, such as detrital K-feldspar, mica and heavy accessory minerals will mostly have been deposited earlier before reaching the sites of mudrock deposition, though this may not always be true of micas. Two Th/K cross-plots for the case-study mudrocks confirm a direct relationship between K and Th (Figs 5 and 6). The K and Th content and Th/K ratio reflect the proportion, type and composition of clay minerals present.

Uranium differs from K and Th in that as well as being associated with the detrital clay fraction it may also be removed from solution to enrich the sediment in 'authigenic' U, i.e. uranium from non-detrital

TABLE 1. Radioelement contents in various minerals (from Schlumberger, 1982; Fertl, 1982; Adams *et al.*, 1959).

Mineral	*K* (%)	*U (ppm)*	*Th (ppm)*
Major			
Quartz	0	0.7	2.0
K-Feldspar[c]	11.8–14	3–7	0.2–5
Muscovite[a]	7.9(a)	2–8	10–25
Biotite[b]	8.5(a)	1–4	0.5–50
Calcite	0	1.5	0
Plagioclase	0	0.5–3	0.2–5
Pyroxene	0	2–25	0.01–40
Hornblende	0	1–30	5–50
Clay Minerals			
Illite	6.7a (3.4–8.3)	1.5?	10–25
Glauconite	4.5a (3.2–5.8)	10?	?
Montmorillonite	1.6a (0–4.9)	2–5	10–24
Chlorite	0.1a (0–0.35)	?	3–5
Kaolinite	0.30 (0–0.6)	1.5–9	6–42
Accessory Minerals			
Zircon	Z	100–6000	50–4000
Sphene	E	100–700	100–600
Apatite	R	5–150	20–150
Monazite	O	500–3000	25,000–200,000
Xenotime		500–35,000	Low
Ilmenite		1–50	?
Magnetite		1–30	0.3–20
Epidote		20–50	50–500
Evaporites			
Halite	0	Negligible	
Anhydrite	0	Negligible	
Gypsum	0	Negligible	
Sylvite	52	Negligible	
Polyhalite	14	Negligible	

[a] Pure muscovite does not contain thorium but muscovite is often closely associated with thorium-bearing accessory minerals in sediments.
[b] The thorium and uranium will be contained in heavy mineral inclusions such as zircon.
[c] Refers to pure K-feldspar—K% probably lower in sedimentary K-feldspar.

TABLE 2. Typical shale composition from Potter *et al.* (1980).

	%	*Range*
Clay minerals	58	<50–>90
Quartz	28	10–80
Feldspar (dom Na)	6	0–30
Carbonates	5	0–>50
Iron oxide	2	
Organic carbon	?	

sources. The possibility of uranium being removed from solution has been of interest for some time as a means of explaining the very high U contents of some organic-rich sediments (Swanson, 1961). Recently several authors have used $^{234}U/^{238}U$ ratios to show that many organic-rich sediments do contain 'authigenic' U derived from sea water and that often there is a general correlation between U and organic carbon (Veeh, 1967; Veeh *et al.*, 1974; Holland, 1984).

The mechanism of uranium fixation in organic-rich sediments has been the subject of some dispute. Degens *et al.* (1977) suggested that the uranium was fixed by living organisms during growth and that anoxic depositional environments were only necessary to preserve the already uranium-enriched

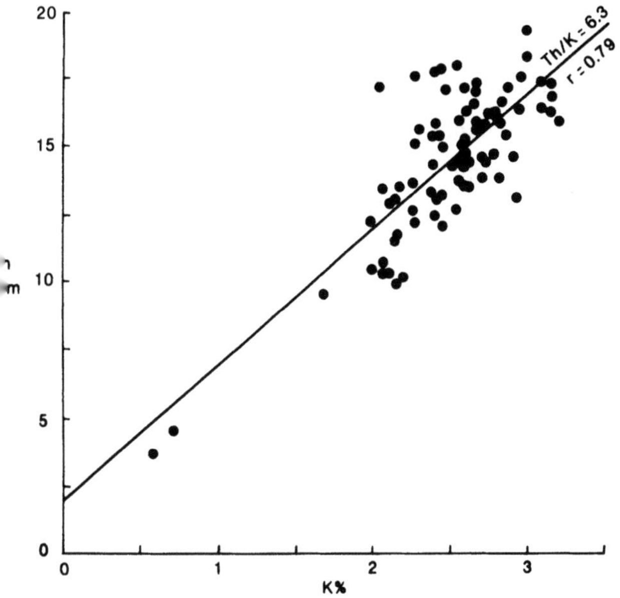

Fig. 5. Thorium/potassium cross-plot for the Lower Toarcian mudstones.

organic material which would otherwise be recycled before sedimentation. However, Mann and Fyfe (1985) only recorded a maximum of 1 ppm U in modern marine algae. Uranium from this source can only account for a small part of the enrichments observed in organic-rich sediments.

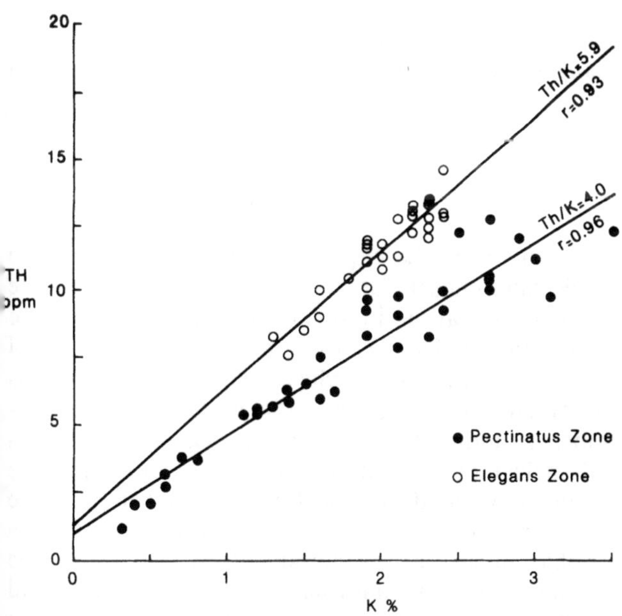

Fig. 6. Thorium/potassium cross-plot for the Kimmeridge Clay Formation.

Reactions at the sediment–water interface are likely to be more important in the fixation of uranium. Experimental work by Kochenov *et al.* (1977) has shown that organic matter plays an active role in the extraction of U from dilute solutions in reducing environments. Uranium can be fixed in the sediment by oxidation–reduction reactions following sorption of uranyl ions by humic acids (Borovec *et al.*, 1979). Sediments enriched in 'authigenic' U tend to be deposited under anoxic conditions which permit both large amounts of organic matter to accumulate and U to be fixed. In detail, the uranium–organic carbon relationship is complex, depending on such factors as the type of organic matter (terrigenous or marine), sedimentation rate and the position of the redox boundary relative to the sediment–water interface. The U/organic carbon ratio varies considerably between different organic-rich shales (Table 3). The U/organic carbon ratios for the Jurassic shales of this study are considerably lower than those for Palaeozoic black shales and the Recent Black Sea, with the exception of some coccolith limestones in the Kimmeridge Clay.

Estimation of 'authigenic' uranium

Adams and Weaver (1958) calculated that the average Th/U ratio of 67 mudrocks of various ages was 3.8 ± 1.1, very close to the crustal average of 3.5 (Adams *et al.*, 1959). The average Th/U ratio of 154 low organic carbon marine and deltaic mudstones measured in the present study was 3.9 ± 0.7, whilst 88% of the samples have Th/U ratios in the range 3–5. It is reasonable, therefore, to take this as the 'normal' Th/U ratio of the detrital clay fraction. A decrease in this ratio should indicate the presence of authigenic uranium, based on the assumption that Th is effectively immobile and entirely in the detrital fraction.

In an organic-rich mudrock the uranium associated with the detrital fraction can be estimated by dividing the measured Th content by 3. The difference between the actual U measured and the calculated detrital U will give a value for the amount of authigenic U in the sample. A Th/U ratio of 3 is assumed in order that only significant amounts of authigenic U are measured. Sediments with a Th/U ratio of greater than 3 will give negative values for authigenic U.

TABLE 3. A comparison of U/organic carbon ratios between Kimmeridge Clay and Lower Toarcian lithologies and other black shales.

Lithology	*U/% organic carbon*[a] $\times 10^{-4}$ *gm/gm*	
Cambrian Alum shale	17	(Holland, 1984)
Devonian black shales	4	(Holland, 1984)
Recent Black Sea:		
coccolith ooze	3.2	(Rhona and Jensu, 1974)
organic-rich laminate	1.2	(Rhona and Jensu, 1974)
Kimmeridge Clay:		(Author's data)
Blackstone oil shale	0.1	
elegans zone bituminous shale	0.2–0.5	
pectinatus zone oil shale	0.2–0.5	
pectinatus zone coccolith limestone	1.0–2.9	
Lower Toarcian:		
Lower Jet Rock member bituminous shale	0.6–2.0 (1.2a)	

[a] U is total U (detrital and 'authigenic') in ppm; organic carbon is TOC expressed in %.

For example, if a sample is found to contain 12 ppm Th and 10 ppm U:

$$U_{det} = \frac{12}{3} = 4$$

$$U_{aut} = U_{act} - U_{det}$$

$$\text{U authigenic} = 10 - 4 = 6 \text{ ppm}$$

U can substitute to a limited extent in the carbonate lattice and up to 2 ppm authigenic U can be accounted for in this way.

We now demonstrate that the authigenic U content of mudrocks can be used as a broad measure of benthic oxygen level, and that combined with traditional palaeontological tools it can help to refine existing depositional models.

RESULTS

Kimmeridge Clay

The outcrops of the Kimmeridge Clay Formation examined in this study occur at the type locality, Kimmeridge Bay, Dorset. The exposed sequence comprises claystones and organic-rich shales interbedded on a 0.5 m scale (Cox and Gallois, 1981). Amounts of preserved organic carbon are exceptionally high, ranging from 1% for some claystones to over 50% for some oil shales (Farrimond *et al*., 1984; Ebukanson and Kinghorn, 1985). Previous depositional models, usually based on the Dorset outcrops, have tended to focus on comparisons with recent analogues. Gallois (1976) suggested that toxic blooms of phytoplankton poisoned the benthic fauna, caused anoxic bottom waters and led to enhanced levels of organic carbon preservation. Tyson *et al.* (1979) suggested, by analogy with the recent Black Sea, that a rising O_2/H_2S interface in an increasingly stratified water column could account for the various lithologies.

Palaeoecology

Sediments from the *elegans* and *pectinatus* ammonite zones were chosen for analysis as they illustrate the range of variation in the depositional system (see Figs 7 and 8). The *elegans* zone sediments consist of cyclic alternations of claystones and bituminous (organic-rich) shales, while the *pectinatus* zone contains, in addition, exceptionally organic-rich shales (oil shales) and coccolith limestones. Comprehensive counts were made of the fauna contained in these beds, and the numerically dominant elements, the trophic nuclei, are summarized in Table 4. The Shannon–Weaver information index (Dodd and Stanton, 1981) is given for each lithology to enable a comparison of the dominance-diversity between beds.

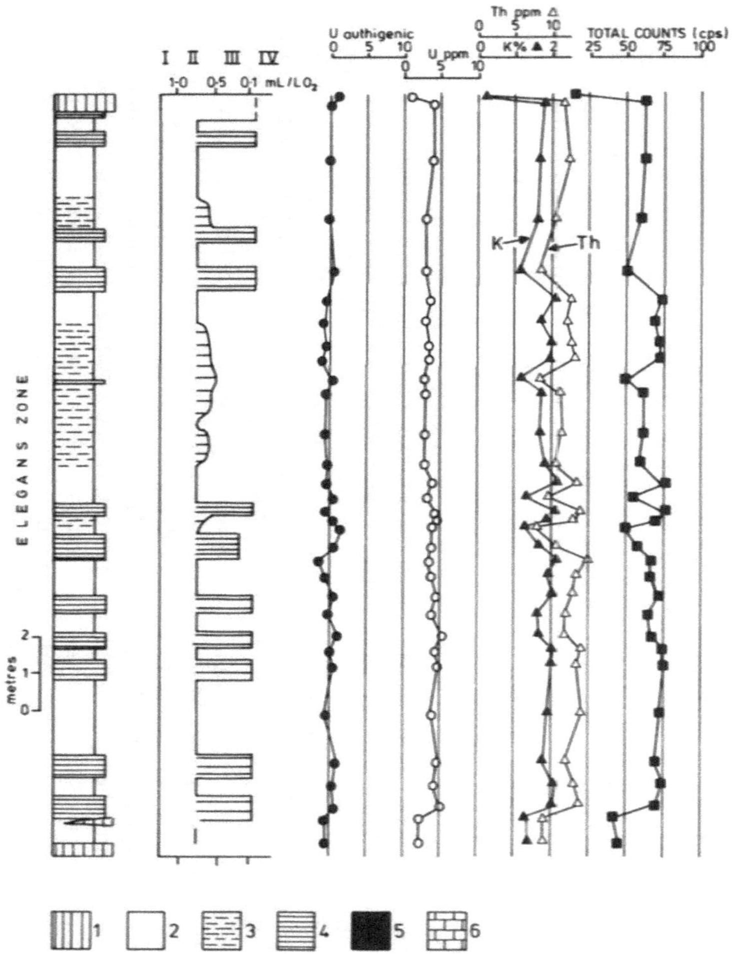

Fig. 7.

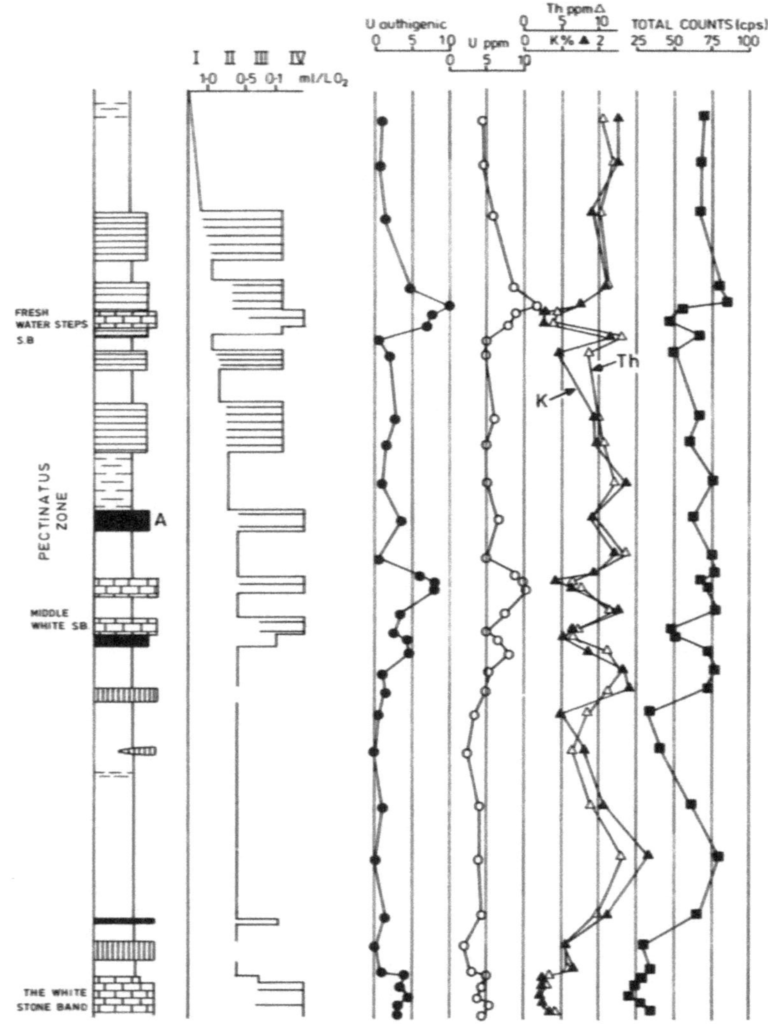

Fig. 8.

Figs 7 and 8. *Lithologies:* 1, diagenetic dolostone (Irwin, 1980); 2, claystone to marly claystone; 3, sub-bituminous shale; 4, bituminous shale; 5, oil shale; 6, coccolith limestone. Lithologies 3, 4 and 5 are distinguished in the field by the fossil content and sedimentary fabric, and in the laboratory by increasing proportions of organic carbon measured by a Perkin–Elmer elemental analyser. *Biofacies:* I, aerobic; II, restricted aerobic; III, dysaerobic; IV, azoic. Oxygen levels are based on recent analogues. The left-hand columns show a log of the *elegans* and *pectinatus* zone sediments at Kimmeridge Bay, Dorset, with the interpreted benthic oxygen levels to the right of it. The right-hand columns illustrate the amounts of K, U and Th determined by gamma-ray spectrometry.

TABLE 4. Trophic nuclei for the *elegans* and lower *pectinatus* zones based on the field analysis of 59 samples with over 6000 specimens. Samples were chosen every metre, or less when the lithological alternations are more frequent.

	Trophic nuclei				*Dom.-div. index*
	Elegans zone				
Lithofacies	*1*	*2*	*3*	*4*	*H*
Marly clay–clay	'*Lucina*' 30.9	*Corbulomima* 30.8	*Protocardia* 13.9	Ammonites 7.0	1.20
Bituminous shale	'*Lucina*' 29.9	*Protocardia* 26.6	*Corbulomima* 19.6	Ammonites 9.6	1.23
	Pectinatus zone				
Normal marly clay (above bed A)	*Discinisca* 29.3	*Quadrinervus* 24.9	Ammonites 22.9	*Protocardia* 7.8	1.55
Restricted marly clay (below bed A)	Ammonites 51.3	*Protocardia* 14.8	*Lingula* 12.3	*Liostrea* 7.4	0.69
Bituminous shale	Ammonites 60.1	*Protocardia* 14.2	*Discinisca* 9.1		1.16
Oil shale	Ammonites 69.3	*Protocardia* 9.5	*Discinisca* 4.9		0.64
Coccolith limestone	Ammonites ~70	Fish ~25			0.23

The results for the *elegans* zone show that both claystones and bituminous shales have a very similar fauna dominated by infaunal, filter-feeding bivalves. The two lithologies can, however, be distinguished by their biofabrics. The claystones contain a homogeneously distributed, disarticulated fauna with moderate fragmentation. In contrast, the bituminous shales have a consistently less fragmented, disarticulated fauna forming shell pavements on discrete bedding planes. One species tends to dominate on any one pavement, so the dominance-diversity is much lower for an individual pavement than the average given in Table 4.

The origin of these shell pavements is clearly of some importance in determining the sedimentary dynamics of these beds. Aigner (1980) has interpreted similar shell pavements from the *autissiodorensis* zone of the lower Kimmeridge Clay as being the allochthonous products of storms. The studies of faunal fragmentation and the concave : convex and left : right ratios of the valves (which are typically 50 : 50) do not support this interpretation. Moreover, shell imbrication and other current-generated structures are extremely rare. The beds are also laterally persistent (Downie, 1955) and, therefore, shell transport distances would have been of the order of tens of kilometres. This is clearly not in accord with the dominantly unfragmented thin-shelled fauna. An alternative, and more realistic, explanation for the *elegans* zone sediments is that the shell pavements represent short-lived periods of colonization by a low-diversity, para-autochthonous fauna.

Examination of the trophic nuclei for the *pectinatus* zone (Table 4) shows an increased proportion of pelagic components (i.e. ammonites) throughout the lithologies, although their overall abundance is not greatly different from that in the *elegans* zone. Coevally, the absolute number and size of the benthic fauna decreases. These trends probably reflect a less oxic benthic environment. Within the bituminous shales the fauna is restricted to discrete horizons, although in insufficient density to form shell pavements. Fauna is extremely rare or absent in the oil shales. The coccolith limestones bands contain a number of moderately burrowed horizons probably caused by a small soft-bodied fauna, but a shelly benthos is lacking, excepting a single oyster horizon in the Freshwater Steps Stone Band. Rare scoured horizons may also occur in the limestones. Marly claystones interbedded with the coccolith limestones and oil shales contain a similar impoverished benthic fauna. However, the thick sequence of marly claystones above the youngest oil shale (Fig. 8) contains an increasingly diverse and numerically abundant fauna indicating a change to more normally oxygenated benthic conditions.

There is no systematic pattern in the occurrence of oil shales and coccolith limestones: for example, a discrete coccolith-poor oil shale is enclosed within the White Stone Band coccolith limestone (Wilson 1980, Fig. 26b). The two lithologies appear to be end-member deposits of the same depositional environment (cf. Tyson *et al.*, 1979) in which subtle variation in environmental parameters may have controlled the productivity of the dominant phytoplankton.

Palaeoecological interpretation

Sediments from the *elegans* zone can be interpreted as having been deposited in an environment which fluctuated between relatively normal benthic oxygen levels (claystone facies) and dominantly anaerobic conditions (bituminous shale facies). Frequent influxes of oxygen allowed a low-diversity infauna to colonize the substrate. Weak currents continued to operate after the influxes of oxygen, exhuming and disarticulating the shells. We attribute the oxygenation events to the breakdown of stratification by vertical mixing of the water column.

The benthic fauna of the lower *pectinatus* sequence is considerably more impoverished than that of the *elegans* zone, and is virtually absent in the oil shales and coccolith limestones. At times of benthic colonization, dissolved oxygen levels were probably in the range of 0.3 to 1.0 ml/litre. This is the equivalent of the upper dysaerobic biofacies of Savrada *et al.* (1984) and Kammer *et al.* (1986), characterized by an impoverished shelly benthos. The composition and abundance of the benthic fauna has been used to estimate benthic oxygen levels during deposition (Figs 5 and 6).

Gamma-ray spectrometry

The triangular composition plot (Fig. 9) shows that sediments from the two zones form distinct compositional fields. The *elegans* zone has very little enrichment in authigenic uranium and many of the values are negative (Fig. 7). In contrast, almost all of the sediments from the *pectinatus* zone are enriched in authigenic uranium over thorium. The highest values are associated with the coccolith limestones and some oil shale beds. The amount is too high to be explained by substitution in the carbonate lattice. Thorium and potassium decrease when replaced by carbonate in the diagenetic dolostones (Irwin *et al.*, 1980) and coccolith limestones and organic matter in the oil shales.

Figure 6 shows that the Th/K ratio decreases from the *elegans* zone to the *pectinatus* zone, suggesting a change in the composition of the detrital clay fraction between the two zones. It is not clear whether this reflects a change in clay mineralogy or a change in the chemical composition of the clay minerals.

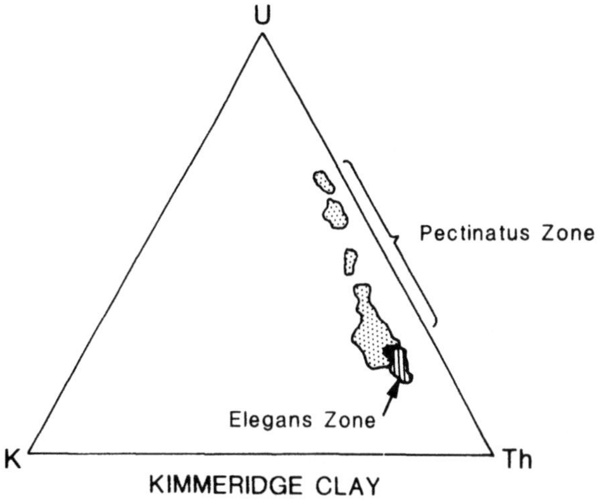

Fig. 9. Triangular composition plot for the Kimmeridge Clay.

Total organic carbon contents

The measured values for preserved total organic carbon (TOC) in Table 5 range from 2.6 to 23%. In the *elegans* zone, TOC contents increase with decreasing benthic palaeo-oxygen levels. In the lower *pectinatus* zone, however, the coccolith limestones—interpreted as the most oxygen deficient—contain similar levels of organic carbon to the better-oxygenated mudstones and bituminous shales. Apart from individual oil shale horizons, the lower *pectinatus* zone sediments have lower TOC levels than the *elegans* zone sediments, despite being deposited in an apparently more restricted environment.

Interpretation and discussion

We argue that the enrichment of the lower *pectinatus* zone sediments in authigenic uranium reflects a dif-

TABLE 5. Total organic carbon contents of various Kimmeridge Clay lithologies.

	Total organic carbon (%)	
Lithology	*Elegans*	*Pectinatus*
Claystones and sub-bituminous shales	4.2–9.7 (6)[a]	3.0–3.7 (4)
Bituminous shales	10.0–13.3 (4)[a]	4.6–5.2 (3)
Oil shale	—	13.6–22.5 (2)
Coccolith limestones[b]	—	2.6–6.6 (6)

[a] Number of samples.
[b] This includes four values from Farrimond *et al.* (1984).

ference in primary depositional environment from the *elegans* zone. This is best accounted for by more prolonged periods of anoxia, which allowed more uranium to be fixed by organic matter. In the *elegans* zone it appears that bottom-water anoxia did not occur for a sufficient length of time for significant uranium fixation to occur. This is further supported by the increased frequency of shell pavements in this zone. The coccolith limestones and oil shales of the lower *pectinatus* zone represent the most extreme periods of bottom-water anoxia. The data cannot, however, show whether the Kimmeridgian seas were ever truly euxinic, i.e. whether or not the O_2/H_2S interface extended above the sediment–water interface for long periods of time (cf. Tyson *et al.*, 1979).

The more restricted aerobic facies of the *pectinatus* zone mudstones would suggest a deeper-water environment than the equivalent mudstones of the *elegans* zone, by the general model of Byers (1977). This depth increase is further substantiated by a decrease in the effectiveness and lower frequency of the oxygenation events at the *pectinatus* zone. Comparison with recent analogues for the maximum depth of vertical mixing of the water column (Ross and Degens, 1974), and allowing for basin size and less rigorous Mesozoic circulation (Hallam, 1986), suggests estimated depths of 50 m for the *elegans* zone and 100 m for the *pectinatus* zone. This would put the former within the influence of storm wave-base and the latter at the very limit of its influence. Such depths do not require the O_2/H_2S interface to rise to any great heights in the water column before nutrient recycling in the euphotic zone could occur.

It is worth speculating as to how large amounts of organic matter were preserved in the *elegans* zone sediments without either inducing prolonged bottom-water anoxia or the precipitation of 'authigenic' uranium. The benthic oxygen level represents a dynamic equilibrium between supply—by circulating currents, and demand—by decaying organic matter (Waples, 1983). The amount of organic matter ultimately preserved is a complex function of primary productivity, water circulation, sedimentation rate and the type of organic matter. The longer organic matter is in contact with sea water, the more uranium can be fixed (Swanson, 1961). High organic sedimentation rates under periodically anoxic waters would allow large amounts of organic carbon to be preserved (Curtis, this volume) while inhibiting uranium fixation. Perhaps also the organic matter, comprised dominantly of marine algal material (Farrimond *et al.*, 1984), was less efficient in fixing uranium than other types of organic matter as discussed by Swanson (1961). Periods of bottom-water anoxia may have been too short for uranium fixation compared to Palaeozoic black shales or the density-stratified Black Sea, as evidenced by the presence of a benthic fauna. If the source of organic matter is assumed to be similar in the *elegans* and *pectinatus* zone sediments, the apparent conflict between lower overall levels of organic carbon and more prolonged periods of anoxia in the *pectinatus* zone can be explained by lower organic input into the sediments through lower surface productivity or lower sedimentation rates causing greater loss of organic carbon by sulphate reduction. Thin-section analysis reveals a decrease in the proportion of silt from the *elegans* to *pectinatus* zones, suggesting a decrease in sedimentation rate. Productivity changes are very difficult to evaluate. The high U/organic carbon values of the coccolith limestones and associated lithologies could be explained by very slow rates of sedimentation during prolonged periods of bottom-water anoxia.

Depositional model

The cyclic alternation between more and less oxygen-depleted lithologies is sufficiently rapid (*c.* 14000 years: Dunn 1974) to discount all but climatic factors as the major driving mechanism for their formation. In the present-day eastern Mediterranean, it is predicted that an increase of less than one degree in surface water temperature could cause stratification of the water column (Mangini and Schlosser, 1986), hence the climatic change may have only been slight in order to effect the transition. The decrease in fragmentation of the fauna in the organic-rich shales is suggestive of decreasing energy levels. An increasingly stratified water column in which oxygen renewal became less efficient is supported by the palaeontological evidence already cited.

Palaeo-water-depth is also an important factor in accounting for the differences between the two zones, although it had no direct effect on the cyclicity.

Figure 10 illustrates the main points of the new depositional model. The *elegans* zone sediments were deposited within reach of storm wave-base and water column stratification was, therefore, frequently broken down even during deposition of the bituminous shales, allowing benthic colonization and inhibiting fixation of 'authigenic' uranium. Since the *pectinatus* zone sediments were deposited at the extreme influence of storm wave-base, stratification was less frequently and less effectively disturbed; consequently benthic oxygen levels were lower, and authigenic uranium shows greater levels of enrichment.

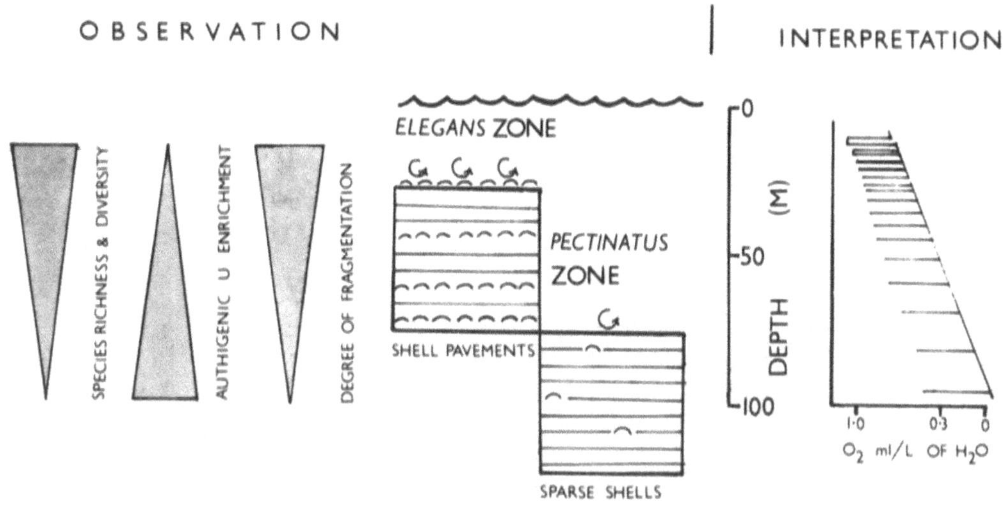

Fig. 10. The main points of the depositional model. The horizontal lines on the graph of benthic oxygen level indicate the frequency and magnitude of the oxygenation 'events'. Climate-induced stratification of the water column was the predominant control on the formation of the different lithologies. Sea level controlled the position of storm wave-base relative to the substrate and thus the frequency of oxygenation events during shale deposition.

The level of preserved organic carbon is controlled both by water column stratification and sedimentation rate. The overall higher levels of organic carbon in the *elegans* zone are a reflection of higher sedimentation rate, probably coupled with higher surface productivity.

The Lower Toarcian Jet Rock of Yorkshire

The Liassic of Yorkshire is an entirely marine succession dominated by shales of several types, usually rich in sideritic concretions, but with some sandstones, siltstones and ironstones. It accumulated in a shallow epicontinental sea towards the western margin of the southern North Sea Basin (Jenkyns, 1985). This second case-study considers deposition in the early part of the Toarcian (*tenuicostatum* and *falciferum* zones) represented on the Yorkshire coast by the Grey Shale and Jet Rock Members of the Whitby Mudstone Formation (Fig. 11). Previous workers are agreed that the Jet Rock Member accumulated under anoxic and suboxic conditions (Hallam, 1967; Hemingway, 1974; Morris, 1979; Pye and Krinsley, 1986; Raiswell and Berner, 1985). Jenkyns (1985) termed the lower *falciferum* zone, or *exaratum* subzone, an 'anoxic event' which can be traced across most of Europe.

Palaeoecology and sedimentology

There is still considerable debate as to whether bottom waters were truly euxinic in the *falciferum* zone. The palaeontological evidence indicates that benthic oxygen levels were below 0.3 ml/litre for prolonged periods of time, since a true shelly benthos is absent from much of the sequence. Two species of crinoid occur, but both were undoubtedly pseudopelagic (Simms, 1986). Pendant bivalves also occur scattered on the bedding planes and probably had a similar mode of life to the crinoids, living attached to driftwood. Raiswell and Berner (1985) concluded that the jet rock may have been semi-euxinic, based on C/S relationships, with the O_2/H_2S boundary fluctuating above and below the sediment–water interface during deposition of the shales. However, benthic oxygen levels were probably too low to allow a shelly benthos to colonize, as shell pavements (*sensu* Kimmeridge Clay) are lacking.

Morris (1979) was able to subdivide the jet rock and grey shale into three facies based on palaeoecological and sedimentological criteria. These facies are:

1. normal marine shale, represented by the lower part of the Grey Shale Member, deposited under normally oxidized bottom waters;
2. restricted marine shale, represented by the upper part of the Grey Shale Member, which was

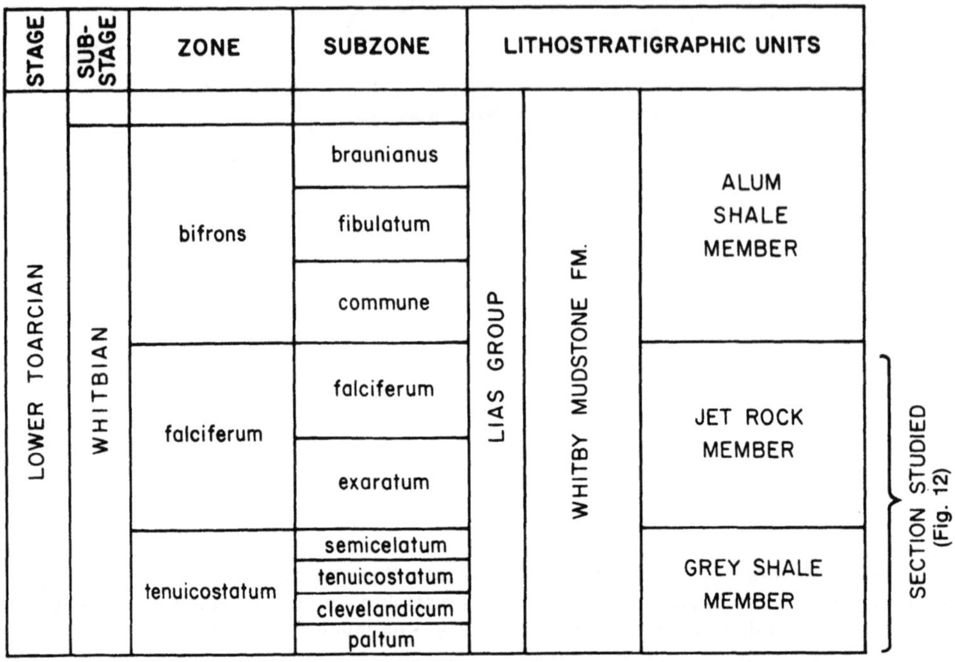

Fig. 11. Lower Toarcian stratigraphy on the Yorkshire coast.

deposited under suboxic or dysaerobic bottom waters;

3. bituminous shale, represented by the Jet Rock Member, which was deposited under anoxic and sometimes euxinic bottom waters.

The fauna of the restricted shale facies is dominated by deposit-feeding bivalves. These rework the top few centimetres of the substrate producing a thixotropic layer which inhibits the settling of the larvae of suspension feeders; this is the trophic group amensalism hypothesis of Rhoads and Young (1970). Therefore, the absence of suspension-feeding bivalves in the upper grey shales does not necessarily imply different oxygen levels from the lower grey shales. This conclusion is in accord with the geochemical and sedimentological evidence discussed below.

Morris's subdivisions have subsequently been refined by Pye (1985) and Pye and Krinsley (1986), who noted an upward transition within the Jet Rock Member from continuously laminated anoxic shales at the base, in the *exaratum* subzone to weakly laminated suboxic shales at the top, in the *falciferum* subzone.

The sediments were deposited during a period of marine transgression (Hallam, 1981) coincident with a short-cycle rise in sea level (Vail *et al.*, 1981). They fine upwards from the base, with the finest sediments occurring in the middle of the Alum Shale Member which probably represents the deepest water conditions (Pye and Krinsley, 1986).

Pye and Krinsley (1986) found that there is little petrological difference between the normal and restricted marine shale facies, both being rich in quartz, micas, chlorite and kaolin. The bituminous shale facies is composed mainly of quartz, kaolin, fine-grained micas, illite–smectite, chlorite, pyrite and calcite.

Previous geochemical studies (Gad *et al.*, 1969; Pye and Krinsley, 1986) again found little difference between the normal and restricted facies, but the bituminous facies could be distinguished on the basis of a higher content of Ca, Fe, P and S. They also noted a gradual transition to a less bituminous facies in the upper half of the Jet Rock Member compared to the sharp restricted-bituminous transition at the top of the Grey Shale Member. There was some evidence of enrichment of Cu, Ni, As, U, V, and Zn in the *exaratum* subzone though there were too few samples to draw any firm conclusions.

Gamma-ray spectrometry

The restricted and normal shale facies of the Grey Shale Member show no enrichment in authigenic uranium and cannot be distinguished from each other (Fig. 12). There is a sharp change at the base of

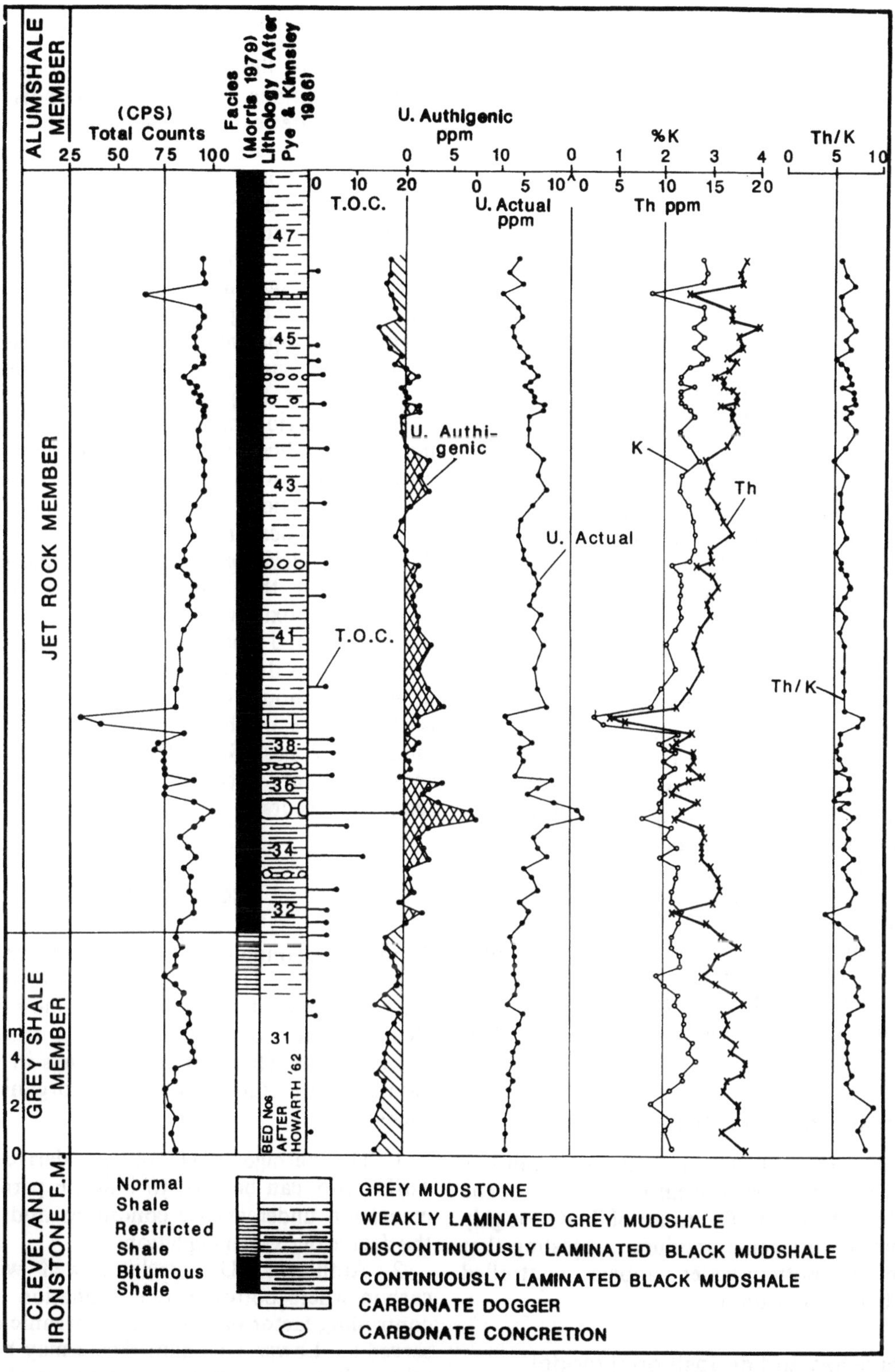

Fig. 12. The distribution of K, U, Th and total organic carbon in the Lower Toarcian of Yorkshire. The cross-hatched area shows sediments enriched in authigenic U.

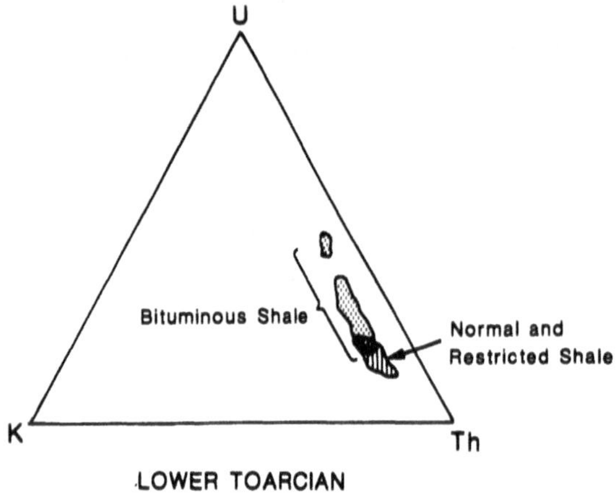

Fig. 13. Triangular composition plot for the Lower Toarcian of Yorkshire.

the Jet Rock Member to sediments enriched in authigenic U. This reaches a peak of 8.2 ppm, 4.5 m above the base, coincidental with the prominent whale stone doggers, and the highest measured total organic carbon content of 20%. The U_a values gradually decrease upward becoming negative 8 m below the top of the unit. The Th and K curves show similar trends, both decreasing upward into the lower Jet Rock Member, as the clay fraction is diluted by organic carbon and carbonate, and then increasing upwards as U_a decreases. The Th/K ratio decreases upward—from an average of 6.3 below the whale stone concretions to 5.5 above—paralleling the decrease in grain size and rise in sea level. The lower part of the Jet Rock Member is clearly distinguished on the composition plot (Fig. 13).

The total gamma-ray curve is rather flat because the decrease and increase in Th and K is largely balanced by an increase and decrease in U. Only the top jet dogger and a sideritic horizon are clearly picked out.

Total organic carbon contents

Total organic carbon (TOC) contents vary from less than 2% in the grey shales member to 20% in the lower Jet Rock Member. The trend of TOC roughly parallels that of authigenic uranium, suggesting that the degree of bottom-water anoxia controlled organic carbon preservation.

Discussion and depositional model

Portable gamma-ray spectrometry has refined our knowledge of the lower *falciferum* zone 'anoxic event'. It has a sharp base and a diffuse top with the most extreme period of anoxia occurring 4.5 m above the base of the Jet Rock Member. It is this level which is most likely to have been deposited under euxinic conditions. The presence of undisrupted lamination indicates oxygen levels permanently below 0.1 ml/litre in the surface sediments. The maximum U_a value of 8.2 ppm is comparable to that of the oil shales and coccolith limestones of the *pectinatus* zone of the Kimmeridge Clay. It lasted for three-quarters of an ammonite zone (*c.* 750 000 years), and occurred during a transgression and sea-level rise but probably before the deepest water conditions were developed. In comparison with the Kimmeridge Clay organic-rich shales, in which oxygenation events were relatively frequent, the Jet Rock represents a prolonged period of stable anoxic conditions. This may either indicate that a considerable portion of the water column was anoxic and thus resistant to the oxygenating effects of storms, or that the sea was considerably deeper. Jenkyns (1985) suggested that during the *falciferum* zone high organic productivity prevailed on the shelf during the transgression, producing anoxic waters which were transported to the continental margins as an oxygen-minimum zone. The new data are consistent with such a model, although it should be noted that the facies sequence is the inverse of that observed for recent oxygen-minimum zones (Stackelberg, 1972) in which the sharp boundary between laminated and bioturbated sediments occurs at the *top* of the zone.

CONCLUSIONS

A portable gamma-ray spectrometer can be used to measure the K, U and Th contents of marine mudrocks in the field, allowing large amounts of radiochemical data to be gathered in a relatively short space of time. There are several ways in which these data can be useful in the study of organic-rich mudrocks:

1. The authigenic U content, derived from the Th/U ratio, can be used to indicate periods of sedimentation under anoxic conditions and to estimate the degree of oxygen depletion.
2. Authigenic U correlates well with organic carbon where bottom-water stagnation alone is the controlling factor on the level of organic carbon preserved (cf. Lower Toarcian). There may be no correlation when factors such as surface productivity and sedimentation rate are important in controlling organic carbon preservation (cf. Kimmeridge Clay).

3. Th and K are contained mainly in the clay fraction of mudrocks. The Th/K ratio tends to decrease with increasing water depth in both the Kimmeridge Clay and the Jet Rock. This is probably related to a decrease in grain size and an increasing distance from the shoreline.

We note that the U/organic carbon ratio in the Jurassic case-studies is generally lower than those recorded for Lower Palaeozoic and Devonian black shales.

Kimmeridge Clay

The central and upper portion of this formation consists of cyclic alternations of relatively organic-poor and organic-rich mudstones. Climatic forcing was the predominant control on this cyclicity: a slight increase in temperature caused stratification and greatly reduced the efficiency of oxygen replenishment to the bottom waters. Frequent storm-induced events supplied oxygen to the substrate, indicating the delicate nature of the stratification. Sea level had no direct effect on the cyclicity but controlled the position of storm wave base and thus the frequency of oxygenation events. The deeper water sequences (the lower *pectinatus* zone) consequently have higher values of 'authigenic' uranium. Euxinic conditions may have occurred intermittently during deposition of the coccolith limestones and some of the oil shales. Surface productivity probably decreased between the two zones.

Lower Toarcian

Enriched values of authigenic uranium indicate that the most intense period of anoxia occurred 4.5 m above the base of the lower Jet Rock member (*exaratum* subzone). The 'anoxic event' was remarkably persistent with no oxygenation events recorded within it. It coincides with a period of rising sea level and transgression, though it does not represent the deepest water conditions.

ACKNOWLEDGEMENTS

We would like to thank Tony Hallam and Mike Simms of the University of Birmingham, Jeremy Leggett and Harry Shaw of Imperial College, and Kevin Pickering of the University of Leicester for criticizing an earlier draft of this paper. We would also like to thank Richard Tyson of University College, London. The final manuscript has benefited considerably from criticisms by Andrew Thickpenny of Lomond Associates. K. Myers acknowledges the support of Fina Exploration Ltd and thanks them for permission the publish the gamma-ray spectrometry data. P. Wignall gratefully acknowledges funding from a CASE award with NERC and the Natural History Museum, London.

REFERENCES

Adams, J. A. S. and Gasperini, P. (1970). Gamma-ray spectrometry of rocks, *In Methods in Geochemistry and Geophysics* 10, Elsevier.

Adams, J. A. S., Osmond, J. K. and Rogers, J. J. W. (1959). The geochemistry of uranium and thorium, *Physical Chemistry of the Earth* **3,** 299–328.

Adams, J. A. S. and Weaver, R. (1958). Thorium-to-uranium ratio as an indicator of sedimentary process: examples of the concept of geochemical facies, *American Association of Petroleum Geologists Bulletin* **42 (2),** 387–430.

Aigner, T. (1980). Biofabrics and stratinomy of the Lower Kimmeridge Clay (Upper Jurassic, Dorset, England), *Nues Jahrbuch für Geologie und Paläontologie Abhandlungen* **159,** 324–338.

Borovec, Z., Kribek, B. and Tolar, V. (1979). Sorption of uranyl by humic acids, *Chemical Geology* **27,** 39–46.

Byers, C. W. (1977). Biofacies patterns in euxinic basins: a general model, *In Deep-water Clastic Environments*, Cook, H. E. and Enos, P. (Eds), S.E.P.M. Special Publication No. 25, pp. 5–17.

Cochraine, J. K., Carey, A. E., Sholkovitz, E. R. and Surprenant, L. D. (1986). The geochemistry of uranium and thorium in coastal marine sediments and pore waters, *Geochimica et Cosmochimica Acta* **50,** 663–680.

Cox, B. M. and Gallois, R. W. (1981). The stratigraphy of the Kimmeridge Clay in the Dorset type area and its correlation with some other Kimmeridgian sequences, *Inst. Geol. Sci. Report* 80/4.

Curtis, C. (1987). Recent data on the variety and consequence of organic–inorganic reactions which influence sedimentary mineral assemblages, this volume.

Degens, E. T., Khoo, K. and Michaelis, W. (1977). Uranium, anomaly in Black Sea sediments, *Nature* **269,** 566–569.

Demaison, G. J. and Moore, G. T. (1980). Anoxic environments and oil source bed genesis. *American Association of Petroleum Geologists Bulletin* **64,** 1179–1209.

Dodd, J. R. and Stanton, R. J. (1981). *Palaeoecology, Concepts and Applications*, New York, Wiley.

Dongarra, G. (1984). Geochemical behaviour of uranium in the supergene environment, *In Uranium Geochemistry, Mineralogy, Geology, Exploration and Resources*, De Viro *et al.* (Eds), Institute of Mining and Metallurgy, pp. 18–22.

Downie, C. (1955). The Kimmeridge oil shale, Unpublished Ph.D. Thesis, University of Sheffield.

Dunn, C. E. (1974). Identification of sedimentary cycles through Fourier analysis of geochemical data, *Chemical Geology* **13,** 217–232.

Ebukanson, E. J. and Kinghorn, R. R. F. (1985). Kerogen facies in the major Jurassic mudrock formations of Southern England and the implications on the depositional environments of their precursors, *Journal of Petroleum Geology* **8 (4),** 435–462.

Farrimond, P. *et al.* (1984) Organic geochemical study of the Upper Kimmeridge Clay of the Dorset type area, *Marine and Petroleum Geology* **1 (4),** 340–354.

Fertl, W. H. (1982). Gamma-ray spectrometry assists in complex formation evaluation, *The Log Analyst* **20 (5),** 23–57.

Gad, M. A., Catt, S. A. and Le Riche, H. H. (1969). Geochemistry of the Whitbian (Upper Lias) sediments of the Yorkshire coast, *Proceedings Yorkshire Geological Society* **37** (1), 105–139.

Gallois, R. W. (1976). Coccolith blooms in the Kimmeridge Clay and the origin of North Sea oil, *Nature* **259,** 473–475.

Glennie, K. W. (Ed.) (1984). *Introduction to the Petroleum Geology of the North Sea*, Oxford, Blackwell Science Publishers.

Guthrie, V. A. and Kleeman, J. D. (1986). Changing uranium distributions during weathering of granite. *Chemical Geology* **54,** 113–126.

Hallam, A. (1967). An environmental study of the Upper Domerian and Lower Toarcian in Great Britain, *Philosophical Transactions Royal Society, London B* **252,** 393–445.

Hallam, A. (1981). A revised sea-level curve for the early Jurassic, *Journal of the Geological Society, London* **138,** 735–743.

Hallam, A. (1986). Role of climate in affecting late Jurassic and early Cretaceous sedimentation in the North Atlantic, *In North Atlantic Paleao-oceanography*, Summerhayes, C. P. (Ed.), Oxford, Blackwell Science Publishers.

Heier, K. S. and Billings, G. K. (1972). Potassium, *In Handbook of Geochemistry*, Wedepohl, (Ed.) Vol. 11-2, part 19-b-n. Berlin, Springer-Verlag.

Hemingway, J. E. (1974). Jurassic, *In The Geology and Mineral Resources of Yorkshire*, Rayner, D. H. and Hemingway, J. E. (Eds), Yorkshire Geological Society, pp. 161–223.

Holland, H. D. (1984). The chemical evolution of the atmosphere and oceans, Princeton Series in Geochemistry, Princeton University Press, Princeton, N.J.

Howarth, M. K. (1962). The Jet Rock Series and the Alum Shale Series of the Yorkshire coast, *Proceedings Yorkshire Geological Society* **33 (4),** 381–422.

Irwin, H. (1980). Early diagenetic carbonate precipitation and pore fluid migration in the Kimmeridge Clay of Dorset, *Sedimentology* **27,** 577–591.

Jenkyns, H. C. (1985). The Early Toarcian and Cenomanian–Turonian anoxic events in Europe: comparisons and contrasts, *Geologische Rundschau* **74,** 505–518.

Kammer, T. W., Brett, C. E., Darwin, R., Boardman, H. and Mapes, R. H. (1986). Ecologic stability of the dysaerobic biofacies during the late Palaeozoic, *Lethiaia* **19,** 109–121.

Killeen, P. G. (1979). Gamma-ray spectrometric methods in uranium exploration—applications and interpretation, *In Geophysics and Geochemistry in the Search for Metallic Ores*, Hood, P. J. (Ed.), Geological Survey of Canada, Economic Geology Report, **31,** 163–229.

Kochenov, A. v., Korolev, K. G., Dubinchuk, V. T. and Medvedev, Y. L. (1977). Experimental data on the conditions of precipitation of uranium from aqueous solutions, translated from *Geokhimiya* **11,** 1711–1716, *Geochemistry International* (1977) p. 82–87.

Langmuir, D. (1978). Uranium solution–mineral equilibria at low temperatures with applications to sedimentary ore deposits. *Geochimica et Cosmochimica Acta* **42,** 547–569.

Langmuir, D. and Herman, J. S. (1980). The mobility of thorium in natural waters at low temperatures, *Geochimica et Cosmochimica Acta* **44,** 1753–1766.

Levinson, A. A. and Coetzee, G. L. (1978). Implications of disequilibrium in exploration for uranium ores in the surficial environment using radiometric techniques—a review, *Minerals Science Engineering* **10** (1), 19–27.

Løvborg, L. (1971). Field determination of uranium and thorium by gamma-ray spectrometry, exemplified by measurements on the Ilimausaq alkaline intrusion, South Greenland, *Economic Geology* **66** (3), 368–384.

Løvborg, L. (1984). The calibration of portable and airborne gamma-ray spectrometers—theory, problems and facilities, Report Riso-M-2456, Riso National Laboratory, Denmark.

Mangini, A. and Schlosser, P. (1986). The formation of eastern Mediterranean sapropels, *Marine Geology* **72,** 115–124.

Mann, H. and Fyfe (1985). Uranium uptake by algae: experimental and natural environments, *Canadian Journal Earth Science* **22,** 1899–1903.

Morris, K. A. (1979). A classification of Jurassic marine shale sequences: an example from the Toarcian (Lower Jurassic) of Great Britain, *Palaeography, Palaeoclimatology, Palaeoecology* **26,** 117–126.

Potter, P. E., Maynard, J. B. and Pryor, W. A. (1980). *The Sedimentology of Shale*, Springer-Verlag, New York.

Pye, K. (1985). Electron microprobe analysis of zoned dolomite rhombs in the Jet Rock Formation (Lower Toarcian) of the Whitby area, U.K., *Geological Magazine* **122,** 279–286.

Pye, K. and Krinsley, D. H. (1986). Microfabric, Mineralogy and early diagenetic history of the Whitby Mudstone Formation (Toarcian), Cleveland Basin, U.K., *Geological Magazine* **123,** 191–203.

Raiswell, R. and Berner, R. A. (1985). Pyrite formation in euxinic and semi-euxinic sediments, *American Journal Science* **285**.

Rhoads, D. C. and Morse, D. C. (1971). Evolutionary and ecological significance of oxygen-deficient marine basins, *Lethaia* **4,** 413–428.

Rhoads, D. C. and Young, D. K. (1970). The influence of deposit-feeding organisms on bottom-sediment stability and trophic structure, *Journal of Marine Research* **28,** 150–178.

Rhona, E. and Joensu, U. (1974). Uranium geochemistry in the Black Sea, *In The Black Sea—Geology, Chemistry and Biology*, D. A. Ross and E. T. Degens, (Eds), *AAPG Mem.* **20,** 570–572.

Ross, D. A. and Degens, E. T. (1974). *The Black Sea—Geology, Chemistry and Biology, American Association of Petroleum Geologists Memoir* **20**.

Savrada, C. E., Boetier, D. J. and Gorsline, D. S. (1984). Development of an oxygen-deficient marine biofacies model: evidence from Santa Monica, San Pedro and Santa Barbara Basins, California continental borderland. *American Association of Petroleum Geologists Bulletin* **68 (9),** 1179–1192.

Schlumberger (1982). Natural gamma-ray spectrometry: essentials of N.G.S. interpretation, Schlumberger.

Silver, L. T., Woodhead, J. A. and Williams, I. S. (1982). Primary mineral distribution and secondary mobilisation of uranium and thorium in radioactive granites, *In Uranium Exploration Methods, I.A.E.A.*, Paris, pp. 355–383.

Simms, M. J. (1986). Contrasting lifestyles in lower Jurassic crinoids: comparison of benthic and pseudopelagic isocrinidia, *Palaeontology* **29,** (3), 475–495.

Stackelberg, U. V. (1972). Faziesverteilung in Sedimenten des Indisch-Pakistanischen Kontinentalrandes (Arabisches Meer): 'Meteor', *Forschungsergebuise, Reihe C*, **9,** 1–173.

Swanson, J. E. (1961). Geology and geochemistry of uranium in marine black shales, a review, US Geological Survey, Prof. paper 356-C.

Tieh, T. T., Ledger, E. B. and Rowe, M. W. (1980). Release of uranium from granitic rocks during in-situ weathering and initial erosion (Central Texas), *Chemical Geology* **29,** 227–248.

Thompson, J. B., Mulins, T. H., Newton, C. R. and Vercontere, T. L. (1985). Alternative biofacies model for dysaerobic communities, *Lethaia* **18,** 167–179.

Tyson, R. V., Wilson, R. C. L. and Downie, C. (1979). A stratified water column environmental model for the type Kimmeridge Clay, *Nature* **277,** 377–380.

Vail, P. R. and Todd, R. G. (1981). Northern North Sea Jurassic unconformities, chronostratigraphy and sealevel changes from seismic stratigraphy, In *Petroleum Geology of the Continental Shelf of North West Europe*, Illing, L. V. and Hobson, G. D. (Eds), Heyden, London, 216–235.

Veeh, H. H. (1967). Deposition of uranium from the ocean, *Earth and Planetary Science Letters* **3,** 145–150.

Veeh, H. H., Calvert, S. E. and Price, N. B. (1974). Accumulation of uranium in sediments and phosphorites on the South-west African shelf, *Marine Chemistry* **2,** 189–202.

Waples, D. W. (1983). Reappraisal of anoxia and organic richness, with emphasis on Cretaceous of North Atlantic, *American Association of Petroleum Geologists Bulletin* **67,** (6), 963–978.

Wilson, R. C. L. (1980). *Changing Sea Level: a Jurassic Case-study*, Open University Press, Milton Keynes.

Chapter 10

Deep-marine Foreland Basin and Forearc Sedimentation: a Comparative Study from the Lower Palaeozoic Northern Appalachians, Quebec and Newfoundland

Kevin T. Pickering, Department of Geology,
University of Leicester, Leicester LE1 7RH, UK

ABSTRACT

Deep-marine foreland basins may develop during continent–arc or continent–continent collision upon relatively thin lithosphere. Plate interactions that generate essentially compressive stresses to develop a thrust imbricate stack may result in lithospheric flexure and the initiation of a foredeep or foreland basin. Such basins can form in remnant forearc and backarc or marginal basins. In ancient orogenic belts, it is often difficult to recognize deep-marine foreland basins and to differentiate them from forearc or backarc basins. This paper presents a general overview of our current understanding on the development of foreland basins, with reference to deep-marine environments, and compares a foreland basin system from the Canadian Appalachians (a remnant forearc in the Quebec Re-entrant) with a Late Ordovician–Early Silurian telescoping remnant marginal basin that because of a reversal in subduction polarity became a forearc associated with considerable sinistral oblique-slip. The ultimate demise of this latter system may have been as a foreland basin-fill that shallows up from deep-marine to shallow-marine and possibly subaerial environments. These Lower Palaeozoic case studies are discussed in the context of plate-tectonic models for the destruction of the Iapetus Ocean.

INTRODUCTION

Foreland basins develop in response to a 'point load' applied to the lithosphere, such as a thick thrust-sheet pile, causing the downbending of the flexed lithosphere. Figure 1, after Beaumont *et al.* (1982), shows the effect of modelling the surface deformation of a continuous visco-elastic lithosphere under a surface load, a case that appears to provide a valid and predictable model for the generation of a foreland basin. Once the load has been applied, as a thick thrust-sheet or imbricate system, the lithosphere shows an elastic response followed by a viscous response, leading to deepening and narrowing of the foreland basin. This model has been successfully applied to explain the overall tectono-stratigraphic evolution of the post-Mid-Jurassic Western Canadian Sedimentary Basin or Alberta Basin by Beaumont (1981), and for the Palaeozoic central Appalachians by Quinlan and Beaumont (1984).

Marine Clastic Sedimentology © J. K. Leggett & G. G. Zuffa (Graham and Trotman, 1987) pp. 190–211

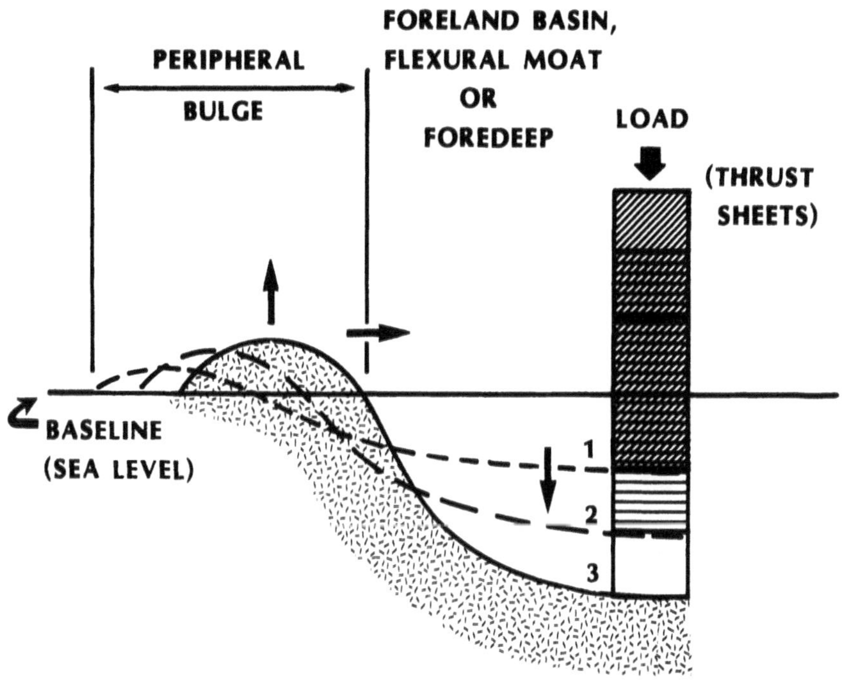

Fig. 1. Cross-sectional profile for the surface deformation of a continuous visco-elastic lithosphere under a surface load, e.g. thrust sheets of a thrust-imbricate fan, duplex or accreted terrane such as an island arc. With time, stress relaxation causes the profile to evolve to stages 2 and 3 as the system tends towards local isostatic equilibrium. After Quinlan and Beaumont (1984).

The consequences for geologists of the above model are that foreland basins show a rapid subsidence phase followed by relatively stable intervals. Basin deepening occurs in pulses and sediment infilling takes place as erosion catches up: typically several such phases occur in response to a chain of discrete thrust-stacking events. Foreland basins are also characterized by episodic and terminal uplift and erosion that appear as major unconformities (Beaumont *et al.*, 1982).

For the visco-elastic lithospheric flexural model, a given driving load applied to a thick lithosphere will tend to produce a relatively wide and shallow basin. In contrast, the same driving load applied to a thin lithosphere tends to give a relatively narrow but deep basin. These different load–response models help to explain why, for example, the Alberta Basin is wide, since it developed on the Canadian Shield lithosphere that is thick and has not undergone a thermal event for at least 500 My, whereas in the case of young (less than about 30 Ma old) and hot lithosphere that is flexurally weak, narrow and deep basins form that may reach a maximum depth of 6 km (Beaumont *et al.*, 1982). Also, in their seminal paper on foreland basins, Beaumont *et al.* (1982) note that the foreland basin model implies that the basin topography experiences little deformation other than regional flexural warping, and that the resulting wedge-shaped basin should have its maximum depth adjacent to the rising and stacking thrust system, or mountains.

An inherent feature of the visco-elastic lithospheric flexure model is that the development of a foreland basin is accompanied by uplift of the basin margin opposite the thrust system to give a peripheral bulge (Fig. 1), a counterpart to the outer trench high that forms in subduction systems. Peripheral bulges, in contrast to the rising thrust system that exerts a significant load on the lithosphere, have relatively little topographic relief, form at least tens of kilometres outboard from the thrust system, and tend to contribute relatively little sediment to the foreland basin infill. The basin half-width, the distance between the centre of the load (thrust system) and the peripheral bulge, increases as a cube power of the lithosphere thickness (Walcott, 1970).

STRATIGRAPHIC IMPLICATIONS

The model for foreland basin development outlined above should, ideally, generate a predictable strati-

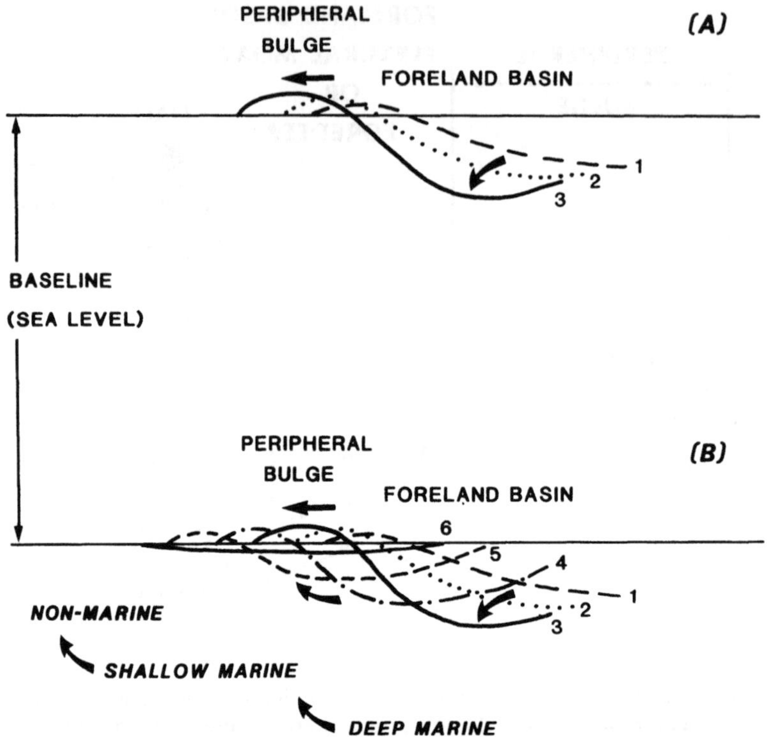

Fig. 2. (A) Cross-sectional profile for the surface deformation of a continuous visco-elastic lithosphere under a dynamic surface load where the load is a foreland-propagating thrust-imbricate fan or duplex. Compare with Fig. 1. With time, the foredeep or foreland basin migrates towards the foreland in front of the advancing thrust stack, from stage 1 to 3. (B) Model as for (A), but showing possible evolution (stages 1 to 3) and demise of foreland basin (stages 4 to 6) if load that is driving subsidence is removed or decreases in magnitude during foreland basin infill, e.g. erosion of thrust stack exceeds rate and magnitude of assembly of thrust stack. In (A), the foreland basin *sensu stricto* may end as a deep-marine system starved of coarse clastic infill, whereas in (B), foreland-propagating depocentres may change from deep-marine, through shallow-marine, to non-marine.

graphy (Fig. 2). Thrust systems that may form 'imbricate fans' or 'duplexes' (terminology of Boyer and Elliott, 1982) contemporaneous with sedimentation tend to migrate towards the basin axis or foredeep. Thus, the peripheral bulge will tend to migrate away from, not towards, the load as in the case of applying a static load to the lithosphere; consequently, while the thrust system is actively developing we might predict a progressively younger (diachronous) influx of clastics into successive foreland basins that develop in front of the migrating thrust system (Fig. 2).

With time, as relative plate motions change together with varying magnitudes of collision, a thrust system coupled to a foreland basin will become inactive. This, in turn, may lead to isostatic readjustment, especially as erosion of the load or thrust system proceeds. The net result, first of collision processes generating a thrust system and deep-marine foreland basin (since such cases concern us in this paper), followed by the demise of that system, is to produce a diachronous basin-fill that ideally shows a rapid deepening stratigraphy that shallows up eventually into subaerial sedimentation. In the case of deep-marine foreland basins, the pre-foreland basin stratigraphy may be represented by a foundered continental margin. Deep-marine turbidite and related depositional systems may abruptly overlie shallow-marine deposits, testifying to the phase of active foreland basin development. Finally, the waning stages of the foreland basin history, or the post-foreland basin phase, may be characterized by a shallowing-up into subaerial sedimentation. Much of the subaerial 'molasse-type' deposits may occupy half-graben basins formed by the reactivation of thrust faults under post-collision extensional, or crustal relaxation, tectonic regimes.

The above idealized stratigraphy is well seen in the SOQUIP seismic line shot in southern Quebec (Laroche, 1983) through the Lower Palaeozoic Appa-

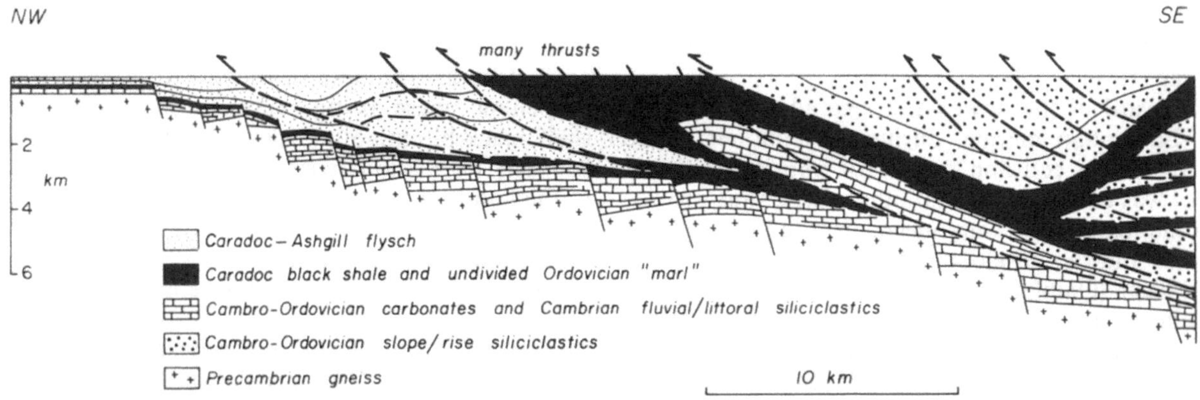

Fig. 3. Simplified interpretation of part of the SOQUIP seismic line fully described by Laroche (1983) and shot approximately across the strike of the Appalachian orogen near Quebec City. Foreland basin flysch lies stratigraphically above Caradoc black shales which overlie the foundered Cambro-Ordovician passive margin carbonate successions. The flysch was subsequently overthrust by nappes containing older rocks originally deposited farther south-east. In this section, the Nicolet River and Lotbiniere Formations form the flysch that was contemporaneous with the Cloridorme Formation. From Hiscott *et al.* (1986).

lachian foreland basin (Fig. 3). Beneath the Ordovician deep-marine foreland basin sedimentation, are the foundered, over-ridden, Cambro-Ordovician passive margin carbonates and siliciclastics of the Laurentian continent. Later, after the Taconic (Middle Ordovician) collision tectonics and thrust-imbrication, Siluro-Devonian, essentially subaerial (continental) deposits rest unconformably upon the highly deformed Ordovician stratigraphy (not seen in Fig. 3 cross-section, but see location map of Laroche, 1983).

Other tectono-stratigraphic collages that may represent, in part, remnant foreland basins include the Late Precambrian Lower Palaeozoic geology of north-west Britain seen, for example, in the MOIST line (Brewer and Smythe, 1983, 1984). Geologically younger, well-documented deep-marine foreland basins that show many of the features outlined above include the Tertiary of the Italian Apennines (Ori *et al.*, 1986; Ricci Lucchi, 1975; Ricci Lucchi and Ori, 1985), and the Tertiary South Pyrenees of Spain (Labaume *et al.*, 1983, 1985; Mutti, 1977, 1985).

There are few case-studies of modern foreland basins in which deep-marine sedimentation is occurring. However, one example involves the north Australian 'Atlantic-type' continental margin being driven northwards to underplate the Banda Arc without any intervening oceanic crust (Audley-Charles, 1986; Bowin *et al.*, 1980; Jacobson *et al.*, 1978). The non-volcanic Outer Banda Arc has been interpreted as the emergent part of the thrust system developed in response to the underplating of the Australian margin beneath the Timor–Babar–Tanimbar–Kai island chain (Audley-Charles, 1986). Farther east and west, this modern foreland basin passes laterally into active subduction zones. Tectonic thickening of the Australian continental margin and shelf has occurred by mainly southerly transported thrust slices being imbricated. The deep-marine sediments, including turbiditic and pelagic deposits, are being deformed in a manner analogous to that found in modern forearc accretionary prisms. Thus, apart from the temporal and spatial association of arc magmatism with forearcs, the overall tectono-stratigraphic record of foreland basins can be somewhat similar to that of active forearc plate margins.

MIDDLE ORDOVICIAN QUEBEC RE-ENTRANT

The Middle Ordovician Taconic Orogeny was one of three major pulses of deformation to affect the Appalachian Orogen (Williams and Hatcher, 1983). The passive 'Atlantic-type' continental margin of Laurentia, lying in equatorial latitudes and fringed by an extensive carbonate platform, experienced diachronous collision, from western Newfoundland in the north to New York State in the south, by an island arc system (Lushs Bight, Tetagouche, Bronson Hill and Long Island arcs). Geological and geophysical evidence along the entire length of the North American Appalachians suggest a south-eastward-dipping Benioff Zone prior to this arc–continent collision

(Malpas and Stevens, 1977; Shanmugam and Lash, 1982; Stanley and Ratcliffe, 1985; Swinden and Strong, 1976).

Seaward of the downward-flexured and foundering Laurentian margin, during arc–continent plate interaction, a collision orogen of imbricated thrust slices was emplaced on thin lithosphere, thus providing the load and driving mechanism for lithospheric flexure and the development of a deep-marine foreland basin (Quinlan and Beaumont, 1984). Contained within the imbricate thrust system were the obducted vestiges of the forearc accretionary prism and volcanic arc (St Julien and Hubert, 1975; Stanley and Ratcliffe, 1985; Williams and Hatcher, 1983). Initial mantle convection-cell processes, causing subduction and forcing down thinned continental crust beneath an overriding arc at this margin, were converted into load-induced subsidence as collision advanced. A useful modern analogue can perhaps be found where the north Australian margin is being forced beneath the Banda Arc (Audley-Charles, 1986; Hamilton, 1979).

The history of this collision and the development of a foreland basin are well seen in the Quebec Reentrant, a major embayment in the Laurentian margin between the St Lawrence Promontory to the north and the New York Promontory to the south (Hiscott *et al.*, 1986; Thomas, 1977). Much of the information in this section of the paper is a summary and synthesis from Pickering and Hiscott (1985), and Hiscott *et al.* (1986). The reader is referred to these papers for further data.

Prior to foreland basin development, and contemporaneous with shelf carbonate sedimentation along the Laurentian margin, there were platform-derived deep-marine turbidite and related systems that can be distinguished from the overlying Taconic-age deep-marine systems primarily on the basis of composition (Hiscott, 1978, 1984). Collision processes have resulted in the pre-foreland basin deep-marine systems being tectonically emplaced above Laurentian shelf carbonates, whereas the foreland basin systems are autochthonous to para-autochthonous with respect to the underlying foundered carbonate margin.

The rate of subsidence of the Laurentian margin varied from a slow and gentle basinwards flexure, tilting and subsidence, as seen in the stratigraphy of Anticosti Island, to a rapid downwarping of the crust in the more outboard region nearer to the rising thrust imbricate system. Sanford *et al.* (1979), in a simplified version of a seismic line across Anticosti Island, show the regional facies changes that developed in response to downbending of the Laurentian margin, with an essentially uninterrupted shallow-marine Ordovician–Silurian stratigraphy on Anticosti Island; the late Caradoc to early Ashgill deepening phase being represented by deep-water shelf, black bituminous shales, 30–170 m thick, of the Macasty Formation, overlain by marls and siltstones of the hundred-metre-thick English Head Member of the Vaureal Formation. Sanford *et al.* (1979) show the Caradoc–Ashgill sedimentary rocks of Anticosti Island passing laterally and thickening over an estimated restored distance of about 20–30 km into a succession of sandstones and mudstones that is at least 4000 m thick. This seismic section ends approximately 15 km from the north Gaspé coastline, where the Caradoc–Ashgill deep-marine Cloridorme Formation, estimated to be about 4000 m thick, crops out (Hiscott *et al.*, 1986). Thus, it appears that the Cloridorme Formation passes laterally into the considerably thinner, fine-grained mudrocks and siltstones seen on Anticosti Island.

The Société Québècoise d'Initiatives Petrolières (SOQUIP) has shot a seismic line, 150 km long, orientated NW–SE, from a location about 50 km south-west of Quebec City on the St Lawrence River to the US border. It shows a Taconic-age (Middle Ordovician south-eastward-dipping thrust system (Fig. 3 after Laroche, 1983). The SOQUIP line shows a lower Cambro-Ordovician foundered Laurentian passive continental margin overlain by thrust-stacked, mainly Ordovician rocks, representing both the deep-marine foreland basin and earlier stratigraphy. Elsewhere, in Newfoundland and south-west of Quebec City, shallow-marine and continental molasse deposits overlie the deep-marine foreland basin successions. The SOQUIP seismic profile, together with a profile described by Sanford *et al.* (1979), reveal a thrust system at least 150 km wide which formed in response to arc–continent collision during Middle Ordovician times in the Quebec Reentrant. Later, Siluro-Devonian (Acadian) reactivation of many of the Taconic-age thrusts, again under crustal compression (seen in the seismic line described by Sanford *et al.* 1979), further telescoped the foreland basin successions.

In the Quebec Re-entrant, the oldest deep-marine systems include small-radius submarine fans that formed in slope basins on the rising thrust system during the Arenig–Llanvirn, such as the Tourelle, Metis and St Modeste Formations 300–500 m thick (Fig. 4) (Hiscott, 1977, 1978, 1980; Hiscott *et al.*, 1986). Deep-marine foreland-basin sedimentation *sensu stricto* began in the Llandeilo–Caradoc with the accumulation of the 2300 m-thick Deslandes Formation (Fig. 4) of essentially basin-floor turbi-

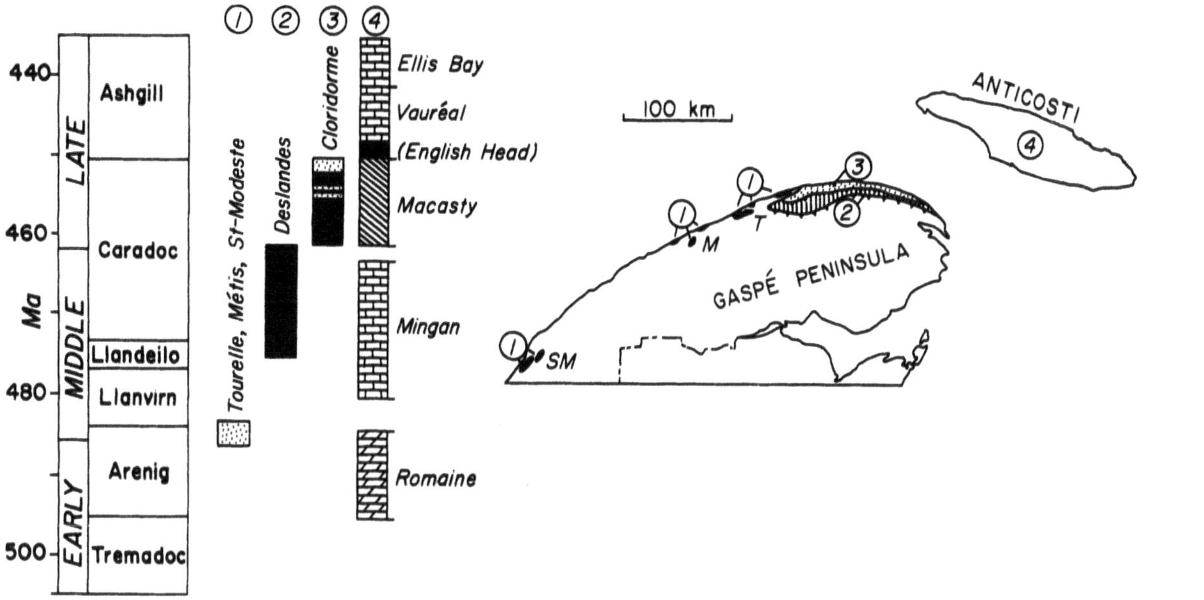

Fig. 4. Ages of Ordovician flysch and platform deposits in the eastern part of the Quebec Re-entrant. Lithologic symbols as for Fig. 3, but do not apply to the map. Location of columns keyed to corresponding numbers on map. Ages from Barnes *et al.* (1981) for columns 1 to 3, and from SOQUIP (1983) for column 4. From Hiscott *et al.* (1986).

dites (Enos, 1969a). Finally, during late Caradoc–Ashgill times, at least 4000 m of basin-floor turbidites and submarine-fan deposits accumulated in the foreland basin as the Cloridorme Formation (Fig. 5) (Enos, 1969a,b; Hiscott *et al.*, 1986; Pickering and Hiscott, 1985). The earlier-deposited Deslandes Formation, as well as the intraslope turbidite systems (such as the Tourelle Formation), were then thrust over the late Caradoc–Ashgill systems; the early phases of thrusting occurring contemporaneously with deposition of the Cloridorme Formation.

The Cloridorme Formation (Fig. 5) comprises approximately 4000 m of flysch without an exposed base. The oldest St Helier Member, up to 1140 m thick, essentially contains Facies Class C, D, E and G deposits (Pickering *et al.*, 1986), including 'megaturbidite beds' that are interpreted as contained, ponded turbidity current beds (Pickering and Hiscott, 1985): this member is believed to represent basin-floor sedimentation. Overlying this, the Pointe-à-la-Frégate Member approximately 580 m thick, correlated with the considerably finer-grained Manche d'Epée Member, 550 m thick, contains similar facies to the St Helier Member, but with more abundant Facies Class C beds occurring in packets in the Pointe-à-la-Frégate Member, suggesting lower fan-lobe and fan-fringe passing west into basin-floor deposits.

The Petite-Vallée Member, 835 m thick, similar in constituent facies to the lower parts of the underlying Pointe-à-la-Frégate Member, appears to represent lower fan-lobe and related environments that pass up into the 475 m thick Mont-St-Pierre Member of mainly Facies Class D, E and G deposits (Pickering *et al.*, 1986). The fine-grained, mud-dominated nature of this member appears to correlate well with the proposed major late Caradoc global sea-level rise (Fortey, 1984). Thus, the Mont-St-Pierre Member is interpreted as an 'abandonment' facies-association that accumulated when the deep-marine clastic system was starved of coarse-grained sediments due to a large rise in sea level. Finally, the Marsoui Member, 1000 m thick, mainly containing Facies Class B, C and D deposits (Pickering *et al.*, 1986), shows features, including channels, that are most consistent with a middle-fan environment (Beeden, 1983).

Palaeocurrent studies suggest that the Cloridorme Formation developed first in response to a westward-prograding (St Helier to Petite Vallée Members), then an eastward-prograding deep-marine clastic system (Marsoui Member), the interfingering occurring during the deposition of the Mont-St-Pierre Member at high sea-level stand (Fig. 5). Figure 6 is a summary history of the evolution of the Arenig–Llanvirn deep-marine intraslope–basin clastic systems, through the Llandeilo to early Ashgill accumulation of the Cloridorme Formation.

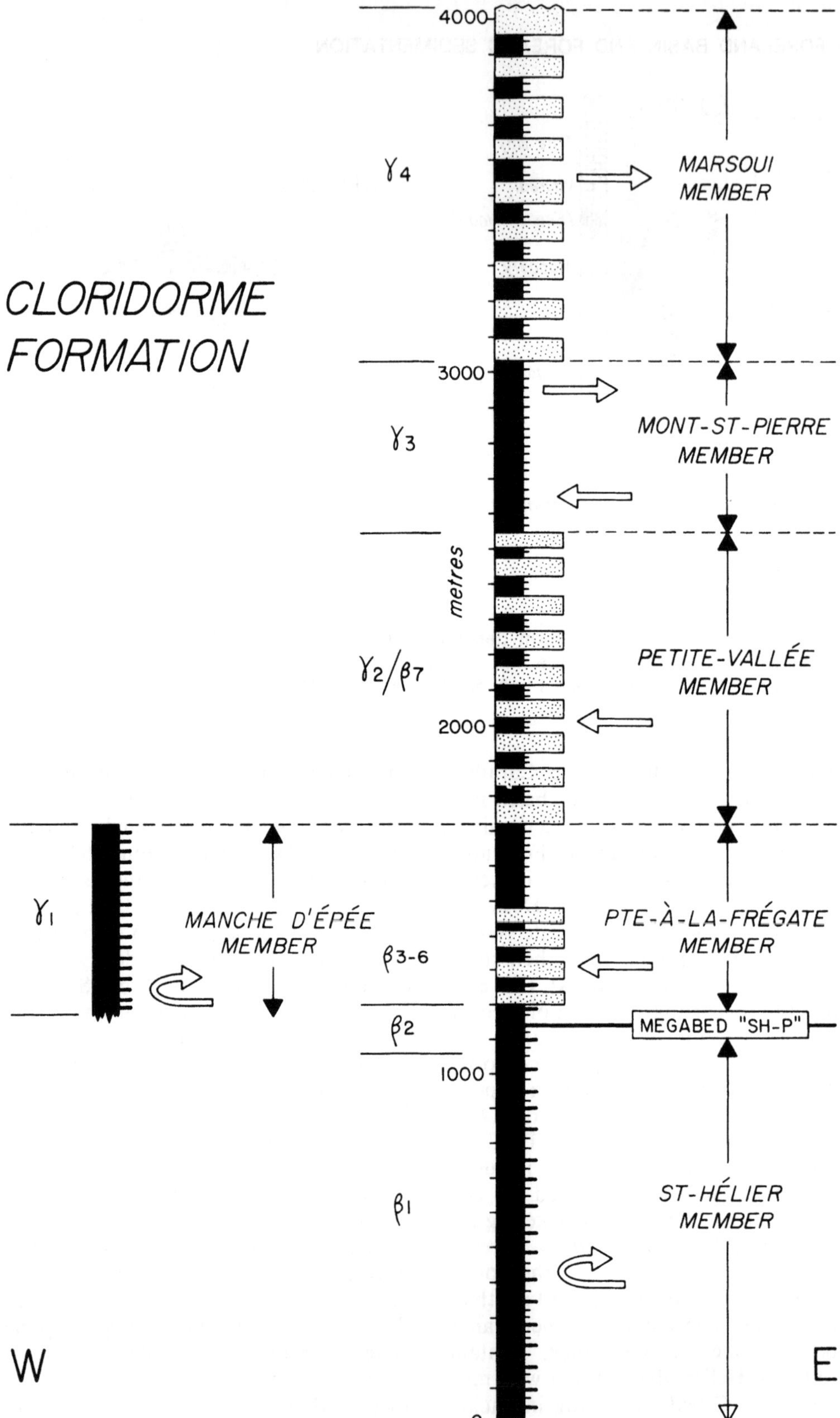

Fig. 5. Informal members of the Cloridorme Formation, and correlation with members of Enos (1969a). Sandstone packets are schematic (stippled). Black—shale; thin ticks—thin-bedded turbidites; thick ticks—megaturbidites (Pickering and Hiscott, 1985). Large arrows crudely summarize palaeoflow. In the St Helier and Manche d'Epée Members, megaturbidites show flow reversals due to reflection from basin margins. From Hiscott *et al.* (1986).

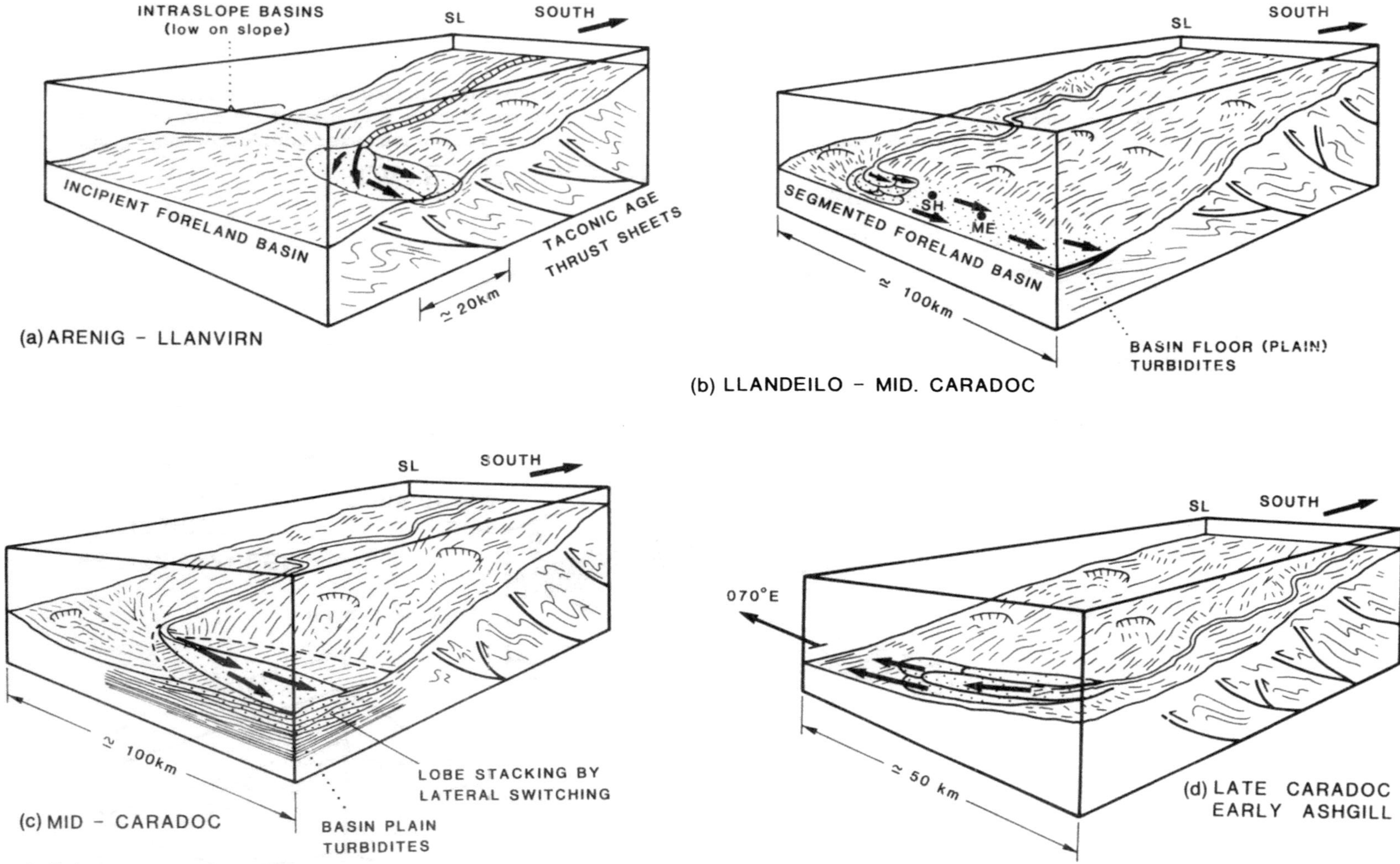

Fig. 6. Palaeogeographic sketches for the evolution of the foreland basin in the Quebec Re-entrant in the following: (a) Arenig–Llanvirn, when small slope basins developed to accommodate small-radius deep-marine clastic systems such as the Tourelle, Metis and St Modeste Formations; (b) Llandeilo–mid-Caradoc, when the deep-marine foreland basin was fully established, with basin-floor turbidite deposition as the Deslandes Formation, St Helier (SH) and Manche d'Epée (ME) Members of the Cloridorme Formation; (c) mid-Caradoc, when sandstone lobes prograded from the east over the basin floor turbidites, as the Pointe-à-la-Fregate and Petite-Vallée Members of the Cloridorme Formation; and (d) late Caradoc–early Ashgill, when a mud blanket of the Mont-St-Pierre Member of the Cloridorme Formation mantled the coarse clastic system, and during which time there was a switch from an eastern to a western source for the sandy turbidites: finally, the uppermost parts of the Cloridorme Formation, in the Marsoui Member, show a return to a sandstone-rich deep-marine clastic system derived from a western source.

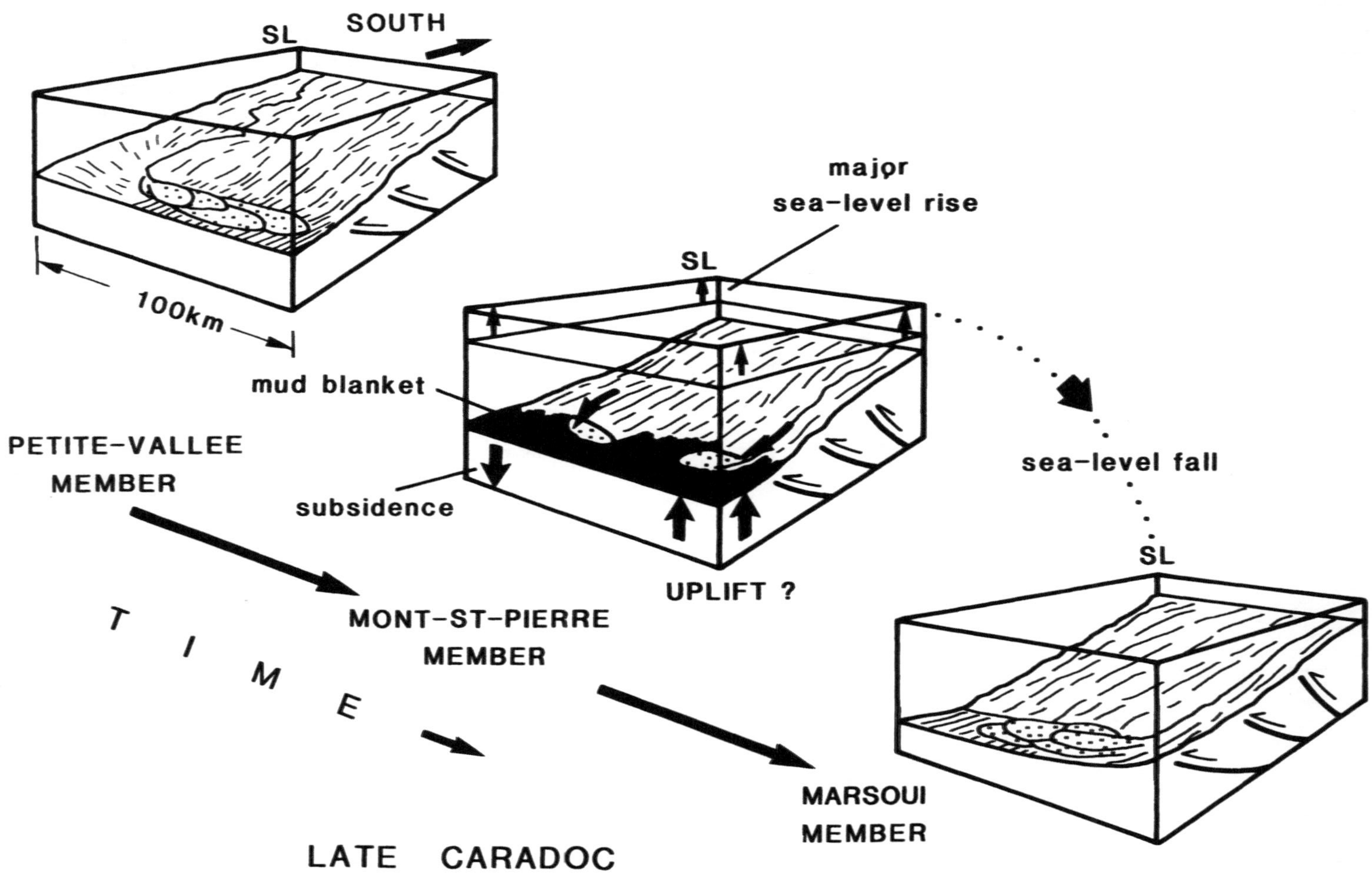

Fig. 7. Palaeogeographic sketches to show in detail the evolution of the Cloridorme Formation during the late Caradoc when there was a change from an eastern to a western source for the clastics, almost certainly tectonically controlled. The mud blanket of the Mont-St-Pierre Member correlates well with the late Caradoc global rise in sea level as documented by Fortey (1984).

Figure 7 summarizes the significance of the late Caradoc major sea-level rise, together with the tectonic influences, in controlling the stratigraphy of the Cloridorme Formation. Towards the late Caradoc, a western source-area for clastic influx into the foreland basin must have become important at the expense of the eastern or south-eastern source: the change almost certainly resulting from tectonic readjustments within the thrust-imbricate system.

The Cloridorme Formation does not show any shallowing-up trends in the youngest Marsoui Member and because of erosion there is no exposed record of the overlying stratigraphy. However, Ashgill shelf carbonates of the White Head Formation overlie the thrust system on the Gaspé Peninsula, suggesting that the shallow seas existed over the former source area, and possibly the filled deep-marine foreland basin. Hiscott *et al.* (1986) estimate that the Cloridorme Formation developed in a 100 km-long segment of a much longer foreland basin, with mean rates of sediment accumulation of about 400 m per million years.

LATE ORDOVICIAN–EARLY SILURIAN, NORTH-CENTRAL NEWFOUNDLAND

The Lower Palaeozoic stratigraphy of central Newfoundland (Figs 8 and 9), involving the Dunnage and Gander zones (Williams, 1978, 1979, 1980, 1984; Willams and Hatcher, 1982, 1983) records protracted accretion and collision events on the western margins of the Iapetus Ocean (Fig. 10). The Lower Palaeozoic history may be summarized as follows:

1. The Middle Cambrian–Middle Ordovician evolution of the eastern seaboard of Laurentia records the development of a passive carbonate continental margin (Hubert *et al.*, 1977).

2. In the Llandeilo to earliest Caradoc, collision between the 'Lushs Bight island arc' (McKerrow, 1983) and Laurentia, above an easterly directed (with respect to present North) subduction zone led to obduction and accretion of a suite of ophiolites such as the Bay of Islands and Coastal Complex ophiolites (Casey and Dewey, 1984; Casey *et al.*, 1985; Church and Stevens, 1971; Dewey and Bird, 1971; Dunning and Krogh, 1985; Karson and Dewey, 1978). Based on age differences within the ophiolites of western Newfoundland, Searle and Stevens (1984) postulate two easterly directed subduction zones, the Coastal Complex being the vestiges of an earlier arc, while the approximately 10 km-thick pile of calc-alkaline rocks of Notre Dame Bay represents a later arc built on ophiolitic basement. The pre-Caradoc history of the Dunnage Zone, east of the Lushs Bight island arc, is best interpreted as a backarc or marginal basin (Swinden and Thorpe, 1984).

Seamounts or chains of seamounts, with MORB-affinity ocean crust, collided with the eastern sides of the Lushs Bight arc where local off-scraping and accretion of the seamount igneous basement occurred above a postulated westward-dipping subduction zone (Pickering and Siveter, in preparation). Geochemical studies of the igneous parts of the Tremadoc–Caradoc (513–454+ Ma; time-scale McKerrow *et al.*, 1985) Summerford Group (Jacobi and Wasowski, 1985), and of the mafic volcanic blocks in the Dunnage melange (Wasowski and Jacobi, 1985), are consistent with a genesis at bathymetric highs astride or slightly off mid-ocean ridges, or mid-plate hot spots (seamounts). The Dunnage melange may have originated as a slope talus around a sea-mount or chain of seamounts; a modern analogue may be the submarine basaltic and carbonate reef rubble, and other sediments, observed on seismic-reflection profiles around Shimada seamount in water depths up to 3000 m (Gardner *et al.*, 1984).

The thick development of Llandeilo basaltic pillow lavas and pipes (e.g. the 800 m-thick Lawrence Head Volcanics) may indicate limited crustal extension behind the colliding Lushs Bight arc. Age relationships within the Lower Exploits Group (Fig. 8), however, are not sufficiently constrained to clarify exactly what igneous–sedimentary events were synchronous with arc–continent collision.

3. During the Caradoc (454–442 Ma), following arc–continent collision between the Lushs Bight arc and Laurentia, red radiolarian-rich cherts overlain by grey bioturbated cherts, then graptolitic black mudstones, were deposited extensively throughout the Dunnage Zone. These Facies Group G deposits (Pickering *et al.*, 1986) correlate with the postulated major sea-level rise in the mid to late Caradoc (Fortey, 1984). Thermal subsidence during the Caradoc lull in arc igneous activity, combined with a major rise in sea level, probably conspired to favour the accumulation of the pelagic and hemipelagic layers. Many parts of the continental margins and shelf areas around the Iapetus Ocean appear to have been sites of enhanced pelagic sedimentation associated with the Caradoc eustatic rise in sea level (Leggett, 1978).

The postulated location of the Southern Uplands somewhere off western Newfoundland in the Caradoc (Elders, 1987), together with other data supporting a possible forearc position for the Caradoc–

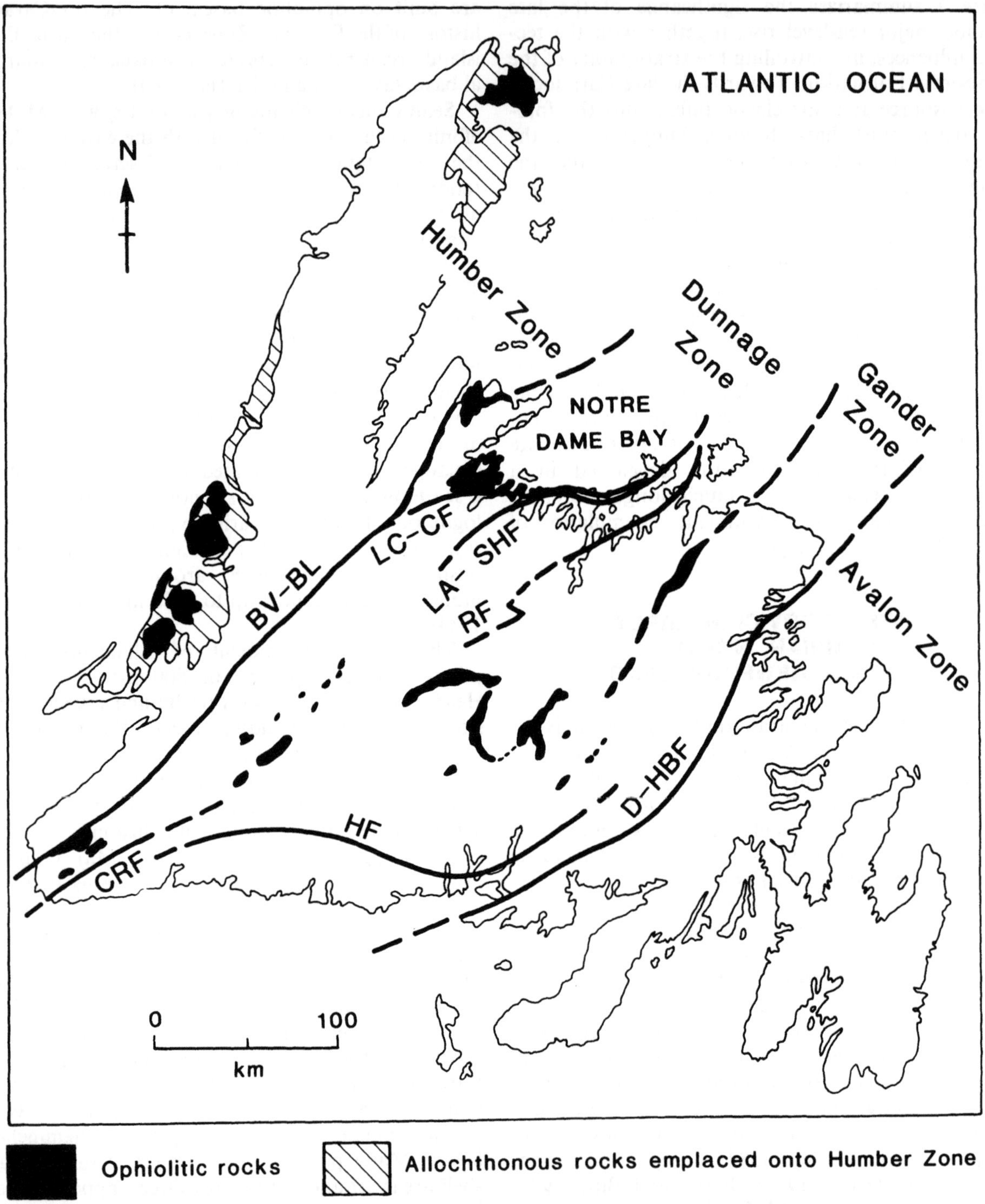

Fig. 8. Map modified from Colmann-Sadd and Swinden (1984) of Newfoundland to show the major tectono-stratigraphic zones (after Williams, 1978), distribution of ophiolitic rocks (black) and allochthonous rocks emplaced onto the Humber Zone (diagonal line legend) during the Taconic Orogeny. BV–BL—Baie Verte–Brompton Line; LC–CF—Lobster Cove–Chanceport Fault; LA–SHF—Lukes Arm–Sops Head Fault; RF—Reach Fault; CRF—Cape Ray Fault; HF—Hermitage Flexure; D–HBF—Dover–Hermitage Bay Fault.

Ashgill of central Newfoundland (Pickering, 1987), suggests that the Caradoc–Ashgill igneous suites in the northern tracts of the Southern Uplands represent arc activity originally generated in the present-day central Newfoundland area. Major strike-slip dismembered the arc to leave relatively small volumes of identifiable arc igneous activity in central Newfoundland. Furthermore, if substantial sinistral shear between plates was occurring in middle–late Ordovician times, then it may be possible to correlate (i) the late Caradoc–Ashgill 'Taconic' thrusting in New York State with the synchronous onset of coarse clastic sedimentation in north-central Newfoundland, and (ii) the late Llandeilo–early Caradoc 'Humberian' thrusting in western Newfoundland with the approximately coeval influx of coarse clastics in the northern tracts of the Southern Uplands. That is, parts of the Dunnage Zone of Newfoundland may have originated as small slope basins off New York State, whereas parts of the northern tracts of the Southern Uplands once may have been located in present-day central Newfoundland.

4. The Ashgill (442–435 Ma) and Llandovery (435–425 Ma) witnessed renewed arc igneous activity (older parts of the Topsails igneous complex—Whalen *et al.*, 1987; rhyodacites in the Point Leamington Formation—Pickering and Saunders (in prep.), with subduction westwards beneath the accreted island arc and Laurentia. Associated with the subduction, an accretionary prism developed involving predominantly E–SE thrust transport directions, with major faults separating different contemporaneous stratigraphies (Arnott, 1983a,b; Arnott *et al.*, 1985; Kusky and Kidd, 1984; Nelson, 1981; Pickering, 1987; Van der Pluijm, 1986). The post-Caradoc calc-alkaline magmatism, however, appears to have been considerably less voluminous compared to the earlier arc activity, although later strike-slip tectonics may have produced a misleading impression of relative abundance levels.

Isotopic dating and petrographic studies of granidiorite suites in pebbles from Scotland suggest that the Southern Uplands was close to the Long Range area of western Newfoundland in the Caradoc (Elders, 1987), therefore major sinistral strike-slip appears to have been important from the late Ordovician to earliest Devonian—for example, to translate the Southern Uplands about 1500 km. The sigmoidal shape of major structural elements such as the Taconic-age Hermitage Flexure (Brown and Colman-Sadd, 1976); the curvature of major terrane-boundary faults that may have been active during sedimentation, such as the Lobster Cove–Chanceport and Lukes Arm–Sops Head faults; the NE–SW, generally westward-dipping, reverse faults associated with similarly orientated folds where stratigraphic thickening occurs, and that may have been actively growing structures during (and controlling) sedimentation, all support a model of major sinistral shear under a predominantly compressive stress regime during the Ashgill and Silurian.

Furthermore, the post-Ashgill igneous suites, such as the younger parts of the Topsails igneous terrane in western Newfoundland (Whalen *et al.*, 1987), and the Springdale Group, are consistent with bimodal suites generated under crustal transtension–transpression in oblique-slip mobile zones. The Californian Borderland, perhaps, provides a modern analogue, where small or incipient ocean basins are developing (Gulf of California) at the same time as transpression (San Andreas Fault in north).

The late Ordovcian–early Silurian stratigraphy of north-central Newfoundland, in Notre Dame Bay, shows overall shallowing-upwards sequences from deep-marine basin and slope to shallow-marine shelf environments (e.g. deep-marine Big Muddy Cove Formation to shallow-marine Goldson Formation—Arnott, 1983a,b). East of the Reach Fault, the Botwood Group (Fig. 9) comprises shallow-marine to non-marine sedimentary and volcanic rocks (Dean, 1978). Thus, by late Ashgill to Llandovery times, the environments of deposition in north-central Newfoundland changed from deep- to shallow-marine or non-marine conditions.

ASHGILL PALAEOGEOGRAPHY, QUEBEC AND NEWFOUNDLAND

The palaeogeography proposed for the Ashgill (Fig. 11) involves the development of two opposing deep-marine foreland basins, either side of the tectonostratigraphic collage created by the accreted Llanvirn–Llandelio island arc(s). Both foreland basins were fully marine and comprised deep-water siliciclastic sedimentation. In the Quebec Re-entrant, the foreland was the downflexed passive margin of Laurentia, whereas in central Newfoundland the foreland was the thinned continental crust of the leading edge of Eastern Avalonia being loaded by the Lushs Bight arc. The Gander Zone represents the vestigial parts of this terrane, the more south-easterly parts probably shearing off along an ancestral Dover–Hermitage Bay Fault to juxtapose against north-west Britain in the Devonian (Emsian).

In central Newfoundland, major sinistral shear in an oblique-slip mobile zone occurred closely associated with thrusting to control sedimentation. In the

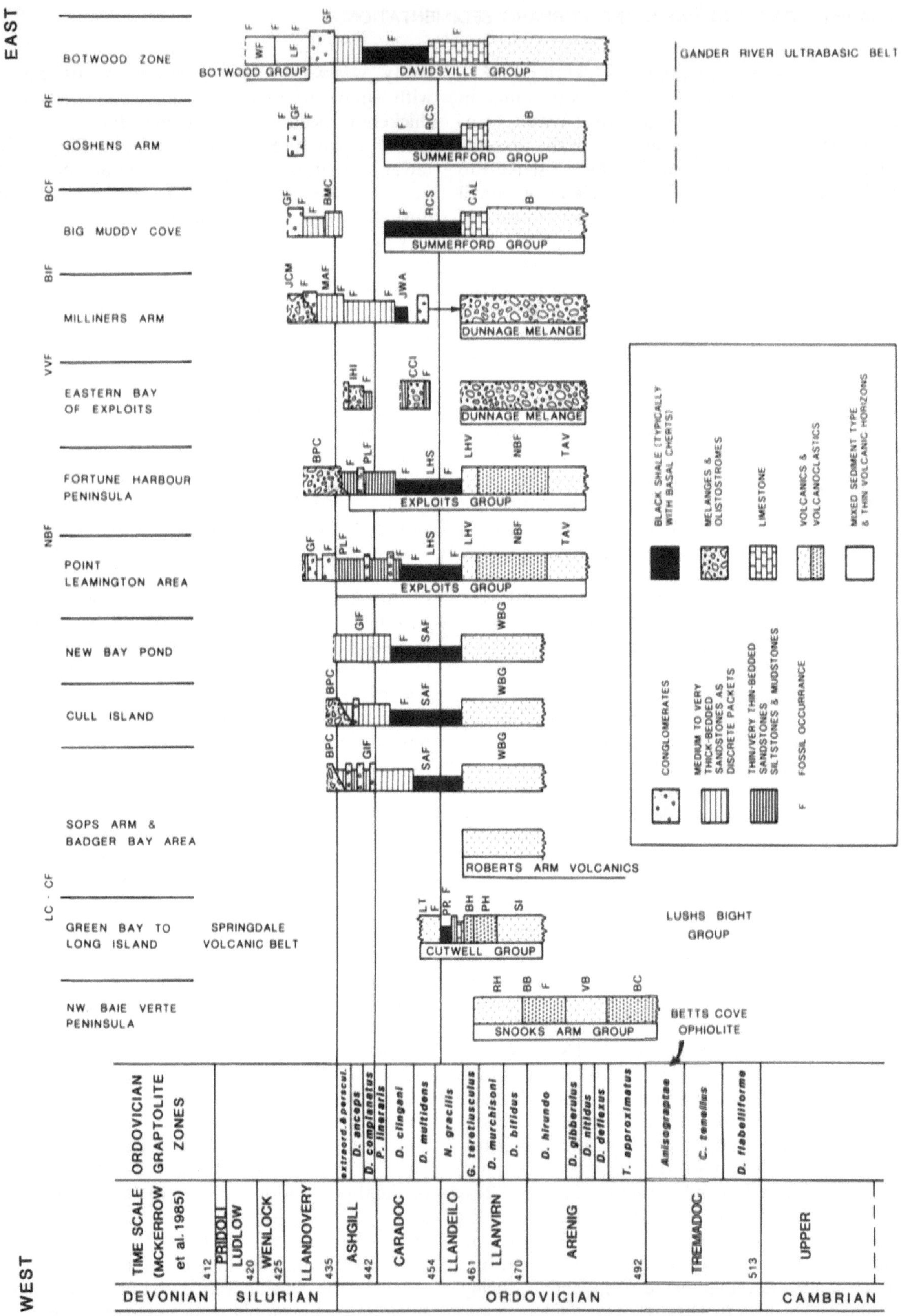
EAST
WEST
BOTWOOD ZONE
BOTWOOD GROUP
DAVIDSVILLE GROUP
GANDER RIVER ULTRABASIC BELT
RF
GOSHENS ARM
SUMMERFORD GROUP
BCF
BIG MUDDY COVE
SUMMERFORD GROUP
BIF
MILLINERS ARM
DUNNAGE MELANGE
VVF
EASTERN BAY OF EXPLOITS
DUNNAGE MELANGE
FORTUNE HARBOUR PENINSULA
EXPLOITS GROUP
NBF
POINT LEAMINGTON AREA
EXPLOITS GROUP
NEW BAY POND
CULL ISLAND
SOPS ARM & BADGER BAY AREA
ROBERTS ARM VOLCANICS
LC - CF
GREEN BAY TO LONG ISLAND
SPRINGDALE VOLCANIC BELT
CUTWELL GROUP
LUSHS BIGHT GROUP
NW. BAIE VERTE PENINSULA
SNOOKS ARM GROUP
BETTS COVE OPHIOLITE
WF LF GF BMC JCM MAF JWA RCS CAL B IHI CCI BPC PLF LHS LHV NBF TAV GIF SAF WBG LT PR BH PH SI RH BB VB BC F
CONGLOMERATES
MEDIUM TO VERY THICK-BEDDED SANDSTONES AS DISCRETE PACKETS
THIN/VERY THIN-BEDDED SANDSTONES SILTSTONES & MUDSTONES
FOSSIL OCCURRANCE
BLACK SHALE (TYPICALLY WITH BASAL CHERTS)
MELANGES & OLISTOSTROMES
LIMESTONE
VOLCANICS & VOLCANOCLASTICS
MIXED SEDIMENT TYPE & THIN VOLCANIC HORIZONS
ORDOVICIAN GRAPTOLITE ZONES
extraord.& perscul.
D. anceps
D. complanatus
P. linearis
D. clingani
D. multidens
N. gracilis
G. teretiusculus
D. murchisoni
D. bifidus
D. hirundo
D. gibberulus
D. nitidus
D. deflexus
T. approximatus
Anisograptae
C. tenellus
D. flabelliforme
TIME SCALE (MCKERROW et al. 1985)
PRIDOLI
LUDLOW
WENLOCK
LLANDOVERY
ASHGILL
CARADOC
LLANDEILO
LLANVIRN
ARENIG
TREMADOC
UPPER
412 420 425 435 442 454 461 470 492 513
DEVONIAN
SILURIAN
ORDOVICIAN
CAMBRIAN

Quebec Re-entrant, there is no evidence for oblique-slip tectonics. The obliquity of collision in central Newfoundland may explain the relatively complex pattern of tectonics and sedimentation compared to the Quebec Re-entrant. Furthermore, the post-Ashgill history of central Newfoundland appears to include bimodal igneous suites that would be consistent with phases of crustal transtension as well as transpression—something that is absent from the Quebec succession.

The proposed Ashgill palaeogeography suggests that the separation of Eastern Avalonia from Laurentia may have involved a marine seaway no more than, say, 500 km in width, floored by interacting continental crust from both continents. A useful analogue may be the continental margin off northern Australia underplating the Banda arc below Timor (Audley-Charles, 1986). This plate-tectonic scenario would be consistent with: (a) palaeomagnetic evidence for Britain south of the Iapetus Suture being in low latitudes, around 20 °S, in the Ashgill (Briden *et al.*, 1984; Piper, 1978, 1979, 1985); (b) palaeontological evidence indicating that the ocean was no longer a barrier to faunal migration by the late Ordovician (Spjeldnaes, 1978; Whittington and Hughes, 1972; Williams, 1973, 1976), and that shelf–shelf contact probably occurred somewhere along the length of the Iapetus Ocean for ostracods with no known pelagic larval stage to interchange between Baltica, Eastern Avalonia and Laurentia (Schallreuter and Siveter, 1985); (c) the structural evidence suggesting extensive thrust-imbricate fan/duplex development in the late Ordovician, contemporaneous with sedimentation in the Dunnage Zone of Newfoundland, (Arnott, 1983a,b; Arnott *et al.*, 1985; Chorlton and Dallmeyer, 1986; Kusky and Kidd, 1984; Nelson, 1981; Pickering, 1987; Van der Pluijm, 1986); and (d) the available structural and stratigraphic data from Newfoundland and Britain to link a common history of plate interaction in the Ashgill, culminating in final closure of the entire system dominated by major sinistral shear in the Emsian (Soper and Hutton, 1984; Soper *et al.*, 1987).

The waning of substantial calc-alkaline igneous activity in both Newfoundland and Britain (e.g. Lake District and Welsh Basin) by late Caradoc times may be related to the initial phases of continent–continent interaction between Laurentia and Avalonia. Murphy and Hutton (1986) develop a palaeogeography for the Irish Caledonides during the Ashgill–Llandovery which is similar to that proposed for the coeval central Newfoundland Appalachians. If, as proposed, the Irish Caledonides were in proximity to the Newfoundland Appalachians in the Ashgill, then the marine foreland basin would have had an along-strike length of at least 500 km.

Although the Ashgill palaeogeography of this paper places Britain south of the Iapetus Suture approximately 2000 km farther north than would be suggested by Cocks and Fortey (1982) or McKerrow and Cocks (1986), it should be stressed that this plate-tectonic reconstruction is consistent with the database used by McKerrow and Cocks (1976, 1977, 1981, 1986). However, unlike the McKerrow and Cocks (1986) reconstruction, this Ashgill palaeogeography attempts to explain the limited post-Caradoc arc-related igneous activity and the timing and nature of the Gander Zone underthrusting (Colman-Sadd and Swinden, 1984), or underplating, in central Newfoundland, in terms of the development of a foreland basin. It incorporates, also, the available palaeomagnetic data for Britain south of the Iapetus Suture, and the faunal data across the Iapetus Ocean.

Fig. 9. Summary stratigraphy of Lower Palaeozoic Dunnage Zone, Notre Dame Bay, Newfoundland (adapted and modified from Arnott *et al.*, 1985). Faults separating successions are: LC–CF—Lobster Cove–Chanceport; NBF—New Bay; VVF—Village Virgin; BIF—Boyds Island; BCF—Byrne Cove; RF—Reach Fault. Stratigraphic units are: BC—Bobby Cove Formation; VB—Venams Bight Basalt; BB—Balsam Bud Cove Formation; RH—Round Harbour Basalt; SI—Stag Island Formation; PH—Pigeon Head Formation; BH—Burnt Head Formation; PP—Parson Point Formation; LT—Long Tickle Formation; WBG—Wild Bight Group (? time equivalent to Frozen Ocean Group); SAF—Shoal Arm Formation; GIF—Gull Island Formation; BPC—Boones Point Complex; TAV—Tea Arm Volcanics; NBF—New Bay Formation; LHV—Lawrence Head Volcanics; LHS—Lawrence Harbour Shale; PLF—Point Leamington Formation; GF—Goldson Formation; CCI—Carters Cove Olistostrome; IHI—Intricate Harbour Olistostrome; JWA—Joe Whites Arm Shale; MAF—Milliners Arm Formation; JCM—Joeys Cove Melange; B—basalts; CAL—Cobbs Arm Limestone; RCS—Rodgers Cove Shale; BMC—Big Muddy Cove Formation; LF—Lawrenceton Formation; WF—Wigwam Formation. The Indian Island Group, in the Botwood Zone, was probably contemporaneous with the uppermost part of the Davidsville Group and the lower part of the Botwood Group (Dean, 1978). Timescale after McKerrow *et al.*, 1985, with graptolite zonation after Williams *et al.*, 1972. For detailed lithostratigraphy, see Dean (1978). Note that pre-Caradoc, mainly volcanoclastic/pyroclastic successions interlayered with arc volcanics are differentiated from post-Caradoc sandstones rich in arc-derived volcanic clasts. After Pickering (1987).

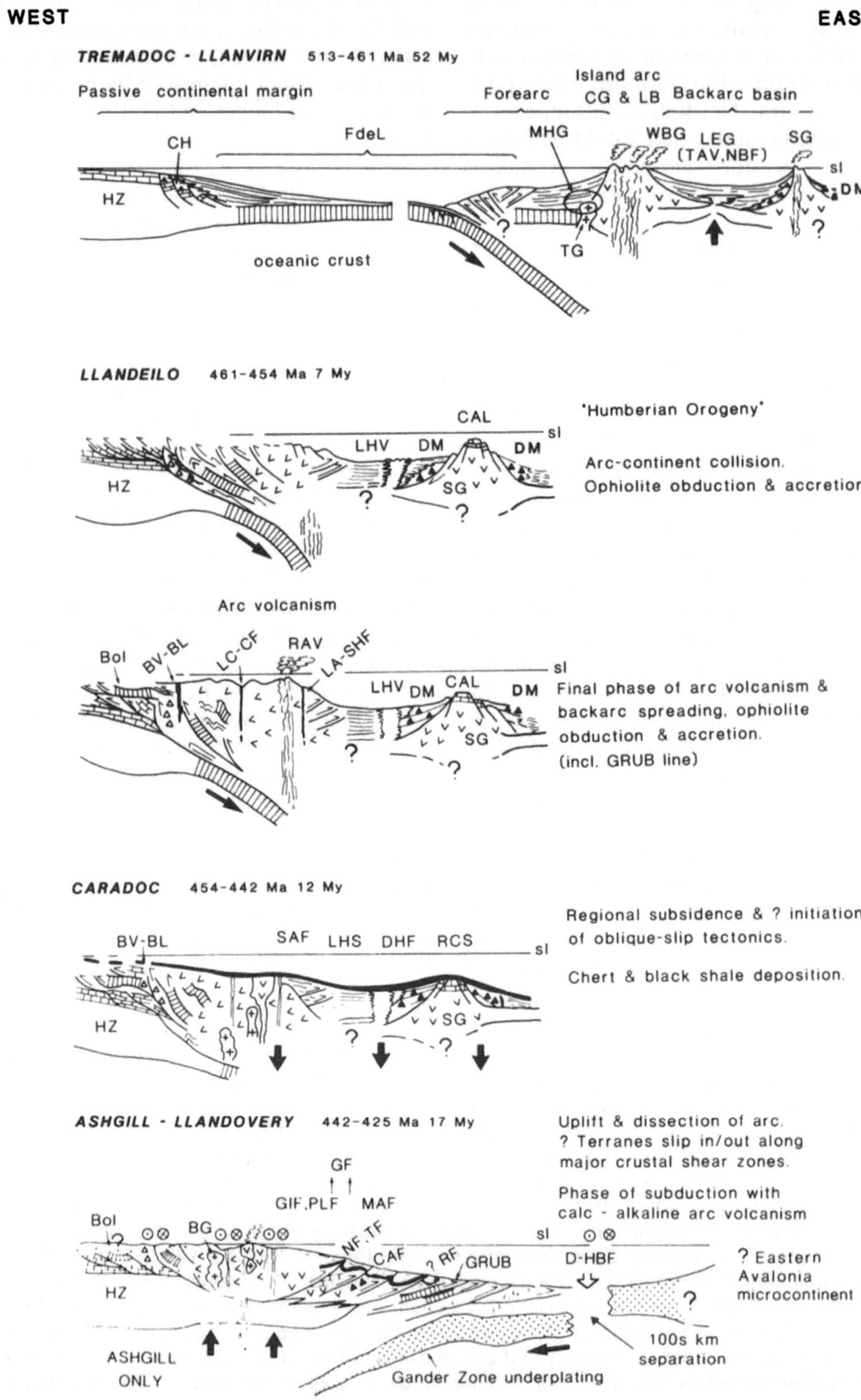
WEST
EAST
TREMADOC - LLANVIRN 513-461 Ma 52 My
Passive continental margin
Forearc
Island arc
CG & LB
Backarc basin
CH
FdeL
MHG
WBG
LEG
(TAV,NBF)
SG
sl
DM
HZ
TG
oceanic crust
LLANDEILO 461-454 Ma 7 My
'Humberian Orogeny'
CAL
sl
LHV
DM
DM
HZ
SG
Arc-continent collision.
Ophiolite obduction & accretion.
Arc volcanism
Bol
BV-BL
LC-CF
RAV
LA-SHF
sl
LHV
DM
CAL
DM
SG
Final phase of arc volcanism & backarc spreading, ophiolite obduction & accretion. (incl. GRUB line)
CARADOC 454-442 Ma 12 My
Regional subsidence & ? initiation of oblique-slip tectonics.
Chert & black shale deposition.
BV-BL
SAF
LHS
DHF
RCS
sl
HZ
SG
ASHGILL - LLANDOVERY 442-425 Ma 17 My
Uplift & dissection of arc.
? Terranes slip in/out along major crustal shear zones.
Phase of subduction with calc - alkaline arc volcanism
GF
GIF,PLF
MAF
Bol
BG
NF,TF
sl
CAF
RF
GRUB
D-HBF
? Eastern Avalonia microcontinent
HZ
100s km separation
ASHGILL ONLY
Gander Zone underplating

Structural evidence from the British Caledonides suggests deformation in the Caradoc–Ashgill, with unconformities (including disconformities) occurring in the following areas: (a) between the Raheen Group above the Tramore Volcanics in the Tramore area; (b) in North Wales, shale successions upon the Snowdon Volcanics; (c) in central Wales around Builth, the Camlo Hill Group upon the Carmel Group; (d) also, the Dholhir Beds and Limestones above the Caradoc volcanics and volcanoclastics, in the Berwyn Hills; (e) in the Welsh Borderland, the succession cuts off in the middle Caradoc; (f) in the Lake District, the Ashgill Shales above the Ashgill Applethwaite Beds; also, the late Caradoc unconformity between the Applethwaite Beds and underlying Stile End Beds; and the Caradoc unconformity above the Borrowdale Volcanics; (g) in the Cross Fell area there are unconformities or disconformities between the Dufton Shale Group and underlying andesitic, rhyolitic and tuffaceous Borrowdale Volcanic Group, and between the Keisley Limestone disconformable on the Dufton Shale Group. It is sometime during the late Caradoc to Ashgill that the interaction between the Eastern Avalonian plate and Laurentia appears to have occurred and, perhaps, these unconformities indicate the beginning of such an event, and/or closure of Tornquist's Sea.

The deep seismic reflection profile across the northern Appalachians, off the north coast of Newfoundland (Keen *et al.*, 1986), has produced an interpretation that reveals a deep crustal structure compatible with the surface structural geology described above (Fig. 12). The stratigraphy and structure of the Dunnage–Gander zones are perhaps most reasonably interpreted as formed during the postulated collisional events. In this interpretation, the west-directed structures would be due to arc–continent collision and associated ophiolite and remnant–arc obduction in the Llanvirn–Llandeilo. The east-directed structures would be due to underplating of the lower crust (?continental affinity) from the east beneath the Grenville crust, leading to the development of a foreland basin from a forearc environment, during continent–continent collision from late Ordovician until at least the middle Silurian. Final closure of 'ensialic' basins in Siluro-Devonian times was manifest as the Acadian Orogeny.

CONCLUSIONS

The stratigraphy in the Quebec Appalachians shows a diachronous northward relocation of the depocentres towards the foreland (Hiscott *et al.*, 1986) as predicted by Quinlan and Beaumont (1984). However, this progressive filling of depocentres migrating towards the foreland is not observed in central Newfoundland, possibly due to relatively poor age constraints on the successions, but probably because the plate-tectonic picture was considerably more complex, being dominated by sinistral

Fig. 10. Summary cartoon of Tremadoc to Llandovery geological evolution of the Dunnage Zone, Newfoundland. Diagrams drawn with information from Arnott (1983a, b), Arnott *et al.* (1985), Colman-Sadd and Swinden (1984), Dean (1978), Jacobi and Wasowski (1985), Nelson (1981), and Watson (1981). HZ—Humber Zone; CH—Cow Head Group; FdeL—Fleur de Lys rocks; MHG—Moretons Harbour Group (shown in forearc, but could be intra-arc); TG—Twillingate Granodiorite; CG and LB—Cutwell Group and Lushs Bight Group (including Pacquet Harbour Group, Snooks Arm Group); WBG—Wild Bight Group (may include Frozen Ocean Group); LEG—Lower Exploits Group, including Tea Arm Volcanics (TAV) and New Bay Formation (NBF); SG—Summerford Group; LHV—Lawrence Head Volcanics; DM—Dunnage Melange; CAL—Cobbs Arm Limestone; BV–BL—Baie Verte–Brompton Line; LC–CF—Lobster Cove–Chanceport Fault; SAF—Shoal Arm Formation; LHS—Lawrence Harbour Shale; DHF—Dark Hole Formation; RCS—Rodgers Cove Shale; RAV—Roberts Arm Volcanic Belt; LA–SHF—Lukes Arm–Sops Head Fault; BG—Burlington Granodiorite; GIF—Gull Island Formation; PLF—Point Leamington Formation; MAF—Milliners Arm Formation; NF—New Bay Fault; TF—Toogood Fault; CAF—Cobbs Arm Fault; BoI—Bay of Islands and associated ophiolites; GF—Goldson Formation. Melanges such as the Boones Point Complex are associated with the Lukes Arm–Sops Head Fault, as olistostromes tectonized by the synsedimentary thrust faulting. Subduction directions are shown, together with backarc spreading centres. Arc–continent collision began in the late Llanvirn; this collision also involved the SG, DM and CAL but the two-dimensional sketches do not adequately show this; throughout this time, over 800 m of tholeiitic basalts (LHV) of the LEG were extruded in a submarine environment, implying continued ?remnant backarc extension. Gander Zone underplating may have occurred towards Upper Llandovery and later. Sea level shown by lines marked sl. During Caradoc, cessation of arc volcanism may have led to thermal contraction of the lithosphere and associated subsidence (chert and black shale deposition), possibly with global rise in sea level. See text for explanation and Fig. 9 for stratigraphy. After Pickering (1987).

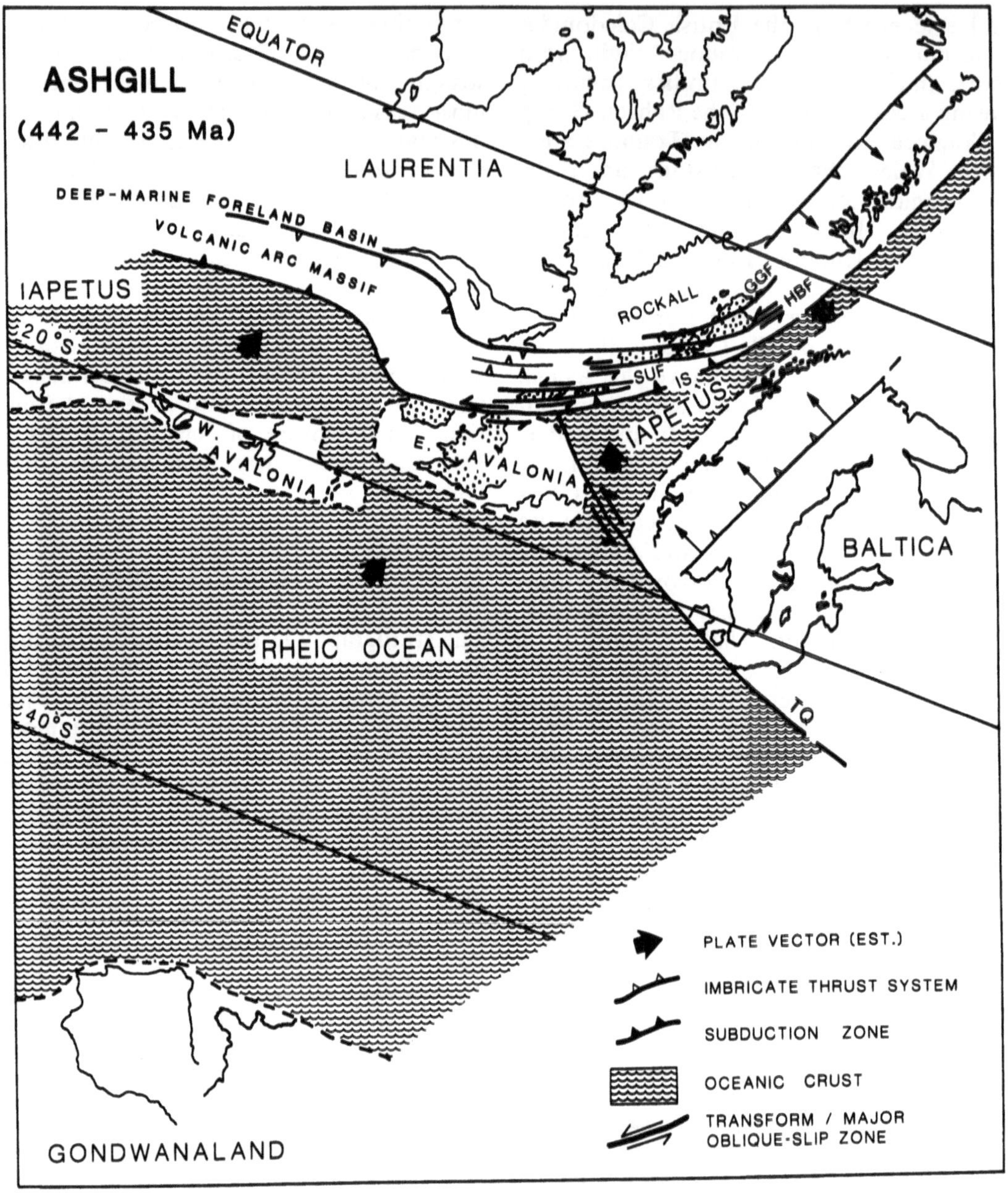

Fig. 11. Possible palaeogeographic reconstruction for Ashgill times, adapted from Pickering *et al.* (in press) as modified from McKerrow and Cocks, 1986 Caradoc palaeogeography. The Avalonian microcontinent may have been considerably more disjointed than shown here. Note the predominance of sinistral oblique-slip tectonics. See text for explanation. IS—Iapetus Suture; GGF—Great Glen Fault; HBF—Highland Boundary Fault; SUF—Southern Uplands Fault; TQ—Tornquist's Line.

oblique-slip. The quality of the data from Newfoundland precludes a resolution of tectonic events into phases of deformation dominated by thrusting or strike-slip.

The Quebec and Newfoundland examples of deep-marine foreland basins evolved as a response to lithospheric flexure induced by an accreted island arc complex acting as a load. Also, in both examples, the pre-foreland basin plate-tectonic setting was that of a forearc accretionary prism. The senses of thrusting prior to, and following, arc–continent collision were similar in orientation. Finally, the Lower Palaeozoic

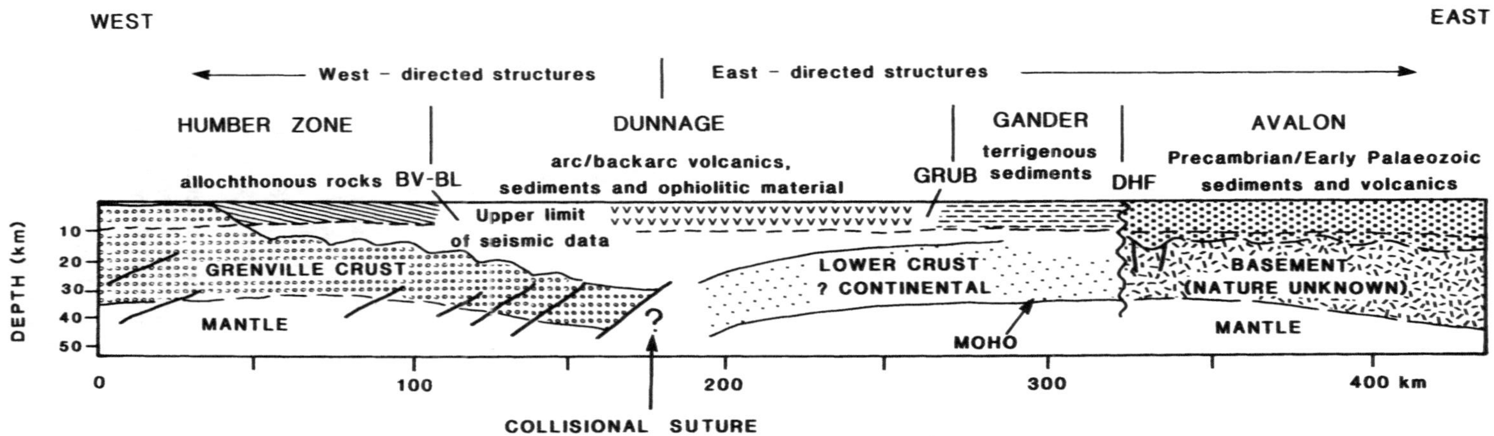

Fig. 12. Simplified interpretation of the deep seismic reflection profile across the Northern Appalachians (west to east off northern Newfoundland), redrawn from Fig. 3 of Keen *et al.* (1986). BV–BL—Baie Verte–Brompton Line; GRUB—Gander River Ultrabasic Belt of ophiolitic rocks; DHF—Dover–Hermitage Bay Fault zone.

stratigraphies in Quebec and Newfoundland suggest that the deep-marine foreland basins were eventually infilled by shallow-marine and non-marine sediments, presumably after the lithosphere had reached a state of flexural rigidity in response to the imposed load (accreted island arc and thrust sheets), and/or the driving subsidence imposed by the load had been effectively reduced by erosion.

References

Arnott, R. J. (1983a). Sedimentology, structure and stratigraphy of north-east New World Island, Newfoundland, Unpublished D.Phil. Thesis, Oxford University.

Arnott, R. J. (1983b). Sedimentology of Upper Ordovician–Silurian sequences on New World Island, Newfoundland: separate fault-controlled basins? *Canadian Journal of Earth Sciences* **20,** 345–354.

Arnott, R. J., McKerrow, W. S. and Cocks, L. R. M. (1985). The tectonics and depositional history of the Ordovician and Silurian rocks of Notre Dame Bay, Newfoundland, *Canadian Journal of Earth Sciences* **22,** 607–618.

Audley-Charles, M. G. (1986). Rates of Neogene and Quaternary tectonic movements in the Southern Banda Arc based on micropalaeontology, *Journal of the Geological Society, London* **143,** 161–175.

Barnes, C. R., Norford, B. S. and Skevington, D. (1981). The Ordovician System in Canada. Publication of the International Union of Geological Sciences, Volume 8, 27 pp.

Beaumont, C. (1981). Foreland basins, *Geophysical Journal of the Royal Astronomical Society* **65,** 291–329.

Beaumont, C., Keen, C. E. and Boutilier, R. (1982). A comparison of foreland and rift margin sedimentary basins, *Philosophical Transactions of the Royal Society London* Series A 305, 295–317.

Beeden, D. R. (1983). Sedimentology of some turbidites and related rocks from the Cloridorme Group, Ordovician, Quebec, M.Sc. Thesis, McMaster University, Hamilton, Canada.

Bowin, C., Purdy, G. M., Johnston, C., Shor, G. G., Lawver, L., Hartong, H. M. S. and Jezek, P. (1980). Arc–continent collision in the Banda Sea region, *Bulletin of the American Association of Petroleum Geologists* **64,** 868–915.

Boyer, S. E. and Elliott, D. (1982). Thrust systems, *Bulletin of the American Association of Petroleum Geologists* **66,** 1196–1230.

Brewer, J. A. and Smythe, D. K. (1983). The Moine and Outer Isles seismic traverse (MOIST), *In* A. W. Bally (Ed.), *Seismic Expression of Structural Styles*, Vol. 3, pp. 3.2.1–23 to 3.2.1–28, American Association of Petroleum Geologists, Studies in Geology Series, No. 15.

Brewer, J. A. and Smythe, D. K. (1984). MOIST and the continuity of crustal reflector geometry along the Caledonian–Appalachian orogen, *Journal of the Geological Society, London* **141,** 105–120.

Briden *et al.* (1984). British palaeomagnetism, Iapetus Ocean, and the Great Glen Fault, *Geology* **12,** 428–431.

Brown, P. A. and Colman-Sadd, S. P. (1976). Hermitage flexure: figment or fact? *Geology* **4,** 561–564.

Casey, J. F. and Dewey, J. F. (1984). Initiation of subduction zones along transform and accreting plate boundaries, triple junction evolution, and forearc spreading centres—implications for ophiolitic geology and obduction, *In* I. G. Gass, S. J. Lippard and A. W. Shelton (Eds), *Ophiolites and Oceanic Lithosphere*, Geological Society of London Special Publication, 13, pp. 269–290.

Casey, J. F., Elthon, D. L., Siroky, F. X., Karson, J. A. and Sullivan, J. (1985). Geochemical and geological evidence bearing on the origin of the Bay of Islands and Coastal Complex ophiolites of western Newfoundland, *Tectonophysics* **116,** 1–40.

Chorlton, L. R. and Dallmeyer, R. D. (1986). Geochronology of early to middle Paleozoic tectonic development in the south-west Newfoundland Gander Zone, *Journal of Geology* **94,** 67–89.

Church, W. R. and Stevens, R. K. (1971). Early Palaeozoic ophiolite complexes of the Newfoundland Appalachians as mantle–oceanic crust sequences, *Journal of Geophysical Research* **76,** 1460–1466.

Cocks, L. R. M. and Fortey, R. A. (1982). Faunal evidence for oceanic separations in the Palaeozoic of Britain, *Journal of the Geological Society, London* **139,** 465–478.

Colman-Sadd, S. P. and Swinden, H. S. (1984). A tectonic window in central Newfoundland? Geological evidence that the Appalachian Dunnage Zone may be allochthonous, *Canadian Journal of Earth Sciences* **21,** 1349–1367.

Dean, P. L. (1978). The volcanic stratigraphy and metallogeny of Notre Dame Bay, Newfoundland, *Memorial University, Newfoundland, Geology Report* **7.**

Dewey, J. F. and Bird, J. M. (1971). Origin and emplacement of the Ophiolite Suite: Appalachian ophiolites in Newfoundland, *Journal of Geophysical Research* **76,** 3179–3206.

Dunning, G. R. and Krogh, T. E. (1985). Geochronology of ophiolites of the Newfoundland Appalachians, *Canadian Journal of Earth Sciences* **22,** 1659–1670.

Elders, C. (1987). The provenance of granite boulders in conglomerates of the northern central belts of the Southern Uplands of Scotland, *Journal of the Geological Society, London*, in press.

Enos, P. (1969a). Cloridorme Formation, Middle Ordovician flysch, northern Gaspe Peninsula, Quebec. *Geological Society of America* Special Paper, 117.

Enos, P. (1969b). Anatomy of a flysch, *Journal of Sedimentary Petrology* **39,** 680–723.

Fortey, R. A. (1984). Global earlier Ordovician transgressions and regressions and their biological implications, *In* D. L. Bruton (Ed.), *Aspects of the Ordovician System*, Palaeontological Contributions from the University of Oslo, No. 295, pp. 37–50, Universitetsforlaget, Oslo.

Gardner, J. V., Dean, W. E. and Blakely, R. J. (1984). Shimada Seamount: an example of recent mid-plate volcanism. *Bulletin Geological Society of America*, **95,** 855–862.

Hamilton, W. (1979). Tectonics of the Indonesian region, *US Geological Survey Professional Paper*, 1078.

Hiscott, R. N. (1977). Sedimentology and regional implications of deep-water sandstones of the Tourelle Formation, Ordovician, Quebec. Ph.D. Thesis, McMaster University, Hamilton, Canada.

Hiscott, R. N. (1978). Provenance of Ordovician deep-water sandstones, Tourelle Formation, Quebec, and implications for initiation of the Taconic orogeny, *Canadian Journal of Earth Sciences* **15,** 1579–1597.

Hiscott, R. N. (1980). Depositional framework of sandy mid-fan complexes of Tourelle Formation, Ordovician, Quebec, *American Association of Petroleum Geologists Bulletin*, **64,** 1052–1077.

Hiscott, R. N. (1984). Ophiolitic source rocks for Taconic-age flysch: trace-element evidence, *Bulletin of the Geological Society of America* **95,** 1261–1267.

Hiscott, R. N., Pickering, K. T. and Beeden, D. R. (1986). Progressive filling of a confined Middle-Ordovician foreland basin associated with the Taconic Orogeny, Quebec, Canada, *In* P. A. Allen and P. Homewood (Eds), *Foreland Basins*. International Association of Sedimentologists Special Publication, 8, pp. 165–181.

Hubert, J. F., Suchecki, R. K. and Callahan, R. K. M. (1977). The Cow Head Breccia: sedimentology of the Cambro-Ordovcian continental margin, Newfoundland. *In* H. E. Cook and P. Enos (Eds), *Deep-water Carbonate Environments*, Society of Economic Paleontologists and Mineralogists Special Publication 25, pp. 125–154.

Jacobi, R. D. and Wasowski, J. J. (1985). Geochemistry and plate-tectonic significance of the volcanic rocks of the Summerford Group, north-central Newfoundland, *Geology* **13,** 126–130.

Jacobson, R. S., Shor, G. G., Kieckheffer, R. M. and Purdy, G. M. (1978). Seismic refraction and reflection studies in the Timor–Aru trough system and Australian continental shelf, *Memoirs of the American Association of Petroleum Geologists* **29,** 209–222.

Karson, J. and Dewey, J. F. (1978). Coastal Complex, western Newfoundland: An Early Ordovician oceanic fracture zone, *Bulletin of the Geological Society of America* **89,** 1037–1049.

Keen, C. E., Keen, M. J., Nichols, B., Reid, G., Stockmal, G. S., Colman-Sadd, S. P., O'Brien, S. J., Miller, H., Quinlan, G., Williams, H. and Wright, J. (1986). Deep seismic-reflection profile across the northern Appalachians, *Geology* **14,** 141–145.

Kusky, T. M. and Kidd, W. S. F. (1984). Middle Ordovician conodonts from the Buchans Group, central Newfoundland, and their significance for regional stratigraphy of the Central Volcanic Belt: Discussion, *Canadian Journal of Earth Sciences* **22,** 484–485.

Labaume, P., Mutti, E., Seguret, M. and Rosell, J. (1983). Megaturbidites carbonatées du bassin turbiditique de l'Eocene inférieur et moyen sud-pyrénéen, *Bulletin of the Geological Society France* **25,** 927–941.

Labaume, P., Seguret, M. and Seyve, C. (1985). Evolution of a turbidite foreland basin and analogy with an accretionary prism: example of the Eocene South-Pyrenean Basin, *Tectonics* **4,** 661–685.

Laroche, P. J. (1983). Appalachians of southern Quebec seen through Seismic Line no. 2001, *In* A. W. Bally (Ed.), *Seismic Expression of Structural Styles*, American Association of Petroleum Geologists, *Studies in Geology Series, N15*, Vol. 3, pp. 3.2.1–7 to 3.2.1–24.

Leggett, J. K. (1978). Eustacy and pelagic regimes in the Iapetus Ocean during the Ordovician and Silurian, *Earth and Planetary Science Letters* **41,** 163–169.
and C. L. Drake (Eds), *Studies in Continental Margin Geology*, American Association of Petroleum Geologists, Memoir 34, pp. 521–533.

McKerrow, W. S. and Cocks, L. R. M. (1976). Progressive faunal migration across the Iapetus Ocean, *Nature* **263,** 304–306.

McKerrow, W. S. and Cocks, L. R. M. (1977). The location of the Iapetus Ocean suture in Newfoundland, *Canadian Journal of Earth Sciences* **14,** 488–495.

McKerrow, W. S. and Cocks, L. R. M. (1981). Stratigraphy of eastern Bay of Exploits, Newfoundland, *Canadian Journal of Earth Sciences* **18,** 751–764.

McKerrow, W. S. and Cocks, L. R. M. (1986). Oceans, island arcs and olistostromes: the use of fossils in distinguishing sutures, terranes and environments around the Iapetus Ocean, *Journal of the Geological Society, London* **143,** 185–191.

McKerrow, W. S., Lambert, R. St-J. and Cocks, L. R. M. (1985). The Ordovician, Silurian and Devonian Periods, *In* N. J. Snelling (Ed.), *The Chronology of the Geological Record*, Geological Society of London Memoir 10, pp. 73–80.

Malpas, J. and Stevens, R. K. (1977). The origin and emplacement of the ophiolite suite with examples from western Newfoundland, *Geotectonics* **11,** 453–466.

Murphy, P. C. and Hutton, D. H. W. (1986). Is the Southern Uplands of Scotland really an accretionary prism? *Geology* **14,** 354–357.

Mutti, E. (1977). Distinctive thin-bedded turbidite facies and related depositional environments in the Eocene Hecho Group (South-central Pyrenees, Spain), *Sedimentology* **24,** 107–131.

Mutti, E. (1985). Hecho turbidite system, Spain, *In* A. H. Bouma, W. R. Normark and N. E. Barnes (Eds), *Submarine Fans and Related Turbidite Systems*, Springer-Verlag, New York, pp. 205–208.

Nelson, K. D. (1981). Melange development in the Boones Point Complex, north-central Newfoundland, *Canadian Journal of Earth Sciences* **18,** 433–442.

Neuman, R. B. (1984). Geology and palaeobiology of islands in the Ordovician Iapetus Ocean: review and implications, *Bulletin of the Geological Society of America* **95,** 1188–1201.

Ori, G. G., Roveri, M. and Vannoni, F. (1986). Plio-Pleistocene sedimentation in the Apenninic–Adriatic foredeep (Central Adriatic Sea, Italy). *In* P. A. Allen and P. Homewood (Eds), *Foreland Basins.* International Association of Sedimentologists Special Publication, **8,** 183–198.

Pickering, K. T. (1987). Wet-sediment deformation in the Upper Ordovician Point Leamington Formation: an active thrust-imbricate system during sedimentation, Notre Dame Bay, north-central Newfoundland, *In* M. E. Jones and R. M. F. Preston (Eds), *Deformation of Sediments and Sedimentary Rocks*, Geological Society of London, Special Publication, 199–225.

Pickering, K. T. and Hiscott, R. N. (1985). Contained (reflected) turbidity currents from the Middle Ordovician Cloridorme Formation, Quebec, Canada: an alternative to the antidune hypothesis, *Sedimentology* **32,** 373–394.

Pickering, K. T., Stow, D. A. V., Watson, M. P. and Hiscott, R. N. (1986). Deep-water facies, processes and models: a review and classification scheme for modern and ancient sediments, *Earth-Science Reviews* **23,** 75–174.

Pickering, K. T. and Saunders, A. D. (in prep.). Evidence for early Ashgill igneous activity during sedimentation of the Point Leamington Formation, Dunnage Zone, north-central Newfoundland.

Pickering, K. T., Siveter, D. J. and Bassett, M. G. (1987). Evidence for late Ordovician closure of the Iapetus Ocean in central Newfoundland in relation to the Northern Appalachian–Caledonide Orogen. *Transactions of the Royal Society of Edinburgh: Earth Sciences* (in press).

Piper, J. D. A. (1978). Palaeomagnetic survey of the (Palaeozoic) Shelve inlier and Berwyn Hills, Welsh Borderlands, *Geophysical Journal of the Royal Astronomical Society* **53,** 355–371.

Piper, J. D. A. (1979). Aspects of Caledonian palaeomagnetism and their tectonic implications. *Earth and Planetary Science Letters* **44,** 176–192.

Piper, J. D. A. (1985). Palaeomagnetism in the Caledonian–Appalachian orogen: a review, *In* D. G. Gee and B. A. Sturt (Eds), *The Caledonian Orogen—Scandinavia and Related Areas*, Wiley, Chichester, pp. 35–49.

Quinlan, G. M. and Beaumont, C. (1984). Appalachian thrusting, lithospheric flexure, and the Palaeozoic stratigraphy of the Eastern Interior of North America, *Canadian Journal of Earth Sciences* **21,** 973–996.

Ricci Lucchi, F. (1975). Miocene palaeogeography and basin analysis in the Periadriatic Apennines, *In* C. Squyres (Ed.), *Geology of Italy*, Petroleum Exploration Society of Libya, Tripoli, pp. 5–111.

Ricci Lucchi, F. (1981). The Miocene Marnoso–Arenacea turbidites, Romagna and Umbria Apennines, *In* F. Ricci Lucchi (Ed.), *Excursion Guidebook with Contributions on Sedimentology of Some Italian Basins*, International Association of Sedimentologists, Second European Regional Meeting, Bologna, Italy, pp. 229–303.

Ricci Lucchi, F. and Ori, G. G. (1985). Synorogenic deposits of a migrating basin system in the north-west Adriatic foreland: examples from Emilia–Romagna region, northern Apennines, *In* P. A. Allen, P. Homewood and G. Williams (Eds), *International Symposium on Foreland Basins, Excursion Guidebook*, International Association of Sedimentologists, pp. 137–176.

Sanford, B. V., Grant, A. C., Wade, J. A. and Barss, M. S. (1979). Geology of Eastern Canada and adjacent areas, *Geological Survey of Canada*, Map 1401A.

Schallreuter, R. E. L. and Siveter, D. J. (1985). Ostracodes across the Iapetus Ocean, *Palaeontology* **28,** 577–598, plates 1–3.

Searle, M. P. and Stevens, R. K. (1984). Obduction processes in ancient, modern and future ophiolites, *In* I. G. Gass, S. J. Lippard and A. W. Shelton (Eds), *Ophiolites and Oceanic Lithosphere*, Geological Society of London Special Publication, 13, pp. 303–319.

Shanmugam, G. and Lash, G. G. (1982). Analagous tectonic evolution of a foredeep basin in the Middle Ordovician, southern Appalachians, *American Journal of Science* **280,** 479–496.

Soper, N. J. and Hutton, D. H. W. (1984). Late Caledonian sinistral displacements in Britain: implications for a three-plate collision model, *Tectonics* **3,** 781–794.

Soper, N. J., Webb, B. C. and Woodcock, N. H. (1987). Late Caledonian transpression in north-west England: timing, geometry and geotectonic significance, *Proceedings Yorkshire Geological Society*, in press.

Spjeldnaes, N. (1978). Faunal provinces and the Proto-Atlantic, *Geological Journal Special Issue* **10,** 139–150.

St Julien, P. and Hubert, C. (1975). Evolution of the Taconian Orogen in the Quebec Appalachians, *American Journal of Science* **275-A,** 337–362.

Stanley, R. S. and Ratcliffe, N. M. (1985). Tectonic synthesis of the Taconian orogeny in western New England, *Bulletin of the Geological Society of America* **96,** 1227–1250.
Swinden, H. S. and Strong, D. F. (1976). A comparison of plate-tectonic models of metallogenesis in the Appalachians, the North American Cordillera, and the East Australian Palaeozoic, *Geological Association of Canada*, Special Paper No. 14, 443–471.
Swinden, H. S. and Thorpe, R. I. (1984). Variations in the style of volcanism and massive sulfide deposition in early to middle Ordovician island-arc sequences of the Newfoundland Central Mobile Belt, *Economic Geology* **79,** 1596–1619.
Thomas, W. A. (1977). Evolution of Appalachian–Ouachita salients and recesses from re-entrants and promontories in the continental margin, *American Journal of Science* **277,** 1233–1278.
Van der Pluijm, B. A. (1986). Geology of eastern New World Island, Newfoundland: an accretionary terrane in the north-eastern Appalachians, *Bulletin of the Geological Society of America* **97,** 932–945.
Walcott, R. I. (1970). Isostatic response to loading of the crust in Canada, *Canadian Journal of Earth Sciences* **7,** 2–13.
Wasowski, J. J. and Jacobi, R. D. (1985). Geochemistry and tectonic significance of the mafic volcanic blocks in the Dunnage melange, north-central Newfoundland, *Canadian Journal of Earth Sciences* **22,** 1248–1256.
Watson, M. P. (1981). Submarine-fan deposits of the Upper Ordovician–Lower Silurian Milliners Arm Formation, New World Island, Newfoundland. D.Phil. Thesis, Oxford University.
Whalen, J. B., Currie, K. L. and Van Breemen, O. (1987). Episodic Ordovician–Silurian plutonism in the Topsails igneous terrane, western Newfoundland, *Philosophical Transactions of the Royal Society, Edinburgh*, in press.
Whittington, H. B. and Hughes, C. P. (1972). Ordovician geography and faunal provinces deduced from trilobite distribution, *Philosophical Transactions of the Royal Society, London, Series B*, **263,** 235–278.
Williams, A. (1973). Distribution of brachiopod assemblages in relation to Ordovician palaeogeography, *Special Papers in Palaeontology* **12,** 241–269.
Williams, A., Strachan, I., Bassett, D. A., Dean, W. T., Ingham, J. K., Wright, A. D. and Whittington, H. B. (1972). A correlation of Ordovician rocks in the British Isles. Geological Society of London Special Report No. 3.
Williams, A. (1973). Distribution of brachiopod assemblages in relation to Ordovician palaeogeography, *Special Papers in Palaeontology* **12,** 241–269.
Williams, A. (1976). Plate tectonics and biofacies evolution as factors in Ordovician correlation, *In* M. G. Bassett (Ed.), *The Ordovician System*, University of Wales Press and National Museum of Wales, Cardiff, pp. 29–66.
Williams, H. (1978). Tectonic lithofacies map of the Appalachian orogen, Memorial University of Newfoundland Map No. 1, scale 1:1 000 000.
Williams, H. (1979). Appalachian orogen of Canada, *Canadian Journal of Earth Sciences* **16,** 792–807.
Williams, H. (1980). Structural telescoping across the Appalachian orogen and the minimum width of the Iapetus Ocean, *In* D. W. Strangway (Ed.), *The Continental Crust of the Earth and Its Mineral Deposits*, Geological Association of Canada, Special Paper 20, pp. 421–440.
Williams, H. (1984). Miogeoclines and suspect terranes of the Caledonian–Appalachian Orogen: tectonic patterns in the North Atlantic region, *Canadian Journal of Earth Sciences* **21,** 887–901.
Williams, H. and Hatcher, R. D. (1982). Suspect terranes and accretionary history of the Appalachian orogen, *Geology* **10,** 530–536.
Williams, H. and Hatcher, R. D. (1983). Appalachian suspect terranes, *In* R. D. Hatcher, H. Williams and I. Zietz (Eds), *Contributions to the Tectonics and Geophysics of Mountain Chains*, Geological Society of America Memoir 158, 33–53.

MIX
Papier aus verantwortungsvollen Quellen
Paper from responsible sources
FSC® C105338

If you have any concerns about our products,
you can contact us on
ProductSafety@springernature.com

In case Publisher is established outside the EU,
the EU authorized representative is:
Springer Nature Customer Service Center GmbH
Europaplatz 3, 69115 Heidelberg, Germany

Printed by Libri Plureos GmbH
in Hamburg, Germany